Antiplane Motions of
Piezoceramics and
Acoustic Wave Devices

Antiplane Motions of Piezoceramics and Acoustic Wave Devices

Jiashi Yang

University of Nebraska, USA

 **World Scientific**

NEW JERSEY · LONDON · SINGAPORE · BEIJING · SHANGHAI · HONG KONG · TAIPEI · CHENNAI

Published by

World Scientific Publishing Co. Pte. Ltd.

5 Toh Tuck Link, Singapore 596224

USA office: 27 Warren Street, Suite 401-402, Hackensack, NJ 07601

UK office: 57 Shelton Street, Covent Garden, London WC2H 9HE

British Library Cataloguing-in-Publication Data
A catalogue record for this book is available from the British Library.

ANTIPLANE MOTIONS OF PIEZOCERAMICS AND ACOUSTIC WAVE DEVICES

ISBN-13 978-981-4291-44-6
ISBN-10 981-4291-44-7

Printed in Singapore.

Preface

Piezoelectric materials exhibit electromechanical coupling. They are strained when placed in an electric field, and become electrically polarized under mechanical loads. Piezoelectric materials have been used for a long time to make various electromechanical devices. Examples include transducers for converting electrical energy to mechanical energy or vice versa, resonators for frequency control and selection with applications in telecommunication and timing, actuators, and acoustic wave sensors.

Due to material anisotropy and multi-field coupling, the three-dimensional equations of linear piezoelectricity are rather complicated. Theoretical analyses based on these equations usually present considerable mathematical challenges. The results obtained are often very lengthy, obscuring the underlying physical picture. In the analysis of piezoelectric devices, relatively few exact solutions from the three-dimensional equations can be obtained. Numerical techniques like the finite element method are usually needed. Another way to simplify the problems so that theoretical analyses are possible is to use approximate, lower-dimensional structural theories of plates, shells, beams and rods. These two approaches are both very useful in general in the modeling and design of piezoelectric devices.

Polarized ceramics are common materials for piezoelectric devices. They are transversely isotropic and allow the so-called shear-horizontal (SH) or antiplane motions. These motions are with one displacement component only along the poling direction depending on two spatial coordinates, with coupling to an electric potential depending on the same two spatial coordinates. For these motions the mathematical problem is relatively simple. Useful problems of practical interest can often be solved analytically, sometimes with elegant results showing the physics involved clearly. Beginning from the discovery of the simple and truly piezoelectric Bleustein–Gulyaev surface waves over a ceramic half space in 1968, many theoretical results on antiplane motions of polarized ceramics have been obtained for device applications. This book collects a series of solutions of antiplane motions of polarized ceramics. Although

v

some static problems are included, the main purpose is to present results on dynamic problems related to acoustic wave devices. Both surface acoustic waves (SAW) and bulk acoustic waves (BAW) are considered. The operating principles of quite a few acoustic wave and electromechanical devices are exhibited through antiplane problems.

Since the structures of the material tensors of crystals with 6mm symmetry are the same as those of polarized ceramics, 6mm crystals are mathematically the same as polarized ceramics within the theory of piezoelectricity. Therefore solutions for polarized ceramics are also applicable to 6mm crystals which include widely used materials like zinc oxide and aluminum nitride. Crystals of 4mm symmetry differ from 6mm crystals only in that $c_{66} = (c_{11}-c_{12})/2$ is no longer valid for 4mm crystals. This, however, makes no difference for antiplane motions. Therefore the results in the present book are also valid for 4mm crystals like dilithium tetraborate and strontium barium niobate.

The results presented in this book are simple and basic, and therefore are useful for educational purposes. The study of antiplane motions of polarized ceramics is an active research area. This book is also useful as a reference for researchers. The problems analyzed in this book can provide motivation and solution techniques for further studies of other useful problems.

The author would like to thank Dr. Zhenghua Qian of Tokyo Institute of Technology for producing Fig. 5.2 and sharing the electronic file of Sec. 4.3, and Professor Jianke Du of Ningbo University for sharing the electronic file of Sec. 6.3.

Jiashi Yang
Lincoln, Nebraska
October, 2009

Contents

Chapter 1
Basic Equations

Although this book is about antiplane motions depending on two spatial variables, in this chapter we begin with the three-dimensional equations of linear piezoelectricity. This is because it is usually beneficial to look at problems from a three-dimensional point of view and that antiplane motions of ceramics are in fact exact solutions of the three-dimensional equations. We first summarize the basic theory of linear piezoelectricity based on the *IEEE Standard on Piezoelectricity* [1], the classical book on piezoelectricity by H. F. Tiersten [2] who also wrote the theoretical part of [1], and Chapter 2 of a relatively recent book by the present author [3]. Then we show that antiplane motions of ceramics are a special class of motions satisfying the three-dimensional equations and derive special equations for antiplane motions. The Cartesian tensor notation, the summation convention for repeated tensor indices, and the convention that a comma followed by an index denotes partial differentiation with respect to the coordinate associated with the index are used. A superimposed dot represents a time derivative.

1.1. Equations of Linear Piezoelectricity

The equation of linear piezoelectricity can be obtained by linearizing the nonlinear electroelastic equations [4,5] under the assumption of infinitesimal deformation and fields. The equations of motion and the charge equation of electrostatic (Gauss's law) are

$$T_{ji,j} + f_i = \rho \ddot{u}_i,$$
$$D_{i,i} = q, \tag{1.1}$$

where $\mathbf{T}$ is the stress tensor, ρ is the mass density, $\mathbf{f}$ is the body force per unit volume, $\mathbf{u}$ is the displacement vector, $\mathbf{D}$ is the electric displacement vector, and q is the body free charge density which is usually zero. Constitutive relations are given by an electric enthalpy function H

$$H(\mathbf{S}, \mathbf{E}) = \frac{1}{2} c_{ijkl}^E S_{ij} S_{kl} - e_{ijk} E_i S_{jk} - \frac{1}{2} \varepsilon_{ij}^S E_i E_j \tag{1.2}$$

1

through

$$T_{ij} = \frac{\partial H}{\partial S_{ij}} = c_{ijkl}^E S_{kl} - e_{kij} E_k,$$

$$(1.3)$$

$$D_i = -\frac{\partial H}{\partial E_i} = e_{ikl} S_{kl} + \varepsilon_{ik}^S E_k,$$

where the strain tensor, **S**, and the electric field vector, **E**, are related to the displacement, **u**, and the electric potential, ϕ, by

$$S_{ij} = (u_{j,i} + u_{i,j})/2, \quad E_i = -\phi_{,i}. \tag{1.4}$$

c_{ijkl}^E, e_{ijk}, and ε_{ij}^S are the elastic, piezoelectric, and dielectric constants. The superscript, E, in c_{ijkl}^E indicates that the independent electric constitutive variable is the electric field **E**. The superscript, S, in ε_{ij}^S indicates that the mechanical constitutive variable is the strain tensor, **S**. The material constants have the following symmetries:

$$c_{ijkl}^E = c_{jikl}^E = c_{klij}^E,$$

$$(1.5)$$

$$e_{kij} = e_{kji}, \quad \varepsilon_{ij}^S = \varepsilon_{ji}^S.$$

We also assume that the elastic and dielectric tensors are positive definite in the following sense:

$$c_{ijkl}^E S_{ij} S_{kl} \geq 0 \quad \text{for any} \quad S_{ij} = S_{ji},$$

$$\text{and} \quad c_{ijkl}^E S_{ij} S_{kl} = 0 \quad \Rightarrow \quad S_{ij} = 0,$$

$$(1.6)$$

$$\varepsilon_{ij}^S E_i E_j \geq 0 \quad \text{for any} \quad E_i,$$

$$\text{and} \quad \varepsilon_{ij}^S E_i E_j = 0 \quad \Rightarrow \quad E_i = 0.$$

With successive substitutions from Eqs. (1.3) and (1.4), Eq. (1.1) can be written as four equations for **u** and ϕ:

$$c_{ijkl} u_{k,lj} + e_{kij} \phi_{,kj} + f_i = \rho \ddot{u}_i,$$

$$(1.7)$$

$$e_{ikl} u_{k,li} - \varepsilon_{ij} \phi_{,ij} = q,$$

where we have neglected the superscripts of the material constants. The linearity of Eq. (1.7) allows the superposition of solutions.

Let the region occupied by a piezoelectric body be V and its boundary surface be S, as shown in Fig. 1.1. Let the unit outward normal of S be **n**.

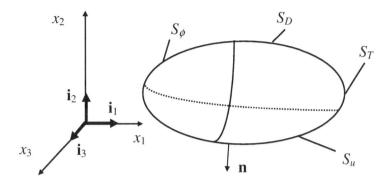

Fig. 1.1. A piezoelectric body and partitions of its boundary surface.

For boundary conditions, we consider the following partitions of S:

$$S_u \cup S_T = S_\phi \cup S_D = S,$$
$$S_u \cap S_T = S_\phi \cap S_D = 0,$$

(1.8)

where S_u is the part of S on which the mechanical displacement is prescribed, and S_T is the part of S where the traction vector is prescribed. S_ϕ represents the part of S which is electroded where the electric potential is no more than a function of time, and S_D is the unelectroded part. For mechanical boundary conditions, we have prescribed displacement $\bar{u}_i$

$$u_i = \bar{u}_i \quad \text{on} \quad S_u,$$

(1.9)

and prescribed traction $\bar{t}_j$

$$T_{ij} n_i = \bar{t}_j \quad \text{on} \quad S_T.$$

(1.10)

Electrically, on the electroded portion of S,

$$\phi = \bar{\phi} \quad \text{on} \quad S_\phi,$$

(1.11)

where $\bar{\phi}$ does not vary spatially. On the unelectroded part of S, the charge condition can be written as

$$D_j n_j = -\bar{\sigma} \quad \text{on} \quad S_D,$$

(1.12)

where $\bar{\sigma}$ is free charge density per unit surface area. In the above formulation, we assume very thin electrodes. The mechanical effects like inertia and stiffness of the electrodes are neglected.

On an electrode, S_ϕ, the total free electric charge, Q_e (a scalar), can be represented by

$$Q_e = \int_{S_\phi} -n_i D_i \, dS \,. \tag{1.13}$$

The electric current flowing out of the electrode is given by

$$I = -\dot{Q}_e \,. \tag{1.14}$$

Sometimes there are two (or more) electrodes on a body, and the electrodes are connected to an electric circuit. In this case, circuit equation(s) will need to be considered.

Consider the following initial-boundary-value problem:

$$
\begin{aligned}
&T_{ji,j} + \rho f_i = \rho \ddot{u}_i \quad \text{in} \quad V, \quad t > t_0, \\
&D_{i,i} = q \quad \text{in} \quad V, \quad t > t_0, \\
&T_{ij} = c_{ijkl} S_{kl} - e_{kij} E_k \quad \text{in} \quad V, \quad t > t_0, \\
&D_i = e_{ijk} S_{jk} + \varepsilon_{ij} E_j \quad \text{in} \quad V, \quad t > t_0, \\
&S_{ij} = (u_{i,j} + u_{j,i})/2 \quad \text{in} \quad V, \quad t > t_0, \\
&E_i = -\phi_{,i} \quad \text{in} \quad V, \quad t > t_0,
\end{aligned}
\tag{1.15}
$$

and

$$
\begin{aligned}
&u_i = \bar{u}_i \quad \text{on} \quad S_u, \quad t > t_0, \\
&T_{ji} n_j = \bar{t}_i \quad \text{on} \quad S_T, \quad t > t_0, \\
&\phi = \bar{\phi} \quad \text{on} \quad S_\phi, \quad t > t_0, \\
&D_i n_i = -\bar{\sigma} \quad \text{on} \quad S_D, \quad t > t_0, \\
&u_i = u_i^0 \quad \text{in} \quad V, \quad t = t_0, \\
&\dot{u}_i = v_i^0 \quad \text{in} \quad V, \quad t = t_0, \\
&\phi = \phi^0 \quad \text{in} \quad V, \quad t = t_0,
\end{aligned}
\tag{1.16}
$$

where u_i^0, v_i^0 and ϕ^0 are initial data. Under Eqs. (1.15) and (1.16), **S**, **E**, **T**, **D** and the velocity fields are unique. The displacement field may have an undetermined static rigid-body displacement if $S_u = 0$. The electric potential may have an undetermined constant if $S_\phi = 0$.

1.2. Cylindrical Coordinates

Quite a few problems in this book are analyzed in cylindrical coordinates. For convenience we summarize the relevant equations below. The cylindrical coordinates (r, θ, z) are defined by

$$x_1 = r\cos\theta, \quad x_2 = r\sin\theta, \quad x_3 = z . \tag{1.17}$$

In cylindrical coordinates, we have the strain-displacement relation

$$S_{rr} = u_{r,r}, \quad S_{\theta\theta} = \frac{1}{r}u_{\theta,\theta} + \frac{u_r}{r}, \quad S_{zz} = u_{z,z},$$

$$2S_{r\theta} = u_{\theta,r} + \frac{1}{r}u_{r,\theta} - \frac{u_\theta}{r}, \quad 2S_{\theta z} = \frac{1}{r}u_{z,\theta} + u_{\theta,z}, \tag{1.18}$$

$$2S_{zr} = u_{r,z} + u_{z,r}.$$

The electric field-potential relation is given by

$$E_r = -\phi_{,r}, \quad E_\theta = -\frac{1}{r}\phi_{,\theta}, \quad E_z = -\phi_{,z} . \tag{1.19}$$

The equations of motion are

$$\frac{\partial T_{rr}}{\partial r} + \frac{1}{r}\frac{\partial T_{\theta r}}{\partial \theta} + \frac{\partial T_{zr}}{\partial z} + \frac{T_{rr} - T_{\theta\theta}}{r} + f_r = \rho\ddot{u}_r,$$

$$\frac{\partial T_{r\theta}}{\partial r} + \frac{1}{r}\frac{\partial T_{\theta\theta}}{\partial \theta} + \frac{\partial T_{z\theta}}{\partial z} + \frac{2}{r}T_{r\theta} + f_\theta = \rho\ddot{u}_\theta, \tag{1.20}$$

$$\frac{\partial T_{rz}}{\partial r} + \frac{1}{r}\frac{\partial T_{\theta z}}{\partial \theta} + \frac{\partial T_{zz}}{\partial z} + \frac{1}{r}T_{rz} + f_z = \rho\ddot{u}_z.$$

The electrostatic charge equation is

$$\frac{1}{r}(rD_r)_{,r} + \frac{1}{r}D_{\theta,\theta} + D_{z,z} = q . \tag{1.21}$$

1.3. Matrix Notation

We now introduce a compact matrix notation [1,2]. This notation consists of replacing pairs of indices, ij or kl, by single indices, p or q, where i, j, k and l take the values of 1, 2 and 3; and p and q take the values of 1, 2, 3, 4, 5 and 6 according to

ij or kl :	11	22	33	23 or 32	31 or 13	12 or 21	
p or q :	1	2	3	4	5	6	(1.22)

Thus

$$c_{ijkl} \to c_{pq}, \quad e_{ikl} \to e_{ip}, \quad T_{ij} \to T_p. \tag{1.23}$$

For the strain tensor, we introduce S_p such that

$$
\begin{aligned}
S_1 &= S_{11}, \quad S_2 = S_{22}, \quad S_3 = S_{33}, \\
S_4 &= 2S_{23}, \quad S_5 = 2S_{31}, \quad S_6 = 2S_{12}.
\end{aligned} \tag{1.24}
$$

The constitutive relations can then be written as

$$
\begin{aligned}
T_p &= c_{pq}^E S_q - e_{kp} E_k, \\
D_i &= e_{iq} S_q + \varepsilon_{ik}^S E_k.
\end{aligned} \tag{1.25}
$$

In matrix form, Eq. (1.25) becomes

$$
\begin{Bmatrix} T_1 \\ T_2 \\ T_3 \\ T_4 \\ T_5 \\ T_6 \end{Bmatrix} =
\begin{pmatrix}
c_{11}^E & c_{12}^E & c_{13}^E & c_{14}^E & c_{15}^E & c_{16}^E \\
c_{21}^E & c_{22}^E & c_{23}^E & c_{24}^E & c_{25}^E & c_{26}^E \\
c_{31}^E & c_{32}^E & c_{33}^E & c_{34}^E & c_{35}^E & c_{36}^E \\
c_{41}^E & c_{42}^E & c_{43}^E & c_{44}^E & c_{45}^E & c_{46}^E \\
c_{51}^E & c_{52}^E & c_{53}^E & c_{54}^E & c_{55}^E & c_{56}^E \\
c_{61}^E & c_{62}^E & c_{63}^E & c_{64}^E & c_{65}^E & c_{66}^E
\end{pmatrix}
\begin{Bmatrix} S_1 \\ S_2 \\ S_3 \\ S_4 \\ S_5 \\ S_6 \end{Bmatrix}
$$

$$
- \begin{pmatrix}
e_{11} & e_{21} & e_{31} \\
e_{12} & e_{22} & e_{32} \\
e_{13} & e_{23} & e_{33} \\
e_{14} & e_{24} & e_{34} \\
e_{15} & e_{25} & e_{35} \\
e_{16} & e_{26} & e_{36}
\end{pmatrix}
\begin{Bmatrix} E_1 \\ E_2 \\ E_3 \end{Bmatrix}, \tag{1.26}
$$

$$
\begin{Bmatrix} D_1 \\ D_2 \\ D_3 \end{Bmatrix} =
\begin{bmatrix}
e_{11} & e_{12} & e_{13} & e_{14} & e_{15} & e_{16} \\
e_{21} & e_{22} & e_{23} & e_{24} & e_{25} & e_{26} \\
e_{31} & e_{32} & e_{33} & e_{34} & e_{35} & e_{36}
\end{bmatrix}
\begin{Bmatrix} S_1 \\ S_2 \\ S_3 \\ S_4 \\ S_5 \\ S_6 \end{Bmatrix}
$$

$$
+ \begin{pmatrix}
\varepsilon_{11}^S & \varepsilon_{12}^S & \varepsilon_{13}^S \\
\varepsilon_{21}^S & \varepsilon_{22}^S & \varepsilon_{22}^S \\
\varepsilon_{31}^S & \varepsilon_{32}^S & \varepsilon_{33}^S
\end{pmatrix}
\begin{Bmatrix} E_1 \\ E_2 \\ E_3 \end{Bmatrix}.
$$

1.4. Constitutive Relations of Polarized Ceramics

In this section we obtain the constitutive relations of polarized ceramics using tensor invariants in the manner of [6]. Polarized ceramics are transversely isotropic. Let $\mathbf{a}$, a constant unit vector, represent the direction of the axis of rotational symmetry or the poling direction of the ceramics. For linear constitutive relations we need a quadratic electric enthalpy function H. For transversely isotropic materials, a quadratic H is a function of the following invariants of degrees one and two (higher degree invariants are not included):

$$I_1 = \mathbf{a} \cdot \mathbf{S} \cdot \mathbf{a}, \quad I_2 = \mathrm{tr}\mathbf{S}, \quad I_3 = \mathbf{a} \cdot \mathbf{E},$$

$$II_1 = \mathbf{a} \cdot \mathbf{S}^2 \cdot \mathbf{a}, \quad II_2 = \mathrm{tr}\mathbf{S}^2, \tag{1.27}$$

$$II_3 = \mathbf{E} \cdot \mathbf{E}, \quad II_4 = \mathbf{a} \cdot \mathbf{S} \cdot \mathbf{E} + \mathbf{E} \cdot \mathbf{S} \cdot \mathbf{a}.$$

A complete quadratic function of the above seven invariants can be written as

$$H = c_1 I_1^2 + c_2 I_2^2 + c_3 I_1 I_2 + c_4 II_1 + c_5 II_2$$

$$+ \varepsilon_1 I_3^2 + \varepsilon_2 II_3 \tag{1.28}$$

$$+ e_1 I_1 I_3 + e_2 I_2 I_3 + e_3 II_4,$$

where c_1, c_2, c_3, c_4 and c_5 are elastic constants, ε_1 and ε_2 are dielectric constants, and e_1, e_2 and c_3 are piezoelectric constants. Differentiation of Eq. (1.28) yields

$$\mathbf{T} = \frac{\partial H}{\partial \mathbf{S}} = \frac{\partial H}{\partial I_1} \mathbf{a} \otimes \mathbf{a} + \frac{\partial H}{\partial I_2} \mathbf{1} + \frac{\partial H}{\partial II_1}(\mathbf{a} \otimes \mathbf{S} \cdot \mathbf{a} + \mathbf{a} \cdot \mathbf{S} \otimes \mathbf{a})$$

$$+ 2\frac{\partial H}{\partial II_2} \mathbf{S} + \frac{\partial H}{\partial II_4}(\mathbf{a} \otimes \mathbf{E} + \mathbf{E} \otimes \mathbf{a}) \tag{1.29}$$

$$= (2c_1 I_1 + c_3 I_2 + e_1 I_3)\mathbf{a} \otimes \mathbf{a} + (2c_2 I_2 + c_3 I_1 + e_2 I_3)\mathbf{1}$$

$$+ c_4(\mathbf{a} \otimes \mathbf{S} \cdot \mathbf{a} + \mathbf{a} \cdot \mathbf{S} \otimes \mathbf{a}) + 2c_5 \mathbf{S} + e_3(\mathbf{a} \otimes \mathbf{E} + \mathbf{E} \otimes \mathbf{a}),$$

and

$$\mathbf{D} = -\frac{\partial H}{\partial \mathbf{E}} = -\frac{\partial H}{\partial I_3} \mathbf{a} - 2\frac{\partial H}{\partial II_3} \mathbf{E} - 2\frac{\partial H}{\partial II_4} \mathbf{S} \cdot \mathbf{a} \tag{1.30}$$

$$= -(2\varepsilon_1 I_3 + e_1 I_1 + e_2 I_2)\mathbf{a} - 2\varepsilon_2 \mathbf{E} - 2e_3 \mathbf{S} \cdot \mathbf{a}.$$

Let $\mathbf{a} = \mathbf{i}_3$, and rearrange Eqs. (1.29) and (1.30) in the form of Eq. (1.25). The following matrices will result:

$$\begin{pmatrix} c_{11} & c_{12} & c_{13} & 0 & 0 & 0 \\ c_{21} & c_{11} & c_{13} & 0 & 0 & 0 \\ c_{31} & c_{31} & c_{33} & 0 & 0 & 0 \\ 0 & 0 & 0 & c_{44} & 0 & 0 \\ 0 & 0 & 0 & 0 & c_{44} & 0 \\ 0 & 0 & 0 & 0 & 0 & c_{66} \end{pmatrix},$$

$$\begin{pmatrix} 0 & 0 & 0 & 0 & e_{15} & 0 \\ 0 & 0 & 0 & e_{15} & 0 & 0 \\ e_{31} & e_{31} & e_{33} & 0 & 0 & 0 \end{pmatrix}, \begin{pmatrix} \varepsilon_{11} & 0 & 0 \\ 0 & \varepsilon_{11} & 0 \\ 0 & 0 & \varepsilon_{33} \end{pmatrix}, \qquad (1.31)$$

where $c_{66} = (c_{11} - c_{12})/2$. The matrices in Eq. (1.31) have the same structures as those of crystals class C_{6v} (or 6mm) [7]. The elements of the matrices in Eq. (1.31) are related to the material constants in Eq. (1.28) by

$$c_1 = c_{11} - 2c_{13} + c_{33} - 4c_{44}, \quad c_2 = c_{12}/2,$$
$$c_3 = c_{13} - c_{12}, \quad c_4 = -c_{11} + c_{12} + 2c_{44},$$
$$c_5 = (c_{11} - c_{12})/2, \qquad\qquad\qquad\qquad (1.32)$$
$$\varepsilon_1 = (\varepsilon_{11} - \varepsilon_{22})/2, \quad \varepsilon_2 = -\varepsilon_{11}/2,$$
$$e_1 = e_{31} + 2e_{15} - e_{33}, \quad e_2 = -e_{31}, \quad e_3 = -e_{15}.$$

With Eq. (1.31), the constitutive relations of ceramics poled in the x_3 direction take the following form:

$$T_{11} = c_{11}u_{1,1} + c_{12}u_{2,2} + c_{13}u_{3,3} + e_{31}\phi_{,3},$$
$$T_{22} = c_{12}u_{1,1} + c_{22}u_{2,2} + c_{13}u_{3,3} + e_{31}\phi_{,3},$$
$$T_{33} = c_{13}u_{1,1} + c_{13}u_{2,2} + c_{33}u_{3,3} + e_{33}\phi_{,3},$$
$$T_{23} = c_{44}(u_{2,3} + u_{3,2}) + e_{15}\phi_{,2}, \qquad\qquad (1.33)$$
$$T_{31} = c_{44}(u_{3,1} + u_{1,3}) + e_{15}\phi_{,1},$$
$$T_{12} = c_{66}(u_{1,2} + u_{2,1}),$$

and

$$D_1 = e_{15}(u_{3,1} + u_{1,3}) - \varepsilon_{11}\phi_{,1},$$
$$D_2 = e_{15}(u_{2,3} + u_{3,2}) - \varepsilon_{11}\phi_{,2}, \qquad\qquad (1.34)$$
$$D_3 = e_{31}(u_{1,1} + u_{2,2}) + e_{33}u_{3,3} - \varepsilon_{33}\phi_{,3}.$$

The equations of motion and charge are

$$c_{11}u_{1,11} + (c_{12} + c_{66})u_{2,12} + (c_{13} + c_{44})u_{3,13} + c_{66}u_{1,22}$$
$$+ c_{44}u_{1,33} + (e_{31} + e_{15})\phi_{,13} = \rho\ddot{u}_1,$$
$$c_{66}u_{2,11} + (c_{12} + c_{66})u_{1,12} + c_{11}u_{2,22} + (c_{13} + c_{44})u_{3,23}$$
$$+ c_{44}u_{2,33} + (e_{31} + e_{15})\phi_{,23} = \rho\ddot{u}_2,$$
$$c_{44}u_{3,11} + (c_{44} + c_{13})u_{1,31} + c_{44}u_{3,22} + (c_{13} + c_{44})u_{2,23} \quad (1.35)$$
$$+ c_{33}u_{3,33} + e_{15}(\phi_{,11} + \phi_{,22}) + e_{33}\phi_{,33} = \rho\ddot{u}_3,$$
$$e_{15}u_{3,11} + (e_{15} + e_{31})u_{1,13} + e_{15}u_{3,22} + (e_{15} + e_{31})u_{2,32}$$
$$+ e_{31}u_{3,33} - \varepsilon_{11}(\phi_{,11} + \phi_{,22}) - \varepsilon_{33}\phi_{,33} = 0.$$

Sometimes a piezoelectric device is heterogeneous, with ceramics poled in different directions in different parts. In this case it is not always possible to orient the x_3 axis along the poling directions unless a few local coordinate systems are introduced. Therefore, material matrices of ceramics poled along other axes are useful. They can be obtained by tensor transformations or effectively from the matrices in Eq. (1.31) by rotating rows and columns properly. For ceramics poled in the x_1 direction, we have

$$\begin{pmatrix} c_{33} & c_{13} & c_{13} & 0 & 0 & 0 \\ c_{13} & c_{11} & c_{12} & 0 & 0 & 0 \\ c_{13} & c_{12} & c_{11} & 0 & 0 & 0 \\ 0 & 0 & 0 & c_{66} & 0 & 0 \\ 0 & 0 & 0 & 0 & c_{44} & 0 \\ 0 & 0 & 0 & 0 & 0 & c_{44} \end{pmatrix}, \quad (1.36)$$

$$\begin{pmatrix} e_{33} & e_{31} & e_{31} & 0 & 0 & 0 \\ 0 & 0 & 0 & 0 & 0 & e_{15} \\ 0 & 0 & 0 & 0 & e_{15} & 0 \end{pmatrix}, \begin{pmatrix} \varepsilon_{33} & 0 & 0 \\ 0 & \varepsilon_{11} & 0 \\ 0 & 0 & \varepsilon_{11} \end{pmatrix}.$$

For ceramics poled in the x_2 direction, we obtain

$$\begin{pmatrix} c_{11} & c_{13} & c_{12} & 0 & 0 & 0 \\ c_{13} & c_{33} & c_{13} & 0 & 0 & 0 \\ c_{12} & c_{13} & c_{11} & 0 & 0 & 0 \\ 0 & 0 & 0 & c_{44} & 0 & 0 \\ 0 & 0 & 0 & 0 & c_{66} & 0 \\ 0 & 0 & 0 & 0 & 0 & c_{44} \end{pmatrix},$$

(1.37)

$$\begin{pmatrix} 0 & 0 & 0 & 0 & 0 & e_{15} \\ e_{31} & e_{33} & e_{31} & 0 & 0 & 0 \\ 0 & 0 & 0 & e_{15} & 0 & 0 \end{pmatrix}, \begin{pmatrix} \varepsilon_{11} & 0 & 0 \\ 0 & \varepsilon_{33} & 0 \\ 0 & 0 & \varepsilon_{11} \end{pmatrix}.$$

1.5. Antiplane Problems

For motions independent of x_3, Eq. (1.35) reduces to:

$$c_{11}u_{1,11} + (c_{12} + c_{66})u_{2,12} + c_{66}u_{1,22} = \rho \ddot{u}_1,$$

$$c_{66}u_{2,11} + (c_{12} + c_{66})u_{1,12} + c_{11}u_{2,22} = \rho \ddot{u}_2,$$

$$c_{44}(u_{3,11} + u_{3,22}) + e_{15}(\phi_{,11} + \phi_{,22}) = \rho \ddot{u}_3,$$

$$e_{15}(u_{3,11} + u_{3,22}) - \varepsilon_{11}(\phi_{,11} + \phi_{,22}) = 0.$$

(1.38)

Equation (1.38) shows that u_1 and u_2 are coupled but they do not interact with the electric field. They form the usual plane-strain problem of linear elasticity. In this book we are interested in the so-called antiplane or shear-horizontal (SH) motions described by u_3 which is coupled to ϕ. In the rest of the book we limit ourselves to

$$u_1 = u_2 = 0, \quad u_3 = u(x_1, x_2, t), \quad \phi = \phi(x_1, x_2, t). \tag{1.39}$$

Corresponding to Eq. (1.39), the nonzero components of the strain tensor S_{ij} and electric field E_i are

$$\begin{Bmatrix} S_5 \\ S_4 \end{Bmatrix} = \begin{Bmatrix} 2S_{31} \\ 2S_{23} \end{Bmatrix} = \nabla u, \quad \begin{Bmatrix} E_1 \\ E_2 \end{Bmatrix} = -\nabla \phi, \tag{1.40}$$

where $\nabla = \mathbf{i}_1 \partial_1 + \mathbf{i}_2 \partial_2$ is the two-dimensional gradient operator. The nontrivial components of the stress tensor T_{ij} and the electric displacement vector D_i are

$$\begin{Bmatrix} T_5 \\ T_4 \end{Bmatrix} = \begin{Bmatrix} T_{31} \\ T_{23} \end{Bmatrix} = c\nabla u + e\nabla \phi, \quad \begin{Bmatrix} D_1 \\ D_2 \end{Bmatrix} = e\nabla u - \varepsilon \nabla \phi, \tag{1.41}$$

where we have denoted the relevant elastic, piezoelectric, and dielectric constants by

$$c = c_{44}, \quad e = e_{15}, \quad \varepsilon = \varepsilon_{11}. \tag{1.42}$$

We will consider source-free problems with $q = 0$ and $f_3 = 0$. The equation of motion and the charge equation can be written as

$$\begin{aligned} c\nabla^2 u + e\nabla^2 \phi &= \rho\ddot{u}, \\ e\nabla^2 u - \varepsilon\nabla^2 \phi &= 0, \end{aligned} \tag{1.43}$$

where $\nabla^2 = \partial_1^2 + \partial_2^2$ is the two-dimensional Laplacian.

1.6. Bleustein's Formulation

Equation (1.43) can be decoupled [8]. We introduce

$$\psi = \phi - \frac{e}{\varepsilon} u. \tag{1.44}$$

Then, in terms of u and ψ,

$$\begin{aligned} T_{31} &= \bar{c}u_{,1} + e\psi_{,1}, \\ T_{23} &= \bar{c}u_{,2} + e\psi_{,2}, \end{aligned} \tag{1.45}$$

$$\begin{aligned} D_1 &= -\varepsilon\psi_{,1}, \\ D_2 &= -\varepsilon\psi_{,2}, \end{aligned} \tag{1.46}$$

and

$$\begin{aligned} v_T^2 \nabla^2 u &= \ddot{u}, \\ \nabla^2 \psi &= 0, \end{aligned} \tag{1.47}$$

where

$$v_T^2 = \frac{\bar{c}}{\rho}, \quad \bar{c} = c + \frac{e^2}{\varepsilon} = c(1 + k^2), \quad k^2 = \frac{e^2}{\varepsilon c}. \tag{1.48}$$

Equation (1.47) is defined over a two-dimensional domain (see Fig. 1.2). Let $\mathbf{n}$ and $\mathbf{s}$ be the unit normal and tangent of the boundary. Typical boundary conditions are the specifications of

$$u \quad \text{or} \quad T_{n3}, \quad {}_{\bullet}$$

$$\psi + \frac{e}{\varepsilon} u \quad \text{or} \quad D_n. \tag{1.49}$$

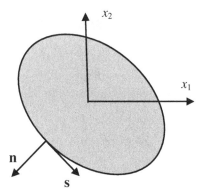

Fig. 1.2. Cross section of a cylindrical body of polarized ceramics.

Although Eq. (1.47) is decoupled, electromechanical coupling still exists in the constitutive relations in Eq. (1.45) and may appear in boundary conditions.

1.7. A Static General Solution in Polar Coordinates

For static problems Eq. (1.43) reduces to

$$c\nabla^2 u + e\nabla^2\phi = 0,$$
$$e\nabla^2 u - \varepsilon\nabla^2\phi = 0. \tag{1.50}$$

When $c\varepsilon + e^2 \neq 0$ which we always assume, Eq. (1.50) is equivalent to

$$\nabla^2 u = 0,$$
$$\nabla^2\phi = 0. \tag{1.51}$$

In polar coordinates, we have

$$\frac{\partial^2 u}{\partial r^2} + \frac{1}{r}\frac{\partial u}{\partial r} + \frac{1}{r^2}\frac{\partial^2 u}{\partial \theta^2} = 0,$$
$$\frac{\partial^2 \phi}{\partial r^2} + \frac{1}{r}\frac{\partial \phi}{\partial r} + \frac{1}{r^2}\frac{\partial^2 \phi}{\partial \theta^2} = 0. \tag{1.52}$$

For problems periodic in θ, by separation of variables, the general solution for u and ϕ is

$$u = A_0 (P_0 + Q_0 \ln r)$$

$$+ \sum_{n=1}^{\infty} (A_n \cos n\theta + B_n \sin n\theta)(P_n r^n + Q_n r^{-n}), \tag{1.53}$$

$$\phi = C_0 (R_0 + S_0 \ln r)$$

$$+ \sum_{n=1}^{\infty} (C_n \cos n\theta + D_n \sin n\theta)(R_n r^n + S_n r^{-n}), \tag{1.54}$$

where A_n, B_n, C_n, D_n, P_n, Q_n, R_n and S_n are undetermined constants. The corresponding stress and electric displacement components are determined from

$$T_{rz} = c \frac{\partial u}{\partial r} + e \frac{\partial \phi}{\partial r},$$

$$T_{\theta z} = c \frac{1}{r} \frac{\partial u}{\partial \theta} + e \frac{1}{r} \frac{\partial \phi}{\partial \theta}, \tag{1.55}$$

$$D_r = e \frac{\partial u}{\partial r} - \varepsilon \frac{\partial \phi}{\partial r},$$

$$D_\theta = e \frac{1}{r} \frac{\partial u}{\partial \theta} - \varepsilon \frac{1}{r} \frac{\partial \phi}{\partial \theta}. \tag{1.56}$$

1.8. A Time-harmonic General Solution in Polar Coordinates

For time-harmonic problems with an $\exp(i\omega t)$ factor where "i" is the imaginary unit, we will drop the factor very often for simplicity. In polar coordinates, from Eq. (1.47) we have

$$v_T^2 \left(\frac{\partial^2 u}{\partial r^2} + \frac{1}{r} \frac{\partial u}{\partial r} + \frac{1}{r^2} \frac{\partial^2 u}{\partial \theta^2} \right) = -\omega^2 u,$$

$$\frac{\partial^2 \psi}{\partial r^2} + \frac{1}{r} \frac{\partial \psi}{\partial r} + \frac{1}{r^2} \frac{\partial^2 \psi}{\partial \theta^2} = 0. \tag{1.57}$$

Consider the possibility of the following fields:

$$u = u(r) \times \begin{Bmatrix} \cos v\theta \\ \sin v\theta \end{Bmatrix}, \quad \psi = \psi(r) \times \begin{Bmatrix} \cos v\theta \\ \sin v\theta \end{Bmatrix}, \tag{1.58}$$

which are consequences of the method of separation of variables. Substitution of Eq. (1.58) into Eq. (1.57) results in

$$\frac{\partial^2 u}{\partial r^2} + \frac{1}{r} \frac{\partial u}{\partial r} + (\xi^2 - \frac{v^2}{r^2})u = 0, \tag{1.59}$$

$$\frac{\partial^2 \psi}{\partial r^2} + \frac{1}{r}\frac{\partial \psi}{\partial r} - \frac{v^2}{r^2}\psi = 0, \tag{1.60}$$

where we have denoted

$$\xi = \frac{\omega}{v_T}. \tag{1.61}$$

Equation (1.59) can be written as Bessel's equation of order v. Equation (1.60) allows a simpler power function solution. The general solution can be written as:

$$u = C_0 J_0(\xi r) + D_0 Y_0(\xi r)$$
$$+ \sum_{m=1}^{\infty} [C_m J_{v_m}(\xi r) + D_m Y_{v_m}(\xi r)]\cos v_m \theta, \tag{1.62}$$

$$\psi = F_0 \ln r + G_0$$
$$+ \sum_{m=1}^{\infty} [F_m r^{v_m} + G_m r^{-v_m}]\cos v_m \theta, \tag{1.63}$$

where C_m, D_m, F_m and G_m are undetermined constants. The electric potential, the stress and the electric displacement components can be obtained as

$$\phi = \psi + \frac{e}{\varepsilon}u$$
$$= \frac{e}{\varepsilon}C_0 J_0(\xi r) + \frac{e}{\varepsilon}D_0 Y_0(\xi r) + F_0 \ln r + G_0$$
$$+ \sum_{m=1}^{\infty} [\frac{e}{\varepsilon}C_m J_{v_m}(\xi r) + \frac{e}{\varepsilon}D_m Y_{v_m}(\xi r)$$
$$+ F_m r^{v_m} + G_m r^{-v_m}]\cos v_m \theta, \tag{1.64}$$

$$T_{rz} = \bar{c}\frac{\partial u}{\partial r} + e\frac{\partial \psi}{\partial r}$$
$$= -\bar{c}\xi C_0 J_1(\xi r) - \bar{c}\xi D_0 Y_1(\xi r) + eF_0 \frac{1}{r}$$
$$+ \sum_{m=1}^{\infty} [\bar{c}\xi C_m J'_{v_m}(\xi r) + \bar{c}\xi D_m Y'_{v_m}(\xi r)$$
$$+ ev_m F_m r^{v_m-1} - ev_m G_m r^{-v_m-1}]\cos v_m \theta, \tag{1.65}$$

$$T_{\theta z} = \bar{c}\frac{1}{r}\frac{\partial u}{\partial \theta} + e\frac{1}{r}\frac{\partial \psi}{\partial \theta}$$

$$= \frac{1}{r}\sum_{m=1}^{\infty} [\bar{c}C_m J_{\nu_m}(\xi r) + \bar{c}D_m Y_{\nu_m}(\xi r) \tag{1.66}$$

$$+ eF_m r^{\nu_m} + eG_m r^{-\nu_m}](-\nu_m)\sin\nu_m\theta,$$

$$D_r = -\varepsilon\frac{\partial \psi}{\partial r}$$

$$\tag{1.67}$$

$$= -\varepsilon F_0\frac{1}{r} + \sum_{m=1}^{\infty} [-\varepsilon\nu_m F_m r^{\nu_m-1} + \varepsilon\nu_m G_m r^{-\nu_m-1}]\cos\nu_m\theta,$$

$$D_\theta = -\varepsilon\frac{1}{r}\frac{\partial \psi}{\partial \theta}$$

$$\tag{1.68}$$

$$= -\varepsilon\frac{1}{r}\sum_{m=1}^{\infty} [eF_m r^{\nu_m} + eG_m r^{-\nu_m}](-\nu_m)\sin\nu_m\theta.$$

For solutions periodic in θ, we must have

$$\nu_m = m, \quad m = 1, 2, 3, \cdots. \tag{1.69}$$

1.9. Boundary Integral Equation Formulation

For time-harmonic motions, with Eq. (1.61), we can write Eq. (1.57) as

$$\nabla^2 u + \xi^2 u = 0,$$

$$\nabla^2 \psi = 0. \tag{1.70}$$

Let u^* and ψ^* be the fundamental solutions of the following differential operators:

$$-\nabla^2 u^* - \xi^2 u^* = \delta,$$

$$-\nabla^2 \psi^* = \delta, \tag{1.71}$$

where δ is the Dirac delta function. For any two functions over a two-dimensional domain A with a boundary curve C, there exist the following Green's identities:

$$\int_A [u(\nabla^2 u^* + \xi^2 u^*) - u^*(\nabla^2 u + \xi^2 u)]dA$$

$$= \int_C \left[u\frac{\partial u^*}{\partial n} - u^*\frac{\partial u}{\partial n}\right]dL, \tag{1.72}$$

$$\int_A [\psi \nabla^2 \psi^* - \psi^* \nabla^2 \psi]dA = \int_C \left[\psi \frac{\partial \psi^*}{\partial n} - \psi^* \frac{\partial \psi}{\partial n} \right]dL , \qquad (1.73)$$

where dA and dL are differential area and line elements. From Eqs. (1.70) through (1.73), we obtain the following boundary integral equations for u and ψ :

$$\frac{1}{2}u(P) = -\int_C \frac{\partial u^*(P,Q)}{\partial n(Q)} u(Q)dL(Q)$$

$$+ \int_C u^*(P,Q)\frac{\partial u(Q)}{\partial n(Q)}dL(Q), \qquad (1.74)$$

$$\frac{1}{2}\psi(P) = -\int_C \frac{\partial \psi^*(P,Q)}{\partial n(Q)} \psi(Q)dL(Q)$$

$$+ \int_C \psi^*(P,Q)\frac{\partial \psi(P,Q)}{\partial n(Q)}dL(Q). \qquad (1.75)$$

The fundamental solutions can be found in [9].

Chapter 2
Static Problems

In this chapter, solutions to a few static problems are presented. The concept of electromechanical coupling coefficient is introduced. We point out that there exists some subtlety between the theory of piezoelectric statics and reality. The theory is about dielectrics and usually involves an electric field. However, real piezoelectric materials more or less have some conductivity. This usually small conductivity tends to neutralize the electric field (relaxation). Therefore, in static problems, an applied voltage is usually needed to maintain an electric field. In dynamic problems, the characteristic relaxation time of conduction competes with the characteristic frequency (or period) of motion. For low-frequency motions, the electric field may in fact be zero due to conduction unless there is an applied voltage. For high-frequency motions an electric field is usually accompanying mechanical fields.

2.1. A Surface Distribution of Electric Potential

Consider a ceramic half-space poled along the x_3 direction (see Fig. 2.1). The surface at $x_2 = 0$ is traction-free and a periodic potential is applied. $x_2 = +\infty$ is mechanically fixed and electrically grounded.

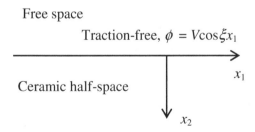

Fig. 2.1. A ceramic half-space.

From Eq. (1.51) the boundary-value problem is:

$$\nabla^2 u = 0, \quad \nabla^2 \phi = 0, \quad x_2 > 0,$$
$$T_{23} = 0, \quad \phi = V \cos \xi x_1, \quad x_2 = 0, \quad (2.1)$$
$$u, \phi \to 0, \quad x_2 \to +\infty,$$

where ξ is considered given and assumed to be positive. Consider the possibility of the following fields:

$$u = A \exp(-\xi x_2) \cos \xi x_1,$$
$$\phi = B \exp(-\xi x_2) \cos \xi x_1, \quad (2.2)$$

which already satisfy the Laplace equations and the boundary conditions at infinity in Eq. (2.1). For the boundary conditions at $x_2 = 0$ we need

$$T_{23} = cu_{,2} + e\phi_{,2} = (-cA\xi - eB\xi) \exp(-\xi x_2) \sin \xi x_1. \quad (2.3)$$

The boundary conditions at $x_2 = 0$ require that

$$c(-\xi)A\cos\xi x_1 + e(-\xi)B\cos\xi x_1 = 0,$$
$$B\cos\xi x_1 = V\cos\xi x_1. \quad (2.4)$$

Equation (2.4) determines

$$A = -\frac{e}{c}V, \quad B = V. \quad (2.5)$$

Hence

$$u = -\frac{e}{c}V \exp(-\xi x_2) \cos \xi x_1,$$
$$\phi = V \exp(-\xi x_2) \cos \xi x_1. \quad (2.6)$$

2.2. Shear of a Plate

Consider an unbounded ceramic plate (see Fig. 2.2). The surfaces of the plate are under a tangential traction p and are electroded. Two cases of shorted and open electrodes will be considered.

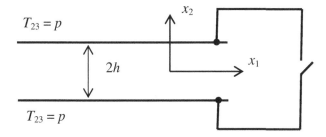

Fig. 2.2. An electroded ceramic plate under mechanical loads.

The boundary-value problem is:

$$\nabla^2 u = 0, \quad \nabla^2 \phi = 0, \quad |x_2| < h,$$

$$T_{23} = p, \quad x_2 = \pm h,$$

$$\phi(x_2 = h) = \phi(x_2 = -h), \quad \text{if the electrodes are shorted,}$$

$$\text{or} \quad D_2(x_2 = \pm h) = 0, \quad \text{if the electrodes are open.}$$

(2.7)

Consider the possibility of the following displacement and potential fields:

$$u = u(x_2), \quad \phi = \phi(x_2). \tag{2.8}$$

The nontrivial components of strain, electric field, stress, and electric displacement are

$$S_4 = 2S_{23} = u_{,2}, \quad E_2 = -\phi_{,2}, \tag{2.9}$$

and

$$T_4 = T_{23} = cu_{,2} + e\phi_{,2},$$

$$D_2 = eu_{,2} - \varepsilon\phi_{,2}. \tag{2.10}$$

The equation of motion and the charge equation reduce to

$$u_{,22} = 0, \quad \phi_{,22} = 0. \tag{2.11}$$

Equation (2.11) implies that all the strain, stress, electric field and electric displacement components are constants.

2.2.1. Shorted electrodes

Since the potential at the two electrodes are equal and E_2 is a constant, we must have

$$E_2 = 0. \tag{2.12}$$

The mechanical boundary conditions require that $T_{23} = p$. Then

$$S_4 = u_{,2} = \frac{p}{c}, \quad D_2 = \frac{e}{c}p. \tag{2.13}$$

The mechanical work done by the surface traction to the plate per unit volume is

$$W_1 = \frac{1}{2}T_4 S_4 = \frac{p^2}{2c}. \tag{2.14}$$

2.2.2. Open electrodes

In this case the boundary conditions require that

$$\begin{aligned} T_4 &= cu_{,2} + e\phi_{,2} = p, \\ D_2 &= eu_{,2} - \varepsilon\phi_{,2} = 0, \end{aligned} \tag{2.15}$$

which implies that

$$S_4 = u_{,2} = \frac{p}{c(1+k^2)}, \quad E_2 = -\phi_{,2} = -\frac{e}{\varepsilon c(1+k^2)}p. \tag{2.16}$$

The work done to the plate per unit volume is

$$W_2 = \frac{1}{2}T_4 S_4 = \frac{p^2}{2c(1+k^2)}. \tag{2.17}$$

2.2.3. Electromechanical coupling coefficient

From Eqs. (2.14) and (2.17), clearly,

$$W_2 < W_1. \tag{2.18}$$

Graphically this is shown in Fig. 2.3. When the electrodes are open, E_2 is nonzero. The plate appears stiffer due to the presence of E_2. This is the so-called piezoelectric stiffening effect. The electromechanical coupling coefficient for the above thickness-shear deformation of a ceramic plate is

$$\frac{W_1 - W_2}{W_1} = 1 - \frac{1}{1+k^2} = \frac{k^2}{1+k^2} = \bar{k}^2. \tag{2.19}$$

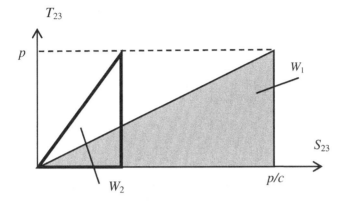

Fig. 2.3. Work done by the surface traction.

2.3. Capacitance of a Plate

Consider the ceramic plate in Fig. 2.4. The surfaces of the plate are electroded. A voltage V is applied across the plate thickness. Two cases of mechanical boundary conditions will be considered.

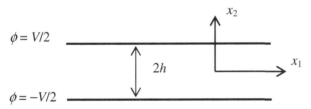

Fig. 2.4. A ceramic plate under a voltage V.

The boundary-value problem is:

$$\nabla^2 u = 0, \quad \nabla^2 \phi = 0, \quad |x_2| < h,$$

$$\phi = \pm V/2, \quad x_2 = \pm h,$$

$$T_{23} = 0, \quad x_2 = \pm h, \quad \text{if the surfaces are traction-free,}$$

$$\text{or} \quad u = 0, \quad x_2 = \pm h, \quad \text{if the surfaces are clamped (fixed).}$$

(2.20)

Equation (2.11) still holds and all the strain, stress, electric field and electric displacement components are constants. In particular,

$$E_2 = -\phi_{,2} = -\frac{V}{2h}.$$

(2.21)

2.3.1. Free surfaces

We have

$$T_4 = cu_{,2} + e\phi_{,2} = 0. \tag{2.22}$$

Then

$$S_4 = u_{,2} = -\frac{e}{c}\frac{V}{2h},$$
$$D_2 = -e\frac{e}{c}\frac{V}{2h} - \varepsilon\frac{V}{2h} = -(1+k^2)\varepsilon\frac{V}{2h}. \tag{2.23}$$

The free charge per unit area on the electrode at $x_2 = h$ is

$$\sigma = -D_2 = (1+k^2)\varepsilon\frac{V}{2h}. \tag{2.24}$$

Hence the capacitance of the plate per unit area is

$$\frac{\sigma}{V} = (1+k^2)\frac{\varepsilon}{2h}. \tag{2.25}$$

Equation (2.25) shows that the effect of piezoelectric coupling enhances the capacitance by a portion of k^2. The electrical energy stored in the plate capacitor per unit area is

$$U_1 = \frac{1}{2}\sigma V = \frac{1}{2}(1+k^2)\varepsilon\frac{V^2}{2h} = \frac{1}{2}\bar{\varepsilon}\frac{V^2}{2h},$$
$$\bar{\varepsilon} = \varepsilon(1+k^2), \tag{2.26}$$

where $\bar{\varepsilon}$ is an effective dielectric constant including piezoelectric effect.

2.3.2. Clamped surfaces

In this case, the strain S_4 is still a constant and u_1 must be a linear function of x_2. The displacement boundary conditions require this linear function to vanish at two points. Hence

$$u = 0, \tag{2.27}$$

which implies that

$$S_4 = u_{,2} = 0, \quad T_4 = e\frac{V}{2h}, \quad D_2 = -\varepsilon\frac{V}{2h}. \tag{2.28}$$

The free charge per unit area on the electrode at $x_2 = h$ is

$$\sigma = -D_2 = \varepsilon\frac{V}{2h}. \tag{2.29}$$

Hence the capacitance of the plate per unit area is

$$\frac{\sigma}{V} = \frac{\varepsilon}{2h}. \tag{2.30}$$

The electric energy stored in the capacitor per unit area is

$$U_2 = \frac{1}{2}\sigma V = \frac{1}{2}\varepsilon\frac{V^2}{2h}. \tag{2.31}$$

2.3.3. Electromechanical coupling coefficient

The electromechanical coupling coefficient for a ceramic plate in thickness-shear is then

$$\frac{U_1 - U_2}{U_1} = \frac{k^2}{1 + k^2} = \bar{k}^2, \tag{2.32}$$

which is the same as Eq. (2.19).

2.4. Capacitance of a Circular Cylindrical Shell

Consider a circular cylindrical ceramic shell with an inner radius R_1 and an outer radius R_2 (see Fig. 2.5). The inner and outer surfaces are electroded, with the electrodes shown by the thick lines in the figure. A voltage V is applied across the thickness. Mechanically the boundary surfaces are either traction-free or fixed. The problem is axisymmetric.

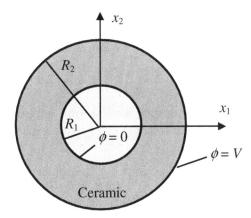

Fig. 2.5. A circular cylindrical ceramic shell as a capacitor.

The boundary-value problem is:

$$\nabla^2 u = 0, \quad \nabla^2 \phi = 0, \quad R_1 < r < R_2,$$
$$\phi = 0, V, \quad r = R_1, R_2,$$
$$T_{rz} = 0, \quad r = R_1, R_2, \quad \text{if the surfaces are traction-free,}$$
$$\text{or} \quad u = 0, \quad r = R_1, R_2, \quad \text{if the surfaces are fixed.} \tag{2.33}$$

The general solution is

$$u = A_1 \ln r + A_2, \quad \phi = B_1 \ln r + B_2, \tag{2.34}$$

where A_1, A_2, B_1 and B_2 are undetermined constants. The stress and electric displacements are

$$T_{rz} = (cA_1 + eB_1)\frac{1}{r}, \quad D_r = (eA_1 - \varepsilon B_1)\frac{1}{r}. \tag{2.35}$$

Consider traction-free surfaces below with

$$(cA_1 + eB_1)\frac{1}{R_1} = 0, \quad (cA_1 + eB_1)\frac{1}{R_2} = 0,$$
$$B_1 \ln R_1 + B_2 = 0, \quad B_1 \ln R_2 + B_2 = V, \tag{2.36}$$

which determines

$$A_1 = -\frac{e}{c}\frac{V}{\ln(R_2 / R_1)},$$
$$B_1 = \frac{V}{\ln(R_2 / R_1)}, \quad B_2 = -\frac{V}{\ln(R_2 / R_1)}\ln R_1. \tag{2.37}$$

Hence

$$\phi = \frac{V}{\ln(R_2 / R_1)}\ln\left(\frac{r}{R_1}\right), \quad u = -\frac{e}{c}\frac{V}{\ln R_2 / R_1}\ln\left(\frac{r}{R_1}\right) + C,$$
$$T_{rz} = 0, \quad D_r = -\bar{\varepsilon}\frac{V}{\ln(R_2 / R_1)}\frac{1}{r}, \quad \bar{\varepsilon} = \varepsilon(1 + k^2), \tag{2.38}$$

where C is an arbitrary constant representing a rigid-body displacement (effectively the A_2 in Eq. (2.34)). The surface charge density on the electrode at $r = R_2$ is given by

$$\sigma = -D_r(r = R_2) = \bar{\varepsilon}\frac{V}{\ln(R_2 / R_1)}\frac{1}{R_2}. \tag{2.39}$$

The capacitance per unit length of the cylinder is

$$C = \frac{2\pi R_2 \sigma}{V} = \frac{2\pi\bar{\varepsilon}}{\ln(R_2 / R_1)}. \tag{2.40}$$

Like in Sec. 2.3.1, Eq. (2.40) shows that the effect of piezoelectric coupling on the capacitance is of the order of k^2 through $\bar{\varepsilon} = \varepsilon(1 + k^2)$ when the surfaces are traction-free.

2.5. A Circular Hole under Axisymmetric Loads

Consider a circular hole of radius R in an unbounded two-dimensional domain (see Fig. 2.6). The hole surface at $r = R$ is electroded, with the electrode shown by the thick line in the figure. On the hole surface we apply a shear stress $T_{rz} = \tau$. We consider the case that the electrode is not connected to other objects. The surface free charge density on the electrode is fixed and is given to be σ. The problem is axisymmetric.

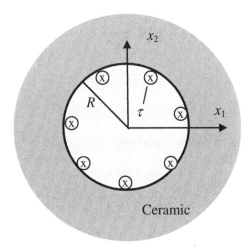

Fig. 2.6. A circular hole under axisymmetric loads.

The boundary-value problem is:

$$\nabla^2 u = 0, \quad \nabla^2 \phi = 0, \quad r > R, \tag{2.41}$$

$$T_{rz} = \tau, \quad D_r = \sigma, \quad r = R,$$
$$T_{rz} \to 0, \quad D_r \to 0, \quad r \to \infty. \tag{2.42}$$

The general solution is

$$u = A_1 \ln r + A_2, \quad \phi = B_1 \ln r + B_2, \tag{2.43}$$

where A_1, A_2, B_1 and B_2 are undetermined constants. A_2 and B_2 represent a rigid-body displacement and a constant in the electric potential and are immaterial to the problem we are considering. The stress and electric displacement are

$$T_{rz} = (cA_1 + eB_1)\frac{1}{r}, \quad D_r = (eA_1 - \varepsilon B_1)\frac{1}{r}, \tag{2.44}$$

which satisfy the boundary conditions at infinity. On the hole surface,

$$(cA_1 + eB_1)\frac{1}{R} = \tau, \quad (eA_1 - \varepsilon B_1)\frac{1}{R} = \sigma, \tag{2.45}$$

which determines

$$A_1 = \frac{1}{c}\left(\tau + \frac{e}{\varepsilon}\sigma\right)R, \quad B_1 = \frac{1}{\bar{\varepsilon}}\left(\frac{e}{c}\tau - \sigma\right)R. \tag{2.46}$$

Hence

$$u = \frac{1}{c}(\tau + \frac{e}{\varepsilon}\sigma)R\ln\frac{r}{R} + C_1,$$

$$\phi = \frac{1}{\bar{\varepsilon}}(-\sigma + \frac{e}{c}\tau)R\ln\frac{r}{R} + C_2, \tag{2.47}$$

$$T_{rz} = \tau\frac{R}{r}, \quad D_r = \sigma\frac{R}{r},$$

where C_1 and C_2 are arbitrary constants.

Now consider the limit of $R \to 0$ and at the same time $\tau \to \infty$ and $\sigma \to \infty$, such that

$$\tau 2\pi R \to F, \quad \sigma 2\pi R \to Q_e. \tag{2.48}$$

Then

$$u = \frac{1}{2\pi c}\left(F + \frac{e}{\varepsilon}Q_e\right)\ln\frac{r}{R} + C_1,$$

$$\phi = \frac{1}{2\pi\bar{\varepsilon}}\left(\frac{e}{c}F - Q_e\right)\ln\frac{r}{R} + C_2, \tag{2.49}$$

$$T_{rz} = \frac{F}{2\pi r}, \quad D_r = \frac{Q_e}{2\pi r}.$$

Equation (2.49) represent the fields of a line force and a line charge at the origin. Mathematically they are the fundamental solutions to the following problems:

$$\overline{c}\nabla^2 u + (-F - \frac{e}{\varepsilon}Q_e)\delta(\mathbf{x}) = 0,$$

$$-\overline{\varepsilon}\nabla^2 \phi = (Q_e - \frac{e}{c}F)\delta(\mathbf{x}),$$

(2.50)

where δ is the Dirac delta function. The stress and electric displacement in Eq. (2.49) are unbounded when $r \to 0$. This is a typical failure of continuum mechanics in problems with a zero characteristic length. Continuum mechanics is valid only when the characteristic length in a problem is much larger than the microstructural characteristic length of matter. Equation (2.49) is valid sufficiently far away from the origin. At a point very close to the origin, the source can no longer be treated as a line source and its size or distribution has to be considered.

2.6. A Circular Hole under Shear

Consider a circular cylindrical hole of radius R in an unbounded ceramic domain. The hole surface is electroded and the electrode is grounded. The hole is under a uniform shear stress τ at $x_2 = \pm\infty$ (see Fig. 2.7). Electrically $x_2 = \pm\infty$ may be shorted or open.

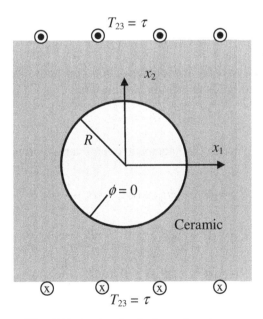

Fig. 2.7. A circular hole under shear.

The boundary-value problem is:

$$\nabla^2 u = 0, \quad \nabla^2 \phi = 0, \quad r > R,$$
$$T_{rz} = 0, \quad \phi = 0, \quad r = R, \tag{2.51}$$
$$u, \phi \to \text{given far fields}, \quad r \to \infty.$$

First we determine the fields in polar coordinates when r is very large. For mechanical fields we have

$$T_{23} = cu_{,2} + e\phi_{,2} = \tau. \tag{2.52}$$

We consider the case when $x_2 = \pm\infty$ are electrically open, such that for large $|x_2|$,

$$D_2 = eu_{,2} - \varepsilon\phi_{,2} = 0. \tag{2.53}$$

Based on Eqs. (2.52) and (2.53), we consider the case when the far fields are given to be

$$u = \frac{\tau}{c}x_2 + C_1 = \frac{\tau}{c}r\sin\theta,$$
$$\phi = \frac{e}{\varepsilon}\frac{\tau}{c}x_2 + C_2 = \frac{e}{\varepsilon}\frac{\tau}{c}r\sin\theta, \tag{2.54}$$

where C_1 and C_2 are arbitrary constants and have been set to zero. In view of the above far-field solution and the general solution in Eqs. (1.53) and (1.54), we look for solutions in the following form:

$$u = (A_1 r + \frac{A_2}{r})\sin\theta, \quad \phi = (B_1 r + \frac{B_2}{r})\sin\theta, \tag{2.55}$$

where A_1, A_2, B_1 and B_2 are undetermined constants. For Eq. (2.55) to match the applied fields at infinity, we must have

$$A_1 = \frac{\tau}{c}, \quad B_1 = \frac{e\tau}{\varepsilon c}. \tag{2.56}$$

The stress and electric displacement components corresponding to Eq. (2.55) are

$$T_{rz} = [(cA_1 + eB_1) - (cA_2 + eB_2)\frac{1}{r^2}]\sin\theta,$$
$$T_{\theta z} = [(cA_1 + eB_1) + (cA_2 + eB_2)\frac{1}{r^2}]\cos\theta,$$
$$\tag{2.57}$$
$$D_r = [(eA_1 - \varepsilon B_1) - (eA_2 - \varepsilon B_2)\frac{1}{r^2}]\sin\theta,$$
$$D_\theta = [(eA_1 - \varepsilon B_1) + (eA_2 - \varepsilon B_2)\frac{1}{r^2}]\cos\theta.$$

At $r = R$ the boundary conditions require that

$$T_{rz}(R) = [(cA_1 + eB_1) - (cA_2 + eB_2)\frac{1}{R^2}]\sin\theta = 0,$$

$$\phi(R) = (B_1 R + \frac{B_2}{R})\sin\theta = 0,$$

(2.58)

which imply that

$$A_2 = \frac{1}{\bar{c}}(1 + 2k^2)R^2\tau, \quad B_2 = -\frac{e}{c\bar{\varepsilon}}R^2\tau.$$

(2.59)

The displacement and potential fields are

$$\phi = \frac{e\tau}{c\bar{\varepsilon}}(r - \frac{R^2}{r})\sin\theta,$$

$$u = \frac{\tau}{\bar{c}}[r + (1 + 2k^2)\frac{R^2}{r}]\sin\theta.$$

(2.60)

2.7. A Circular Cylinder in an Electric Field

Consider an infinite circular cylinder of radius R in a uniform electric field $\mathbf{E}^0 = E^0\mathbf{i}_1$ (see Fig. 2.8). The problem is symmetric about $x_2 = 0$ and is antisymmetric about $x_1 = 0$.

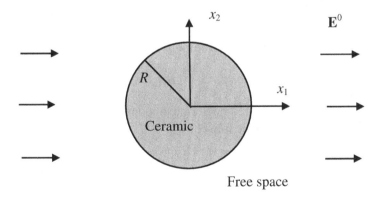

Fig. 2.8. A circular cylinder in a uniform electric field.

The boundary-value problem is

$$\nabla^2 u = 0, \quad \nabla^2 \phi = 0, \quad r < R,$$

$$\nabla^2 \phi = 0, \quad r > R,$$

u and ϕ are bounded, $\quad r = 0,$

$$T_{rz}(r = R) = 0,$$

$$\phi(r = R^-) = \phi(r = R^+), \quad D_r(r = R^-) = D_r(r = R^+),$$

$$\mathbf{E} \to \mathbf{E}^0, \quad r \to \infty.$$

(2.61)

For large r, the fields are given to be

$$\phi = -E^0 x_1 = -E^0 r \cos\theta,$$

$$E_r = E_0 \cos\theta, \quad E_\theta = -E_0 \sin\theta.$$

(2.62)

In view of the far-field solution and the general solution in Eqs. (1.53) and (1.54), we look for solutions in the following form for the field in the free space:

$$\phi = (C_1 r + C_2 \frac{1}{r})\cos\theta,$$

$$E_r = -(C_1 - C_2 \frac{1}{r^2})\cos\theta, \quad E_\theta = (C_1 + C_2 \frac{1}{r^2})\sin\theta,$$

(2.63)

where C_1 and C_2 are undetermined constants. For the electric field in Eq. (2.63) to be equal to the applied field for large r, we must have

$$C_1 = -E^0.$$

(2.64)

Inside the cylinder we look for solutions in the following form:

$$u = (A_1 r + \frac{A_2}{r})\cos\theta, \quad \phi = (B_1 r + \frac{B_2}{r})\cos\theta,$$

(2.65)

where A_1, A_2, B_1 and B_2 are undetermined constants. For the boundedness of u and ϕ at the origin, we must have

$$A_2 = 0, \quad B_2 = 0,$$

(2.66)

and hence

$$u = A_1 r \cos\theta = A_1 x_1, \quad \phi = B_1 r \cos\theta = B_1 x_1.$$

(2.67)

The stress and electric displacement fields in the cylinder are

$$T_{rz} = (cA_1 + eB_1)\cos\theta, \quad T_{\theta z} = -(cA_1 + eB_1)\sin\theta,$$

$$D_r = (eA_1 - \varepsilon B_1)\cos\theta, \quad D_\theta = -(eA_1 - \varepsilon B_1)\sin\theta.$$

(2.68)

We note that Eq. (2.67) leads to uniform strain, stress, electric field and electric displacement inside the cylinder. At $r = R$, the traction-free condition and the continuity of ϕ and D_r require that

$$T_{rz}(r = R) = (cA_1 + eB_1)\cos\theta = 0,$$

$$\phi(r = R^-) = RB_1\cos\theta$$

$$= (-E_0 R + \frac{C_2}{R})\cos\theta = \phi(r = R^+), \tag{2.69}$$

$$D_r(r = R^-) = (eA_1 - \varepsilon B_1)\cos\theta$$

$$= \varepsilon_0(E_0 + \frac{C_2}{R^2})\cos\theta = D_r(r = R^+),$$

which determines

$$C_2 = \frac{\bar{\varepsilon} - \varepsilon_0}{\bar{\varepsilon} + \varepsilon_0}E^0 R^2, \quad B_1 = -\frac{2\varepsilon_0}{\bar{\varepsilon} + \varepsilon_0}E^0, \quad A_1 = \frac{e}{c}\frac{2\varepsilon_0}{\bar{\varepsilon} + \varepsilon_0}E^0. \tag{2.70}$$

Then the electric potential and field in the free space are given by

$$\phi = (-r + \frac{\bar{\varepsilon} - \varepsilon_0}{\bar{\varepsilon} + \varepsilon_0}\frac{R^2}{r})E^0\cos\theta,$$

$$E_r = (1 + \frac{\bar{\varepsilon} - \varepsilon_0}{\bar{\varepsilon} + \varepsilon_0}\frac{R^2}{r^2})E^0\cos\theta, \tag{2.71}$$

$$E_\theta = (-1 + \frac{\bar{\varepsilon} - \varepsilon_0}{\bar{\varepsilon} + \varepsilon_0}\frac{R^2}{r^2})E_0\sin\theta,$$

and the fields inside the cylinder are

$$\phi = -\frac{2\varepsilon_0}{\bar{\varepsilon} + \varepsilon}E^0 r\cos\theta, \quad u = \frac{e}{c}\frac{2\varepsilon_0}{\bar{\varepsilon} + \varepsilon_0}E^0 r\cos\theta,$$

$$T_{rz} = 0, \quad T_{\theta z} = 0, \tag{2.72}$$

$$D_r = \bar{\varepsilon}\frac{2\varepsilon_0}{\bar{\varepsilon} + \varepsilon_0}E^0\cos\theta, \quad D_\theta = -\bar{\varepsilon}\frac{2\varepsilon_0}{\bar{\varepsilon} + \varepsilon_0}E^0\sin\theta.$$

2.8. A Screw Dislocation

Consider a screw dislocation at $\theta = \pi$ in a polar coordinate system (see Fig. 2.9).

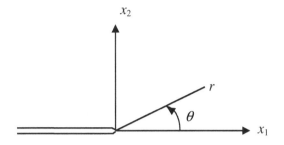

Fig. 2.9. A screw dislocation.

The boundary-value problem is:

$$\nabla^2 u = 0, \quad r > 0, \quad -\pi < \theta < \pi,$$
$$\nabla^2 \phi = 0, \quad r > 0, \quad -\pi < \theta < \pi,$$
$$u(r,\pi) - u(r,-\pi) = \delta,$$
$$\phi(r,\pi) - \phi(r,-\pi) = V.$$

(2.73)

We look for a solution in the following form:

$$u(r,\theta) = u(\theta), \quad \phi(r,\theta) = \phi(\theta).$$

(2.74)

Substitute Eq. (2.74) into the Laplace equations as follows:

$$\nabla^2 u = \frac{\partial^2 u}{\partial r^2} + \frac{1}{r}\frac{\partial u}{\partial r} + \frac{1}{r^2}\frac{\partial^2 u}{\partial \theta^2} = \frac{1}{r^2}\frac{\partial^2 u}{\partial \theta^2} = 0,$$
$$\nabla^2 \phi = \frac{1}{r^2}\frac{\partial^2 \phi}{\partial \theta^2} = 0.$$

(2.75)

The relevant solution is

$$u = A_1\theta + A_2, \quad \phi = B_1\theta + B_2,$$

(2.76)

where A_1, A_2, B_1 and B_2 are undetermined constants. From the boundary conditions in Eq. (2.73),

$$A_1 = \frac{\delta}{2\pi}, \quad B_1 = \frac{V}{2\pi}.$$

(2.77)

Hence

$$u = \frac{\delta}{2\pi}\theta + A_2, \quad \phi = \frac{V}{2\pi}\theta + B_2,$$

(2.78)

and

$$2S_{rz} = 0, \quad 2S_{\theta z} = \frac{1}{r}\frac{\delta}{2\pi},$$

$$E_r = 0, \quad E_\theta = -\frac{1}{r}\frac{V}{2\pi},$$

$$T_{rz} = cu_{,r} + e\phi_{,r} = 0,$$

$$T_{\theta z} = c\frac{1}{r}u_{,\theta} + e\frac{1}{r}\phi_{,\theta} = c\frac{\delta}{2\pi r} + e\frac{V}{2\pi r},$$

$$D_r = eu_{,r} - \varepsilon\phi_{,r} = 0,$$

$$D_\theta = e\frac{1}{r}u_{,\theta} - \varepsilon\frac{1}{r}\phi_{,\theta} = e\frac{\delta}{2\pi r} - \varepsilon\frac{V}{2\pi r}.$$

(2.79)

The singularity of the fields at the origin is another indication of the failure of continuum mechanics in problems with a zero characteristic length. The solution is valid sufficiently far away from the origin only.

2.9. A Crack

Consider a semi-infinite crack at $\theta = \pi$ in a polar coordinate system as shown in Fig. 2.10.

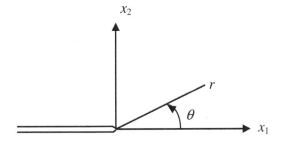

Fig. 2.10. A semi-infinite crack.

The boundary-value problem is:

$$\nabla^2 u = 0, \quad r > 0, \quad -\pi < \theta < \pi,$$

$$\nabla^2 \phi = 0, \quad r > 0, \quad -\pi < \theta < \pi,$$

$$T_{\theta z}(r,-\pi) = 0, \quad T_{\theta z}(r,\pi) = 0,$$

$$D_\theta(r,-\pi) = 0, \quad D_\theta(r,\pi) = 0.$$

(2.80)

Physically, the above boundary conditions indicate traction-free and unelectroded crack faces. We look for a solution in the following form:

$$u(r,\theta) = u(r)\sin\frac{\theta}{2}, \quad \phi(r,\theta) = \phi(r)\sin\frac{\theta}{2}. \tag{2.81}$$

Substitute Eq. (2.81) into the Laplace equations as follows:

$$\nabla^2 u = \frac{\partial^2 u}{\partial r^2} + \frac{1}{r}\frac{\partial u}{\partial r} + \frac{1}{r^2}\frac{\partial^2 u}{\partial \theta^2}$$

$$= \left(\frac{\partial^2 u}{\partial r^2} + \frac{1}{r}\frac{\partial u}{\partial r} - \frac{1}{4}\frac{u}{r^2}\right)\sin\frac{\theta}{2} = 0, \tag{2.82}$$

$$\nabla^2 \phi = \left(\frac{\partial^2 \phi}{\partial r^2} + \frac{1}{r}\frac{\partial \phi}{\partial r} - \frac{1}{4}\frac{\phi}{r^2}\right)\sin\frac{\theta}{2} = 0.$$

The relevant solution is

$$u = A\sqrt{r}\sin\frac{\theta}{2}, \quad \phi = B\sqrt{r}\sin\frac{\theta}{2}, \tag{2.83}$$

where A and B are undetermined constants. The corresponding electromechanical fields are

$$2S_{rz} = \frac{A}{2\sqrt{r}}\sin\frac{\theta}{2}, \quad 2S_{\theta z} = \frac{A}{2\sqrt{r}}\cos\frac{\theta}{2},$$

$$E_r = -\frac{B}{2\sqrt{r}}\sin\frac{\theta}{2}, \quad E_\theta = -\frac{B}{2\sqrt{r}}\cos\frac{\theta}{2},$$

$$T_{rz} = \frac{cA+eB}{2\sqrt{r}}\sin\frac{\theta}{2}, \quad T_{\theta z} = \frac{cA+eB}{2\sqrt{r}}\cos\frac{\theta}{2}, \tag{2.84}$$

$$D_r = \frac{eA-\varepsilon B}{2\sqrt{r}}\sin\frac{\theta}{2}, \quad D_\theta = \frac{eA-\varepsilon B}{2\sqrt{r}}\cos\frac{\theta}{2}.$$

The boundary conditions in Eq. (2.80) are already satisfied and impose no more restrictions on the above fields. Note the singularity of the fields at the origin. The singularity is again an indication of the failure of continuum mechanics in problems with a zero characteristic length. The solution is valid sufficiently far away from the crack tip only.

Chapter 3
Simple Dynamic Problems

This chapter is on a few simple dynamic problems. This includes the propagation of plane waves in unbounded regions, their reflection and refraction at an interface, and their scattering by a circular cylinder. Fields associated with a moving dislocation and a moving crack are also presented.

3.1. Plane Wave Propagation

First, consider waves in a region without a boundary. The waves are governed by the following equations only without boundary conditions:

$$v_T^2 \nabla^2 u = \ddot{u},$$
$$\nabla^2 \psi = 0. \tag{3.1}$$

We are interested in the following plane waves:

$$u = Af(\mathbf{n} \cdot \mathbf{x} - vt),$$
$$\psi = Bf(\mathbf{n} \cdot \mathbf{x} - vt), \tag{3.2}$$

where A, B, $\mathbf{n}$ and v are constants, and f is an arbitrary function. $\mathbf{n} \cdot \mathbf{x} - vt$ is the phase of the wave. v is the phase velocity. $\mathbf{n} \cdot \mathbf{x} - vt = $ constant determines a wave front which is a plane with a normal $\mathbf{n}$. Differentiating Eq. (3.2), we obtain

$$u_{,l} = Af'n_l, \quad u_{,ll} = Af''n_l n_l = Af'', \quad \ddot{u} = Af''v^2,$$
$$\psi_{,k} = Bf'n_k, \quad \psi_{,kk} = Bf''n_k n_k = Bf''. \tag{3.3}$$

Substitution of Eq. (3.3) into Eq. (3.2) yields the following linear equations for A and B:

$$v_T^2 Af'' = v^2 Af'',$$
$$Bf'' = 0. \tag{3.4}$$

For nontrivial solutions of A and/or B, the determinant of the coefficient matrix of Eq. (3.4) has to vanish:

35

$$\begin{vmatrix} v_T^2 f'' - v^2 f'' & 0 \\ 0 & f'' \end{vmatrix} = 0 . \tag{3.5}$$

We are interested in the case when $f'' \neq 0$. Equation (3.5) implies that $B = 0$ and

$$v = v_T . \tag{3.6}$$

Then the plane waves take the form

$$u = Af(\mathbf{n} \cdot \mathbf{x} - v_T t) = Af(n_1 x_1 + n_2 x_2 - v_T t),$$
$$\psi = 0. \tag{3.7}$$

For sinusoidal waves, f is a sine or cosine function. In this case, we use complex functions with the real parts representing the physical fields of interest and write

$$u = A\exp[i\xi(n_1 x_1 + n_2 x_2 - v_T t)],$$
$$\psi = 0. \tag{3.8}$$

The corresponding electric potential and electric displacement are

$$\phi = \frac{e}{\varepsilon} u, \quad D_1 = D_2 = 0 . \tag{3.9}$$

3.2. Reflection at a Boundary

In a semi-infinite medium, a wave solution needs to satisfy boundary conditions in addition to the differential equations in Eq. (3.1). As an example, consider a wave incident upon a plane boundary as shown in Fig. 3.1. The boundary is traction-free and is unelectroded. The electric field in the free space is neglected.

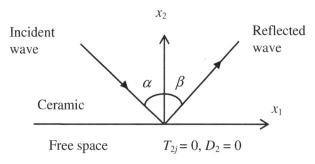

Fig. 3.1. Incident and reflected waves at a plane boundary.

The incident and reflected waves together must satisfy the following equations and boundary conditions:

$$v_T^2 \nabla^2 u = \ddot{u}, \quad \nabla^2 \psi = 0, \quad x_2 > 0,$$
$$T_{2j} = 0, \quad D_2 = 0, \quad x_2 = 0. \tag{3.10}$$

The incident wave can be written as

$$u = A \exp[i\xi(x_1 \sin\alpha - x_2 \cos\alpha - v_T t)],$$
$$\psi = 0, \tag{3.11}$$

where, comparing Eq. (3.11) with Eq. (3.8), we identify

$$n_1 = \sin\alpha, \quad n_2 = -\cos\alpha. \tag{3.12}$$

The incident wave represented by Eq. (3.11) is considered known. It already satisfies the differential equations in Eq. (3.10). The electric displacement and the stress component needed for the boundary conditions are

$$D_2 = 0,$$
$$T_{23} = \bar{c} u_{,2} + e \psi_{,2} \tag{3.13}$$
$$= \bar{c}(-i\xi \cos\alpha) A \exp[i\xi(x_1 \sin\alpha - x_2 \cos\alpha - v_T t)].$$

Similarly, we write the reflected wave as

$$u = B \exp[i\xi(x_1 \sin\beta + x_2 \cos\beta - v_T t)],$$
$$\psi = 0, \tag{3.14}$$

which satisfies the differential equations in Eq. (3.10) and is considered unknown. For boundary conditions, we need

$$D_2 = 0,$$
$$T_{23} = \bar{c}(i\xi \cos\beta) B \exp[i\xi(x_1 \sin\beta + x_2 \cos\beta - v_T t)]. \tag{3.15}$$

The incident and reflected waves already satisfy the governing equations individually and so does their sum. At the boundary, the sum of the incident and reflected waves together has to satisfy the boundary conditions in Eq. (3.10). $D_2 = 0$ is trivially satisfied. We are left with

$$\bar{c}(-i\xi \cos\alpha) A \exp[i\xi(x_1 \sin\alpha - v_T t)]$$
$$+ \bar{c}(i\xi \cos\beta) B \exp[i\xi(x_1 \sin\beta - v_T t)] = 0, \tag{3.16}$$

for any x_1 and any t. This implies that

$$\beta = \alpha, \quad B = A, \tag{3.17}$$

and thus determines the reflected wave. At the boundary surface, the total traction is zero. The total displacement is

$$u = A\exp[i\xi(x_1\sin\alpha - v_T t)] + B\exp[i\xi(x_1\sin\beta - v_T t)]$$
$$= 2A\exp[i\xi(x_1\sin\alpha - v_T t)]. \tag{3.18}$$

Thus at a traction-free boundary the total displacement is twice that of the incident wave.

3.3. Reflection and Refraction at an Interface

At the interface between two semi-infinite media, an incident wave is reflected and refracted. As an example, consider two semi-infinite spaces as shown in Fig. 3.2.

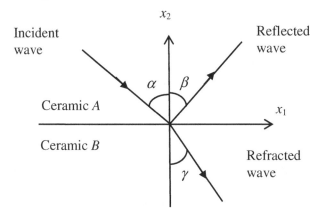

Fig. 3.2. Incident, reflected and refracted waves at an interface.

The incident and reflected waves together must satisfy the following equations:

$$\overline{c}_A\nabla^2 u = \rho_A\ddot{u}, \quad \nabla^2\psi = 0, \quad x_2 > 0. \tag{3.19}$$

The refracted wave must satisfy

$$\overline{c}_B\nabla^2 u = \rho_B\ddot{u}, \quad \nabla^2\psi = 0, \quad x_2 < 0. \tag{3.20}$$

The interface continuity conditions are (assuming $e_A/\varepsilon_A = e_B/\varepsilon_B$)

$$u(x_2 = 0^+) = u(x_2 = 0^-),$$
$$T_{23}(x_2 = 0^+) = T_{23}(x_2 = 0^-),$$
$$\psi(x_2 = 0^+) = \psi(x_2 = 0^-),$$
$$D_2(x_2 = 0^+) = D_2(x_2 = 0^-).$$
(3.21)

The incident wave can be written as

$$u = A\exp[i\xi_A(x_1 \sin\alpha - x_2 \cos\alpha - v_A t)],$$

$$\psi = 0, \quad v_A^2 = \frac{\bar{c}_A}{\rho_A}, \quad D_2 = 0,$$
(3.22)

$$T_{23} = \bar{c}_A(-i\xi_A \cos\alpha)A\exp[i\xi_A(x_1 \sin\alpha - x_2 \cos\alpha - v_A t)],$$

which is considered known. The reflected wave can be written as

$$u = B\exp[i\xi_A(x_1 \sin\beta + x_2 \cos\beta - v_A t)],$$

$$\psi = 0, \quad D_2 = 0,$$
(3.23)

$$T_{23} = \bar{c}_A(i\xi_A \cos\beta)B\exp[i\xi_A(x_1 \sin\beta + x_2 \cos\beta - v_A t)],$$

which is considered unknown. We write the refracted wave as

$$u = C\exp[i\xi_B(x_1 \sin\gamma - x_2 \cos\gamma - v_B t)],$$

$$\psi = 0, \quad v_B^2 = \frac{\bar{c}_B}{\rho_B}, \quad D_2 = 0,$$
(3.24)

$$T_{23} = \bar{c}_B(-i\xi_B \cos\gamma)C\exp[i\xi_B(x_1 \sin\gamma - x_2 \cos\gamma - v_B t)],$$

which is also unknown.

Equations (3.19), (3.20), and (3.21)$_{3,4}$ are already satisfied. From Eq. (3.21)$_{1,2}$ we have

$$
\begin{aligned}
&A\exp[i\xi_A(x_1 \sin\alpha - v_A t)] + B\exp[i\xi_A(x_1 \sin\beta - v_A t)] \\
&\quad = C\exp[i\xi_B(x_1 \sin\gamma - v_B t)], \\
&\bar{c}_A(-i\xi_A \cos\alpha)A\exp[i\xi_A(x_1 \sin\alpha - v_A t)] \\
&\quad + \bar{c}_A(i\xi_A \cos\beta)B\exp[i\xi_A(x_1 \sin\beta - v_A t)] \\
&\quad = \bar{c}_B(-i\xi_B \cos\gamma)C\exp[i\xi_B(x_1 \sin\gamma - v_B t)].
\end{aligned}
$$
(3.25)

Equation (3.25) can be satisfied if the following two sets of relations are true:

$$\exp[i\xi_A(x_1 \sin\alpha - v_A t)]$$
$$= \exp[i\xi_A(x_1 \sin\beta - v_A t)] \tag{3.26}$$
$$= \exp[i\xi_B(x_1 \sin\gamma - v_B t)],$$

and

$$A + B = C,$$
$$\bar{c}_A \xi_A A \cos\alpha - \bar{c}_A \xi_A B \cos\beta = \bar{c}_B \xi_B C \cos\gamma. \tag{3.27}$$

Next we analyze Eqs. (3.26) and (3.27) separately. For Eq. (3.26) to be true for all t and all x_1, we have

$$\xi_A v_A = \xi_B v_B,$$
$$\xi_A \sin\alpha = \xi_A \sin\beta = \xi_B \sin\gamma. \tag{3.28}$$

Hence ξ_B is determined from

$$\frac{\xi_B}{\xi_A} = \frac{v_A}{v_B} = \sqrt{\frac{\bar{c}_A \rho_B}{\rho_A \bar{c}_B}}, \tag{3.29}$$

and β and γ are given by

$$\beta = \alpha, \quad \frac{\sin\gamma}{\sin\alpha} = \frac{\xi_A}{\xi_B} = \frac{v_B}{v_A}, \tag{3.30}$$

which is Snell's law. Then from Eq. (3.27) we solve for B and C:

$$\frac{B}{A} = \frac{\rho_A \sin 2\alpha - \rho_B \sin 2\gamma}{\rho_A \sin 2\alpha + \rho_B \sin 2\gamma},$$
$$\frac{C}{A} = \frac{2\rho_A \sin 2\alpha}{\rho_A \sin 2\alpha + \rho_B \sin 2\gamma}. \tag{3.31}$$

Thus the reflected and refracted waves are fully determined. As a special case consider normal incidence with $\alpha = 0$. From Eq. (3.30) $\beta = \gamma = 0$. Equation (3.31) implies that

$$\frac{B}{A} = \frac{\rho_A v_A - \rho_B v_B}{\rho_A v_A + \rho_B v_B},$$
$$\frac{C}{A} = \frac{2\rho_A v_A}{\rho_A v_A + \rho_B v_B}, \tag{3.32}$$

where $\rho_A v_A$ or $\rho_B v_B$ is the acoustic impedance. When $\rho_A v_A = \rho_B v_B$, the incident wave does not feel the interface and is not reflected, and total transmission occurs.

3.4. Scattering by a Circular Cylinder

In this section we examine the scattering of plane waves by a circular cylinder (see Fig. 3.3). To be simple let the cylinder be a rigid conductor which is grounded.

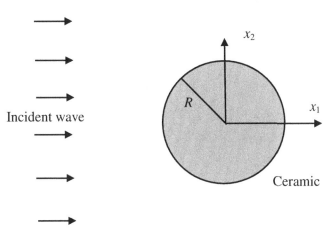

Fig. 3.3. Plane waves incident upon a circular cylinder.

The incident and scattered waves together must satisfy the following equations and boundary conditions:

$$v_T^2 \nabla^2 u = \ddot{u}, \quad \nabla^2 \psi = 0, \quad r > R,$$
$$u = 0, \quad \psi = 0, \quad r = R.$$
(3.33)

Denoting the incident wave by

$$u^I = A \exp[i\xi(x_1 - v_T t)],$$
$$\psi^I = 0,$$
(3.34)

which is considered known. The unknown scattered wave u^S and ψ^S satisfies

$$v_T^2 \nabla^2 u^S = \ddot{u}^S, \quad \nabla^2 \psi^S = 0, \quad r > R,$$
$$u^S + A \exp[i\xi(x_1 - v_T t)] = 0, \quad \psi^S = 0, \quad r = R.$$
(3.35)

In addition, the scattered wave must be outgoing at infinity (radiation condition). Clearly $\psi^S \equiv 0$. u^S has a time dependence of $\exp[i\xi(-v_T t)]$ which is dropped for convenience. Then

$$\nabla^2 u^S + \xi^2 u^S = 0, \quad r > R,$$
$$u^S + A\exp(i\xi x_1) = 0, \quad r = R. \tag{3.36}$$

In polar coordinates Eq. $(3.36)_1$ takes the following form:

$$\frac{\partial^2 u^S}{\partial r^2} + \frac{1}{r}\frac{\partial u^S}{\partial r} + \frac{1}{r^2}\frac{\partial^2 u^S}{\partial \theta^2} + \xi^2 u^S = 0. \tag{3.37}$$

Let

$$u^S(r,\theta) = u^S(r)\cos n\theta, \quad n = 0, 1, 2,\dots. \tag{3.38}$$

Then Eq. (3.37) becomes

$$\frac{\partial^2 u^S}{\partial r^2} + \frac{1}{r}\frac{\partial u^S}{\partial r} + \left(\xi^2 - \frac{n^2}{r^2}\right)u^S = 0, \tag{3.39}$$

or

$$\frac{\partial^2 u^S}{\xi^2 \partial r^2} + \frac{1}{\xi^2 r}\frac{\partial u^S}{\partial r} + \left(1 - \frac{n^2}{\xi^2 r^2}\right)u^S = 0. \tag{3.40}$$

Equation (3.40) is Bessel's equation of order n. For scattered waves diverging toward $r = \infty$, we have

$$u^S = \sum_{n=0}^{\infty} C_n H_n^{(1)}(\xi r)\cos n\theta, \tag{3.41}$$

where C_n are undetermined constants. $H_n^{(1)}$ is the Hankel function of the first kind of order n. Substituting Eq. (3.41) into the boundary condition in Eq. $(3.36)_2$, we arrive at the following equation that determines C_n:

$$\sum_{n=0}^{\infty} C_n H_n^{(1)}(\xi R)\cos n\theta + A\exp(i\xi R\cos\theta) = 0. \tag{3.42}$$

Equation (3.42) can be manipulated further analytically but it will not be pursued here.

3.5. A Moving Dislocation

Consider a screw dislocation moving at a constant speed V along the x_1 axis [10]. Figure 3.4 shows the dislocation when $t = 0$.

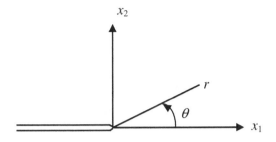

Fig. 3.4. A moving dislocation at $t = 0$.

The fields associated with the dislocation are governed by

$$v_T^2 \nabla^2 u = \ddot{u}, \quad \nabla^2 \psi = 0. \tag{3.43}$$

We choose a moving coordinate system (x,y) to follow the dislocation as follows:

$$x = x_1 - Vt, \quad y = x_2. \tag{3.44}$$

In the moving frame (x,y), Eq. (3.43) becomes

$$\beta^2 \frac{\partial^2 u}{\partial x^2} + \frac{\partial^2 u}{\partial y^2} = 0, \quad \frac{\partial^2 \psi}{\partial x^2} + \frac{\partial^2 \psi}{\partial y^2} = 0, \tag{3.45}$$

where

$$\beta^2 = 1 - \frac{V^2}{v_T^2}. \tag{3.46}$$

Consider Eq. (3.45)$_1$ first. Introduce a new coordinate system by

$$\xi = x, \quad \eta = \beta y. \tag{3.47}$$

Then Eq. (3.45)$_1$ can be written as

$$\frac{\partial^2 u}{\partial \xi^2} + \frac{\partial^2 u}{\partial \eta^2} = 0. \tag{3.48}$$

In polar coordinates defined by

$$\xi = r\cos\theta, \quad \eta = r\sin\theta, \tag{3.49}$$

Equation (3.48) takes the following form:

$$\frac{\partial^2 u}{\partial r^2} + \frac{1}{r}\frac{\partial u}{\partial r} + \frac{1}{r^2}\frac{\partial^2 u}{\partial \theta^2} = 0. \tag{3.50}$$

For a screw dislocation we are interested in solutions to Eq. (3.50) that are r-independent. Denoting the displacement discontinuity at $\theta = \pi$ by b, a constant, we have the displacement field of the dislocation as

$$u = \frac{b}{2\pi}\theta = \frac{b}{2\pi}\tan^{-1}\frac{\eta}{\xi} = \frac{b}{2\pi}\tan^{-1}\frac{\beta y}{x} = \frac{b}{2\pi}\tan^{-1}\frac{\beta x_2}{x_1 - Vt}. \tag{3.51}$$

If we understand a dislocation as a pure mechanical concept as described by Eq. (3.51), the electric field associated with a dislocation is not unique, depending on how we prescribe electrical boundary/continuity conditions at the dislocation. One simple situation is that we also allow a discontinuity of ψ at the dislocation. Then, similarly to Eq. (3.51), we have the following solution to Eq. (3.45)$_2$:

$$\psi = \frac{h}{2\pi}\tan^{-1}\frac{y}{x} = \frac{h}{2\pi}\tan^{-1}\frac{x_2}{x_1 - Vt}, \tag{3.52}$$

where h is a constant.

3.6. A Moving Crack

Consider a semi-infinite crack moving at a constant speed V along the x_1 axis. At $t = 0$ the crack occupies the negative x_1 axis (see Fig. 3.5).

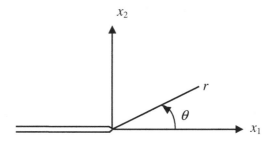

Fig. 3.5. A semi-infinite moving crack at $t = 0$.

The fields associated with the crack are governed by

$$v_T^2\nabla^2 u = \ddot{u}, \quad \nabla^2\psi = 0. \tag{3.53}$$

Similar to the previous section, we first choose a moving coordinate system (x, y) to follow the crack tip:

$$x = x_1 - Vt, \quad y = x_2, \tag{3.54}$$

which changes Eq. (3.53) to

$$\beta^2 \frac{\partial^2 u}{\partial x^2} + \frac{\partial^2 u}{\partial y^2} = 0, \quad \frac{\partial^2 \psi}{\partial x^2} + \frac{\partial^2 \psi}{\partial y^2} = 0, \tag{3.55}$$

where

$$\beta^2 = 1 - \frac{V^2}{v_T^2}. \tag{3.56}$$

Then, for Eq. $(3.55)_1$, like in the previous section, we introduce two new coordinate systems sequentially by

$$\xi = x, \quad \eta = \beta y, \tag{3.57}$$

and

$$\xi = r \cos \theta, \quad \eta = r \sin \theta, \tag{3.58}$$

which changes Eq. $(3.55)_1$ to the following form:

$$\frac{\partial^2 u}{\partial r^2} + \frac{1}{r} \frac{\partial u}{\partial r} + \frac{1}{r^2} \frac{\partial^2 u}{\partial \theta^2} = 0. \tag{3.59}$$

For a crack, the solution of interest to Eq. (3.59) is:

$$u = A\sqrt{r} \sin \frac{\theta}{2}, \quad -\pi \le \theta \le \pi, \tag{3.60}$$

where A is a constant.

For Eq. $(3.55)_2$, in polar coordinates defined by

$$x = R \cos \Theta, \quad y = R \sin \Theta, \tag{3.61}$$

a solution of interest is given by

$$\psi = B\sqrt{R} \sin \frac{\Theta}{2}, \quad -\pi \le \Theta \le \pi, \tag{3.62}$$

where B is a constant.

Chapter 4
Surface and Interface Waves

This chapter is on waves propagating near the surface of a half-space or the interface between two half-spaces. Surface waves in piezoelectrics have been used extensively to make SAW devices. In addition to Rayleigh waves, SH surface waves can also propagate in certain piezoelectrics. These waves are truly piezoelectric in the sense that the existence of these waves relies on piezoelectric coupling. They do not have elastic counterparts.

4.1. Surface Waves over a Half-space

In this section, we discuss the well-known Bleustein–Gulyaev wave [8,11]. Some Japanese researchers also contributed to the discovery of the wave [12]. Consider a ceramic half-space as shown in Fig. 4.1. The surface at $x_2 = 0$ is traction-free and may be electroded or unelectroded.

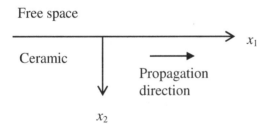

Fig. 4.1. A ceramic half-space.

The governing equations for the ceramic half-space are

$$v_T^2 \nabla^2 u = \ddot{u}, \quad \nabla^2 \psi = 0, \quad \phi = \psi + \frac{e}{\varepsilon} u, \quad x_2 > 0. \quad (4.1)$$

Consider the possibility of solutions in the following form:

$$u = A\exp(-\xi_2 x_2)\cos(\xi_1 x_1 - \omega t),$$

$$\psi = B\exp(-\xi_1 x_2)\cos(\xi_1 x_1 - \omega t),$$

$$T_{23} = -[\bar{c}A\xi_2\exp(-\xi_2 x_2) + eB\xi_1\exp(-\xi_1 x_2)]\cos(\xi_1 x_1 - \omega t), \quad (4.2)$$

$$\phi = [B\exp(-\xi_1 x_2) + \frac{e}{\varepsilon}A\exp(-\xi_2 x_2)]\cos(\xi_1 x_1 - \omega t),$$

$$D_2 = \varepsilon\xi_1 B\exp(-\xi_1 x_2)\cos(\xi_1 x_1 - \omega t),$$

where A and B are undetermined constants, and ξ_2 should be positive for decaying behavior away from the surface. Equation $(4.2)_2$ already satisfies Eq. $(4.1)_2$. For Eq. $(4.2)_1$ to satisfy Eq. $(4.1)_1$ we must have

$$\bar{c}(\xi_1^2 - \xi_2^2) = \rho\omega^2, \quad (4.3)$$

which determines

$$\xi_2^2 = \xi_1^2 - \frac{\rho\omega^2}{\bar{c}} = \xi_1^2(1 - \frac{v^2}{v_T^2}) > 0, \quad v^2 = \frac{\omega^2}{\xi_1^2}. \quad (4.4)$$

4.1.1. A half-space with an electroded surface

First we consider the case when the surface of the half-space is electroded and the electrode is grounded. The corresponding boundary conditions are

$$T_{23} = 0, \quad x_2 = 0,$$

$$\phi = 0, \quad x_2 = 0, \quad (4.5)$$

$$u, \phi \to 0, \quad x_2 \to +\infty,$$

or, in terms of u and ψ,

$$T_{23} = \bar{c}u_{,2} + e\psi_{,2} = 0, \quad x_2 = 0,$$

$$\phi = \psi + \frac{e}{\varepsilon}u = 0, \quad x_2 = 0, \quad (4.6)$$

$$u, \psi \to 0, \quad x_2 \to +\infty.$$

Substituting Eq. (4.2) into Eqs. (4.6)$_{1,2}$:

$$\bar{c}A\xi_2 + eB\xi_1 = 0,$$

$$\frac{e}{\varepsilon}A + B = 0. \tag{4.7}$$

For nontrivial solutions,

$$\begin{vmatrix} \bar{c}\xi_2 & e\xi_1 \\ e/\varepsilon & 1 \end{vmatrix} = \bar{c}\xi_2 - \frac{e^2}{\varepsilon}\xi_1 = 0, \tag{4.8}$$

or

$$\xi_2 = \bar{k}^2\xi_1, \tag{4.9}$$

where

$$\bar{k}^2 = \frac{e^2}{\varepsilon\bar{c}}. \tag{4.10}$$

Substitution of Eq. (4.3) into Eq. (4.9) yields

$$\bar{c}(\xi_1^2 - \bar{k}^4\xi_1^2) = \rho\omega^2, \tag{4.11}$$

from which the surface wave speed can be determined as

$$v^2 = \frac{\omega^2}{\xi_1^2} = \frac{\bar{c}}{\rho}(1 - \bar{k}^4) = v_T^2(1 - \bar{k}^4) < v_T^2. \tag{4.12}$$

When $\bar{k} = 0$, we have $\xi_2 = 0$, and the wave is no longer a surface wave.

4.1.2. A half-space with an unelectroded surface

If the surface of the half-space is unelectroded, electric fields can also exist in the free space of $x_2 < 0$. Denoting the electric potential in the free space by $\hat{\phi}$, we have

$$\nabla^2\hat{\phi} = 0, \quad x_2 < 0. \tag{4.13}$$

The boundary and continuity conditions are

$$T_{23} = 0, \quad \phi = \hat{\phi}, \quad D_2 = \hat{D}_2, \quad x_2 = 0,$$

$$u_3, \phi \to 0, \quad x_2 \to +\infty, \tag{4.14}$$

$$\hat{\phi} \to 0, \quad x_2 \to -\infty,$$

or, in terms of u, ψ, and $\hat{\phi}$,

$$T_{23} = \bar{c}u_{3,2} + e\psi_{,2} = 0, \quad x_2 = 0,$$

$$\phi = \psi + \frac{e}{\varepsilon}u = \hat{\phi}, \quad x_2 = 0,$$

$$-\varepsilon\psi_{,2} = -\varepsilon_0\hat{\phi}_{,2}, \quad x_2 = 0, \tag{4.15}$$

$$u, \psi \to 0, \quad x_2 \to +\infty,$$

$$\hat{\phi} \to 0, \quad x_2 \to -\infty.$$

From Eq. (4.13), in the free space,

$$\hat{\phi} = C\exp(\xi_1 x_2)\cos(\xi_1 x_1 - \omega t),$$

$$D_2 = -\varepsilon_0\hat{\phi}_{,2} = -\varepsilon_0\xi_1 C\exp(\xi_1 x_2)\cos(\xi_1 x_1 - \omega t). \tag{4.16}$$

In Eq. (4.16), C is an undetermined constant. Substituting Eqs. (4.2) and (4.16) into Eqs. $(4.15)_{1,2,3}$:

$$\bar{c}(-A\xi_2) + e(-B\xi_1) = 0,$$

$$-\varepsilon(-\xi_1 B) = -\varepsilon_0 C\xi_1, \tag{4.17}$$

$$\frac{e}{\varepsilon}A + B = C.$$

For nontrivial solutions,

$$\begin{vmatrix} \bar{c}\xi_2 & e\xi_1 & 0 \\ 0 & \varepsilon & \varepsilon_0 \\ e/\varepsilon & 1 & -1 \end{vmatrix} = -\bar{c}\xi_2\varepsilon + \frac{e^2}{\varepsilon}\varepsilon_0\xi_1 - \bar{c}\xi_2\varepsilon_0 = 0, \tag{4.18}$$

or

$$\xi_2 = \bar{k}^2\xi_1\frac{1}{1+\varepsilon/\varepsilon_0}. \tag{4.19}$$

Substitution of Eq. (4.3) into Eq. (4.19) yields the surface wave speed

$$v^2 = \frac{\omega^2}{\xi_1^2} = \frac{\bar{c}}{\rho}\left[1 - \frac{\bar{k}^4}{(1+\varepsilon/\varepsilon_0)^2}\right] = v_T^2\left[1 - \frac{\bar{k}^4}{(1+\varepsilon/\varepsilon_0)^2}\right] < v_T^2. \tag{4.20}$$

We point out that if the electric field in the free space is neglected and $D_2 = 0$ is prescribed as a boundary condition on the surface of the half-space, no surface wave solution can be obtained.

4.2. A Half-space with a Thin Film

In this section, mass sensitivity of the Bleustein–Gulyaev wave in the previous section is discussed [13].

4.2.1. Governing equations

Consider a ceramic half-space with a thin mass layer as shown in Fig. 4.2. The mass layer is very thin compared to the wavelength we are considering. Only the inertial effect of the mass layer is considered. The stiffness and the dielectric effects of the mass layer are neglected.

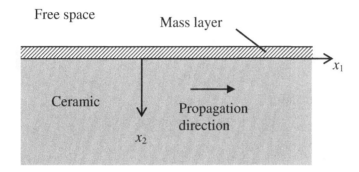

Fig. 4.2. A ceramic half-space with a thin mass layer.

The governing equations for the fields in the ceramic half-space are

$$v_T^2 \nabla^2 u = \rho \ddot{u}, \quad \nabla^2 \psi = 0, \quad \phi = \psi + \frac{e}{\varepsilon} u, \quad x_2 > 0 . \tag{4.21}$$

Fields in the ceramic half-space have the same form as those in Eq. (4.2):

$$u = A \exp(-\xi_2 x_2) \cos(\xi_1 x_1 - \omega t),$$

$$\psi = B \exp(-\xi_1 x_2) \cos(\xi_1 x_1 - \omega t),$$

$$T_{23} = -[\bar{c} A \xi_2 \exp(-\xi_2 x_2) + e B \xi_1 \exp(-\xi_1 x_2)] \cos(\xi_1 x_1 - \omega t), \tag{4.22}$$

$$\phi = [B \exp(-\xi_1 x_2) + \frac{e}{\varepsilon} A \exp(-\xi_2 x_2)] \cos(\xi_1 x_1 - \omega t),$$

$$D_2 = \varepsilon \xi_1 B \exp(-\xi_1 x_2) \cos(\xi_1 x_1 - \omega t),$$

where A and B are undetermined constants, and

$$\overline{c}(\xi_1^2 - \xi_2^2) = \rho\omega^2, \tag{4.23}$$

which determines

$$\xi_2^2 = \xi_1^2 - \frac{\rho\omega^2}{\overline{c}} = \xi_1^2(1 - \frac{v^2}{v_T^2}) > 0, \quad v^2 = \frac{\omega^2}{\xi_1^2}. \tag{4.24}$$

ξ_2 should be positive for decaying behavior away from the surface. Equation $(4.22)_{1,2}$ satisfy Eq. $(4.21)_{1,2}$.

4.2.2. A half-space with an electroded surface

First we consider the case when the surface of the half-space is electroded and the electrode is grounded. The corresponding boundary conditions are

$$T_{23} = \rho'h'\ddot{u}, \quad x_2 = 0,$$
$$\phi = 0, \quad x_2 = 0, \tag{4.25}$$

or, in terms of u and ψ,

$$T_{23} = \overline{c}u_{,2} + e\psi_{,2} = \rho'h'\ddot{u}, \quad x_2 = 0,$$
$$\phi = \psi + \frac{e}{\varepsilon}u = 0, \quad x_2 = 0. \tag{4.26}$$

Substituting Eq. (4.22) into Eq. (4.26):

$$-(\overline{c}A\xi_2 + eB\xi_1) = -\rho'h'\omega^2 A,$$
$$\frac{e}{\varepsilon}A + B = 0. \tag{4.27}$$

For nontrivial solutions,

$$\begin{vmatrix} \overline{c}\xi_2 - \rho'h'\omega^2 & e\xi_1 \\ e/\varepsilon & 1 \end{vmatrix} = \overline{c}\xi_2 - \rho'h'\omega^2 - \frac{e^2}{\varepsilon}\xi_1 = 0, \tag{4.28}$$

or

$$\xi_2 - \frac{\rho'h'\omega^2}{\overline{c}} = \overline{k}^2\xi_1, \quad \overline{k}^2 = \frac{e^2}{\varepsilon\overline{c}}. \tag{4.29}$$

Substitution of Eq. (4.24) into Eq. (4.29) yields

$$\sqrt{\xi_1^2 - \frac{\rho\omega^2}{\overline{c}}} - \frac{\rho'h'\omega^2}{\overline{c}} = \overline{k}^2\xi_1, \tag{4.30}$$

or, in terms of the wave speed $v = \omega / \xi_1$,

$$\sqrt{1 - \frac{v^2}{v_T^2}} = \bar{k}^2 + \xi_1 h' \frac{\rho'}{\rho} \frac{v^2}{v_T^2}, \tag{4.31}$$

which determines the surface speed v. In the special case when the mass layer is not present, i.e., $\rho' h' = 0$, Eq. (4.31) reduces to the corresponding Bleustein–Gulyaev wave with its speed determined by (see Eq. (4.12)):

$$\sqrt{1 - \frac{v^2}{v_T^2}} = \bar{k}^2. \tag{4.32}$$

Equation (4.31) shows that the mass layer inertia lowers the wave speed. The difference between the wave speeds in Eqs. (4.31) and (4.32) can be used to make mass sensors for measuring $\rho' h'$. Note that Eq. (4.31) determines a dispersive wave and Eq. (4.32) determines a nondispersive wave.

4.2.3. A half-space with an unelectroded surface

If the surface of the half-space is unelectroded, the electric field in the free space of $x_2 < 0$ is described by $\hat{\phi}$ and is governed by (see Eq. (4.13)):

$$\nabla^2 \hat{\phi} = 0, \quad x_2 < 0. \tag{4.33}$$

The relevant fields in the free space have the same form as those in Eq. (4.16):

$$\hat{\phi} = C \exp(\xi_1 x_2) \cos(\xi_1 x_1 - \omega t),$$
$$D_2 = -\varepsilon_0 \hat{\phi}_{,2} = -\varepsilon_0 \xi_1 C \exp(\xi_1 x_2) \cos(\xi_1 x_1 - \omega t). \tag{4.34}$$

The boundary and continuity conditions are

$$T_{23} = \rho' h' \ddot{u}, \quad \phi = \hat{\phi}, \quad D_2 = \hat{D}_2, \quad x_2 = 0, \tag{4.35}$$

or, in terms of u, ψ, and $\hat{\phi}$,

$$T_{23} = \bar{c} u_{3,2} + e \psi_{,2} = \rho' h' \ddot{u}, \quad x_2 = 0,$$
$$\phi = \psi + \frac{e}{\varepsilon} u = \hat{\phi}, \quad x_2 = 0, \tag{4.36}$$
$$-\varepsilon \psi_{,2} = -\varepsilon_0 \hat{\phi}_{,2}, \quad x_2 = 0.$$

Substituting Eqs. (4.22) and (4.34) into Eq. (4.36):

$$\bar{c}(-A\xi_2) + e(-B\xi_1) = -\rho'h'\omega^2 A,$$

$$-\varepsilon(-\xi_1 B) = -\varepsilon_0 C\xi_1, \tag{4.37}$$

$$\frac{e}{\varepsilon}A + B = C.$$

For nontrivial solutions,

$$\begin{vmatrix} \bar{c}\xi_2 - \rho'h'\omega^2 & e\xi_1 & 0 \\ 0 & \varepsilon & \varepsilon_0 \\ e/\varepsilon & 1 & -1 \end{vmatrix}$$

$$= \begin{vmatrix} \bar{c}\xi_2 & e\xi_1 & 0 \\ 0 & \varepsilon & \varepsilon_0 \\ e/\varepsilon & 1 & -1 \end{vmatrix} + \begin{vmatrix} -\rho'h'\omega^2 & e\xi_1 & 0 \\ 0 & \varepsilon & \varepsilon_0 \\ 0 & 1 & -1 \end{vmatrix} \tag{4.38}$$

$$= -\bar{c}\xi_2\varepsilon + \frac{e^2}{\varepsilon}\varepsilon_0\xi_1 - \bar{c}\xi_2\varepsilon_0 - \rho'h'\omega^2 \begin{vmatrix} \varepsilon & \varepsilon_0 \\ 1 & -1 \end{vmatrix} = 0.$$

In terms of the wave speed $v = \omega/\xi_1$, Eq. (4.38) becomes

$$\sqrt{1 - \frac{v^2}{v_T^2}} = \frac{\bar{k}^2}{1 + \varepsilon/\varepsilon_0} + \xi_1 h' \frac{\rho'}{\rho}\frac{v^2}{v_T^2}. \tag{4.39}$$

When $\rho'h' = 0$, Eq. (4.39) determines the following speed of the corresponding Bleustein–Gulyaev wave for an unelectroded half-space (see Eq. (4.20)):

$$\sqrt{1 - \frac{v^2}{v_T^2}} = \frac{\bar{k}^2}{1 + \varepsilon/\varepsilon_0}. \tag{4.40}$$

4.3. An FGM Half-space

In this section we study surfaces over a half-space of nonhomogeneous materials [14]. Other terminologies for these materials are functionally graded (or gradient) materials or simply FGM. Consider a nonhomogeneous half-space as shown in Fig. 4.3. The material parameters vary gradually along the x_2 direction. The surface of the half-space may be electroded or unelectroded.

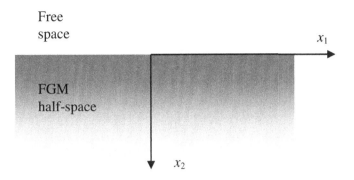

Fig. 4.3. An FGM half-space.

4.3.1. Governing equations

Since the material properties are functions of x_2, the relevant constitutive relations take the following form:

$$T_{31} = c(x_2)u_{,1} + e(x_2)\phi_{,1},$$
$$T_{32} = c(x_2)u_{,2} + e(x_2)\phi_{,2}, \tag{4.41}$$

$$D_1 = e(x_2)u_{,1} - \varepsilon(x_2)\phi_{,1},$$
$$D_2 = e(x_2)u_{,2} - \varepsilon(x_2)\phi_{,2}. \tag{4.42}$$

The equations of motion and charge have variable coefficients:

$$c_{,2}u_{,2} + c\nabla^2 u + e_{,2}\phi_{,2} + e\nabla^2\phi = \rho(x_2)\ddot{u},$$
$$e_{,2}u_{,2} + e\nabla^2 u - \varepsilon_{,2}\phi_{,2} - \varepsilon\nabla^2\phi = 0. \tag{4.43}$$

We consider the case when all material properties have the same exponential variation along the x_2 direction:

$$c = c^0\exp(\alpha x_2), \quad e = e^0\exp(\alpha x_2),$$
$$\varepsilon = \varepsilon^0\exp(\alpha x_2), \quad \rho = \rho^0\exp(\alpha x_2), \tag{4.44}$$

where α is a constant. The parameters with a superscript "0" are the values of the material properties at the surface. Substitution of Eq. (4.44) and into Eq. (4.43) yields the following equations with constant coefficients:

$$c^0\left(\alpha u_{,2} + \nabla^2 u\right) + e^0\left(\alpha\phi_{,2} + \nabla^2\phi\right) = \rho^0\ddot{u},$$

$$e^0\left(\alpha u_{,2} + \nabla^2 u\right) - \varepsilon^0\left(\alpha\phi_{,2} + \nabla^2\phi\right) = 0.$$

(4.45)

The electrical potential in the free space of $x_2 < 0$ is denoted by ϕ_0 and is governed by

$$\nabla^2\phi_0 = 0.$$

(4.46)

At $x_2 = 0$, the traction-free boundary condition is

$$T_{23}(x_1, 0^+, t) = 0.$$

(4.47)

If the surface is unelectroded, the following continuity conditions apply:

$$\phi_0(x_1, 0^-, t) = \phi(x_1, 0^+, t),$$

$$D_2(x_1, 0^-, t) = D_2(x_1, 0^+, t).$$

(4.48)

If the surface is electroded and the electrode is grounded, we have

$$\phi\left(x_1, 0^+, t\right) = 0.$$

(4.49)

When $x_2 \to \pm\infty$, we have:

$$u, \ \phi \to 0, \quad x_2 \to +\infty,$$

$$\phi_0 \to 0, \quad x_2 \to -\infty.$$

(4.50)

4.3.2. Fields in different regions

We look for solutions in the following form:

$$u = W(x_2)\exp[i\xi(x_1 - vt)],$$

$$\phi = \Phi(x_2)\exp[i\xi(x_1 - vt)].$$

(4.51)

Substitution of Eq. (4.51) into Eq. (4.45) yields the following ordinary differential equations for W and Φ:

$$c^0\left(W'' + \alpha W' - \xi^2 W\right) + e^0\left(\Phi'' + \alpha\Phi' - \xi^2\Phi\right) = -\rho^0\xi^2 v^2 W,$$

$$e^0\left(W'' + \alpha W' - \xi^2 W\right) - \varepsilon^0\left(\Phi'' + \alpha\Phi' - \xi^2\Phi\right) = 0,$$

(4.52)

where a prime represents a differentiation with respect to x_2. We write Eq. $(4.52)_2$ as

$$\left(\Phi'' + \alpha\Phi' - \xi^2\Phi\right) = \frac{e^0}{\varepsilon^0}\left(W'' + \alpha W' - \xi^2 W\right),$$ (4.53)

and substitute the expression into Eq. $(4.52)_1$. This results in

$$W'' + \alpha W' + \left(v^2 / v_T^2 - 1\right)\xi^2 W = 0,$$ (4.54)

where

$$v_T^2 = \frac{c^0 + (e^0)^2 / \varepsilon^0}{\rho^0}.$$ (4.55)

Consider a trial solution in the form of $\exp(rx_2)$. Then Eq. (4.54) leads to

$$r^2 + \alpha r + (v^2 / v_T^2 - 1)\xi^2 = 0.$$ (4.56)

The two roots are given by

$$r = \frac{-\alpha \pm \sqrt{\alpha^2 - 4(v^2 / v_T^2 - 1)\xi^2}}{2}.$$ (4.57)

For exponentially decaying behavior from the surface, we need r to be real and negative. This can happen when

$$\Delta = \alpha^2 - 4\left(v^2 / v_T^2 - 1\right)\xi^2 > 0.$$ (4.58)

There may also exist modes that are decaying and oscillating at the same time from the surface. Those modes correspond to complex values of r whose real parts are negative. We are not considering those modes. When Eq. (4.58) is satisfied, we can at least find one real and negative root for r from Eq. (4.57). Then we can determine the corresponding solution for the electric potential from Eq. (4.53). The electric potential for the free space can be obtained from Eq. (4.46). In summary, these fields are

$$u = A_1\exp(rx_2)\exp\left[i\xi(x_1 - vt)\right],$$

$$\phi = \left(A_2\exp(sx_2) + \frac{e^0}{\varepsilon^0}A_1\exp(rx_2)\right)\exp\left[i\xi(x_1 - vt)\right], \quad x_2 > 0, \quad (4.59)$$

$$\phi_0 = A_0\exp(\xi x_2)\exp\left[i\xi(x_1 - vt)\right], \quad x_2 < 0, \quad (4.60)$$

where A_0, A_1 and A_2 are arbitrary constants, and

$$s = \frac{-\alpha \pm \sqrt{\alpha^2 + 4\xi^2}}{2}.$$ (4.61)

We also need s to be real and negative.

4.3.3. An unelectroded surface

Substituting Eqs. (4.59) and (4.60) and the corresponding components of the stress and electrical displacement into Eqs. (4.47) and (4.48), we obtain the following linear homogeneous algebraic equations for A_0, A_1 and A_2:

$$\left(c^0 + \frac{(e^0)^2}{\varepsilon^0} \right) rA_1 + e^0 sA_2 = 0,$$

$$\frac{e^0}{\varepsilon^0} A_1 + A_2 = A_0,$$ (4.62)

$$-\varepsilon^0 sA_2 = -\varepsilon_0 \xi A_0.$$

For nontrivial solutions the determinant of the coefficients matrix has to vanish, which leads to the following dispersion relation:

$$\frac{\alpha + \sqrt{\alpha^2 + 4\xi^2(1 - v^2/v_T^2)}}{\alpha + \sqrt{\alpha^2 + 4\xi^2}}$$

$$+ \frac{\varepsilon^0}{\varepsilon_0} \frac{\alpha + \sqrt{\alpha^2 + 4\xi^2(1 - v^2/v_T^2)}}{2\xi} = \frac{(e^0)^2}{\varepsilon^0 c^0 \left(1 + \frac{(e^0)^2}{\varepsilon^0 c^0} \right)},$$ (4.63)

When $\alpha = 0$, i.e., the material is homogeneous, Eq. (4.63) reduces to

$$\left(1 + \frac{\varepsilon^0}{\varepsilon_0} \right) \sqrt{1 - \frac{v^2}{v_T^2}} = \frac{(e^0)^2}{\varepsilon^0 c^0 \left(1 + \frac{(e^0)^2}{\varepsilon^0 c^0} \right)},$$ (4.64)

which is the speed of the Bleustein–Gulyaev wave (see Eq. (4.20)). As an example, consider a half-space of PZT-5H. Phase velocity versus wave number for different values of α is shown in Fig. 4.4.

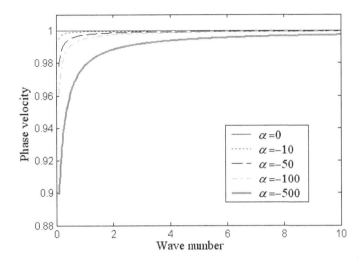

Fig. 4.4. v / v_T versus ξ (unelectroded surface).

4.3.4. An electroded surface

In this case, the second and the third equations in Eq. (4.62) should be replaced by

$$(e^0 / \varepsilon^0) A_1 + A_2 = 0. \qquad (4.65)$$

Then, the phase speed is determined from

$$\frac{\alpha + \sqrt{\alpha^2 + 4\xi^2 (1 - v^2 / v_T^2)}}{\alpha + \sqrt{\alpha^2 + 4\xi^2}} = \frac{(e^0)^2}{\varepsilon^0 c^0 \left(1 + \dfrac{(e^0)^2}{\varepsilon^0 c^0} \right)}. \qquad (4.66)$$

When $\alpha = 0$, Eq. (4.66) reduces to

$$\sqrt{1 - \frac{v^2}{v_T^2}} = \frac{(e^0)^2}{\varepsilon^0 c^0 \left(1 + \dfrac{(e^0)^2}{\varepsilon^0 c^0} \right)}, \qquad (4.67)$$

which is Eq. (4.12). Dispersion curves for PZT-5H are shown in Fig. 4.5 for different α.

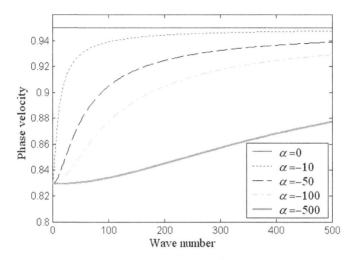

Fig. 4.5. v/v_T versus ξ (electroded surface).

4.4. A Half-space in Contact with a Fluid

A vibrating elastic body in contact with a fluid changes its resonant frequencies. This effect has been used to make fluid sensors for measuring fluid viscosity/density. In this section we discuss waves over a half-space in contact with a fluid [13,15].

4.4.1. Governing equations and fields

Consider a ceramic half-space in contact with a half-space of a dielectric fluid (see Fig. 4.6).

Fluid

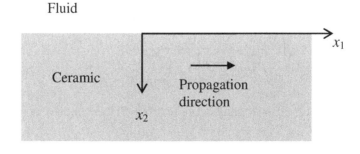

Fig. 4.6. A ceramic half-space in contact with a fluid.

For the ceramic half-space the governing equations are

$$v_T^2 \nabla^2 u = \ddot{u}, \quad \nabla^2 \psi = 0, \quad \phi = \psi + \frac{e}{\varepsilon} u, \quad x_2 > 0. \tag{4.68}$$

The relevant fields are from Eq. (4.2)

$$u = A\exp(-\xi_2 x_2)\exp[i(\xi_1 x_1 - \omega t)],$$

$$\psi = B\exp(-\xi_1 x_2)\exp[i(\xi_1 x_1 - \omega t)],$$

$$T_{23} = -[\bar{c}A\xi_2 \exp(-\xi_2 x_2) + eB\xi_1 \exp(-\xi_1 x_2)]\exp[i(\xi_1 x_1 - \omega t)], \tag{4.69}$$

$$\phi = [B\exp(-\xi_1 x_2) + \frac{e}{\varepsilon} A\exp(-\xi_2 x_2)]\exp[i(\xi_1 x_1 - \omega t)],$$

$$D_2 = \varepsilon \xi_1 B\exp(-\xi_1 x_2)\exp[i(\xi_1 x_1 - \omega t)],$$

where A and B are undetermined constants, $\xi_1 > 0$, and ξ_2 should have a positive real part for decaying behavior away from the surface. Equation $(4.69)_2$ already satisfies Eq. $(4.68)_2$. For Eq. $(4.69)_1$ to satisfy Eq. $(4.68)_1$ we must have

$$\bar{c}(\xi_1^2 - \xi_2^2) = \rho\omega^2, \tag{4.70}$$

which determines

$$\xi_2^2 = \xi_1^2 - \frac{\rho\omega^2}{\bar{c}} = \xi_1^2(1 - \frac{v^2}{v_T^2}), \quad v^2 = \frac{\omega^2}{\xi_1^2}. \tag{4.71}$$

For the fluid, let $v_3(x_1, x_2, t)$ be the velocity field. The governing equation for v_3 and the relevant stress component for the interface continuity condition are

$$\mu(v_{3,11} + v_{3,22}) = \rho' \frac{\partial v_3}{\partial t}, \tag{4.72}$$

$$T_{23} = \mu v_{3,2},$$

where μ is the viscosity of the fluid and ρ' is the fluid density. The following fields are allowed by Eq. $(4.72)_1$:

$$v_3 = C\exp(\lambda x_2)\exp[i(\xi_1 x_1 - \omega t)],$$

$$T_{23} = \mu\lambda C\exp(\lambda x_2)\exp[i(\xi_1 x_1 - \omega t)], \tag{4.73}$$

where C is an arbitrary constant, λ has a positive real part and is determined by

$$\mu(\lambda^2 - \xi_1^2) = \rho'i\omega. \tag{4.74}$$

4.4.2. A half-space with an electroded surface

First consider the case when the surface of the half-space is electroded and the electrode is grounded. The corresponding boundary and continuity conditions are

$$
\begin{aligned}
\dot{u}(x_2 = 0^+) &= v_3(x_2 = 0^-), \\
T_{23}(x_2 = 0^+) &= T_{23}(x_2 = 0^-), \\
\phi &= 0, \quad x_2 = 0.
\end{aligned} \tag{4.75}
$$

Substituting Eqs. (4.69) and (4.73) into Eq. (4.75):

$$
\begin{aligned}
-i\omega A &= C, \\
-(\bar{c}A\xi_2 + eB\xi_1) &= \mu\lambda C, \\
\frac{e}{\varepsilon}A + B &= 0.
\end{aligned} \tag{4.76}
$$

For nontrivial solutions,

$$
\begin{vmatrix} i\omega & 0 & 1 \\ \bar{c}\xi_2 & e\xi_1 & \mu\lambda \\ e/\varepsilon & 1 & 0 \end{vmatrix} = \begin{vmatrix} \bar{c}\xi_2 & e\xi_1 \\ e/\varepsilon & 1 \end{vmatrix} - \mu\lambda \begin{vmatrix} i\omega & 0 \\ e/\varepsilon & 1 \end{vmatrix} = 0, \tag{4.77}
$$

or

$$\xi_2 = \bar{k}^2\xi_1 + \frac{i\omega\mu\lambda}{\bar{c}}, \tag{4.78}$$

which determines the surface wave frequency or speed. In terms of the wave speed $v = \omega / \xi_1$, Eq. (4.78) becomes

$$\sqrt{1 - \frac{v^2}{v_T^2}} = \bar{k}^2 + \frac{i\mu\xi_1 v}{\bar{c}}\sqrt{1 - i\frac{\rho'v}{\mu\xi_1}}. \tag{4.79}$$

Equation (4.79) can be used to design fluid sensors for measuring μ or ρ'. In the special case when the fluid is not present, Eq. (4.79) reduces to the speed of the Bleustein–Gulyaev wave. In the case of a low-density fluid, in the limit of small ρ', Eq. (4.79) can be approximated by

$$\sqrt{1 - \frac{v^2}{v_T^2}} \cong \bar{k}^2 + \frac{i\omega\mu}{\bar{c}}\left(1 - \frac{1}{2}i\frac{\rho'v}{\mu\xi_1}\right) = \bar{k}^2 + \frac{i\mu\xi_1 v}{\bar{c}} + \frac{1}{2}\frac{\rho'}{\rho}\frac{v^2}{v_T^2}, \qquad (4.80)$$

which exhibits the damping effect of viscosity and the frequency shift due to fluid inertia.

4.4.3. A half-space with an unelectroded surface

If the surface of the half-space is unelectroded, the electric potential $\hat{\phi}$ in the fluid is governed by

$$\nabla^2\hat{\phi} = 0, \quad x_2 < 0. \qquad (4.81)$$

The solution is (see Eq. (4.16)):

$$\hat{\phi} = D\exp(\xi_1 x_2)\cos(\xi_1 x_1 - \omega t),$$
$$D_2 = -\varepsilon'\hat{\phi}_{,2} = -\varepsilon_0\xi_1 D\exp(\xi_1 x_2)\cos(\xi_1 x_1 - \omega t), \qquad (4.82)$$

where D is an undetermined constant and ε' is the dielectric constant of the fluid. The boundary and continuity conditions are

$$-i\omega u_3(x_2 = 0^+) = v_3(x_2 = 0^-),$$
$$T_{23}(x_2 = 0^+) = T_{23}(x_2 = 0^-), \qquad (4.83)$$
$$\phi = \hat{\phi}, \quad D_2 = \hat{D}_2, \quad x_2 = 0.$$

Substituting Eqs. (4.69), (4.73) and (4.82) into Eq (4.83):

$$-i\omega A = C$$
$$\bar{c}(-A\xi_2) + e(-B\xi_1) = \mu\lambda C,$$
$$-\varepsilon(-\xi_1 B) = -\varepsilon' D\xi_1, \qquad (4.84)$$
$$\frac{e}{\varepsilon}A + B = D.$$

For nontrivial solutions,

$$\begin{vmatrix} i\omega & 0 & 1 & 0 \\ \overline{c}\xi_2 & e\xi_1 & \mu\lambda & 0 \\ 0 & \varepsilon & 0 & \varepsilon' \\ e/\varepsilon & 1 & 0 & -1 \end{vmatrix}$$

$$= \begin{vmatrix} \overline{c}\xi_2 & e\xi_1 & 0 \\ 0 & \varepsilon & \varepsilon' \\ e/\varepsilon & 1 & -1 \end{vmatrix} + i\omega \begin{vmatrix} e\xi_1 & \mu\lambda & 0 \\ \varepsilon & 0 & \varepsilon' \\ 1 & 0 & -1 \end{vmatrix} \qquad (4.85)$$

$$= \begin{vmatrix} \overline{c}\xi_2 & e\xi_1 & 0 \\ 0 & \varepsilon & \varepsilon' \\ e/\varepsilon & 1 & -1 \end{vmatrix} + i\omega\mu\lambda(\varepsilon + \varepsilon') = 0,$$

or

$$\xi_2 = \frac{\overline{k}^2}{1 + \varepsilon/\varepsilon'}\xi_1 + \frac{i\omega\mu\lambda}{\overline{c}}. \qquad (4.86)$$

In terms of the wave speed $v = \omega/\xi_1$, Eq. (4.86) becomes

$$\sqrt{1 - \frac{v^2}{v_T^2}} = \frac{\overline{k}^2}{1 + \varepsilon/\varepsilon'} + \frac{i\mu\xi_1 v}{\overline{c}}\sqrt{1 - i\frac{\rho'v}{\mu\xi_1}}. \qquad (4.87)$$

4.5. Interface Waves

Consider two half-spaces as shown in Fig. 4.7. The ceramics are both poled along the $\pm x_3$ directions. We are interested in the possibility of a wave traveling near the interface between the two ceramics [16].

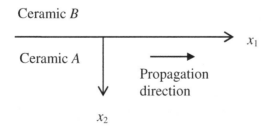

Fig. 4.7. Two semi-infinite ceramic half-spaces.

4.5.1. Governing equations and fields

The governing equations are

$$\overline{c}_A \nabla^2 u_A = \rho_A \ddot{u}_A, \quad \nabla^2 \psi_A = 0, \quad \phi_A = \psi_A + \frac{e_A}{\varepsilon_A} u_A, \quad x_2 > 0,$$

$$\overline{c}_B \nabla^2 u_B = \rho_B \ddot{u}_B, \quad \nabla^2 \psi_B = 0, \quad \phi_B = \psi_B + \frac{e_B}{\varepsilon_B} u_B, \quad x_2 < 0,$$

$$(4.88)$$

where the subscripts indicate fields in ceramic A and ceramic B. For an interface wave, we require that

$$u_A, \psi_A \to 0, \quad x_2 \to +\infty,$$

$$u_B, \psi_B \to 0, \quad x_2 \to -\infty.$$

$$(4.89)$$

For $x_2 > 0$, the solutions to Eq. (4.88)$_1$ that satisfy Eq. (4.89)$_1$ can be written as

$$u_A = U_A \exp(-\eta_A x_2)\cos(\xi x_1 - \omega t),$$

$$\psi_A = \Psi_A \exp(-\xi x_2)\cos(\xi x_1 - \omega t),$$

$$(4.90)$$

where

$$\eta_A^2 = \xi^2 - \frac{\rho_A \omega^2}{\overline{c}_A} = \xi^2 \left(1 - \frac{v^2}{v_A^2}\right) > 0,$$

$$(4.91)$$

and

$$v = \frac{\omega}{\xi}, \quad v_A^2 = \frac{\overline{c}_A}{\rho_A}.$$

$$(4.92)$$

Similarly, for $x_2 < 0$, the solutions to Eq. (4.88)$_2$ that satisfy Eq. (4.89)$_2$ are

$$u_B = U_B \exp(\eta_B x_2)\cos(\xi x_1 - \omega t),$$

$$\psi_B = \Psi_B \exp(\xi x_2)\cos(\xi x_1 - \omega t),$$

$$(4.93)$$

where

$$\eta_B^2 = \xi^2 - \frac{\rho_B \omega^2}{\overline{c}_B} = \xi^2 \left(1 - \frac{v^2}{v_B^2}\right) > 0,$$

$$(4.94)$$

$$v_B^2 = \frac{\overline{c}_B}{\rho_B}.$$

$$(4.95)$$

For interface continuity conditions, we consider two situations separately.

4.5.2. An electroded interface

When the interface is a grounded electrode we have, at $x_2 = 0$,

$$u_A = u_B, \quad T_A = T_B,$$

$$\psi_A + \frac{e_A}{\varepsilon_A} u_A = 0, \quad \psi_B + \frac{e_B}{\varepsilon_B} u_B = 0, \tag{4.96}$$

where $T = T_{23}$. Substitution of Eqs. (4.90) and (4.93) into Eq. (4.96) results in a system of linear, homogenous equations for U_A, Ψ_A, U_B and Ψ_B. For nontrivial solutions, the determinant of the coefficient matrix has to vanish, which yields

$$\bar{c}_A \left(1 - \frac{v^2}{v_A^2}\right)^{1/2} + \bar{c}_B \left(1 - \frac{v^2}{v_B^2}\right)^{1/2} = \frac{e_A^2}{\varepsilon_A} + \frac{e_B^2}{\varepsilon_B}, \tag{4.97}$$

which determines the speed of the interface wave. As a special case, when medium B is free space with

$$\bar{c}_B = 0, \quad \bar{e}_B = 0, \quad \varepsilon_B = \varepsilon_0, \tag{4.98}$$

Eq. (4.97) reduces to

$$v^2 = v_A^2 \left[1 - \left(\frac{e_A^2}{\varepsilon_A \bar{c}_A} \right)^2 \right], \tag{4.99}$$

which is the speed of the Bleustein–Gulyaev wave given by Eq. (4.12).

4.5.3. An unelectroded interface

When the interface is unelectroded, the continuity conditions at $x_2 = 0$ are

$$u_A = u_B,$$

$$T_A = T_B,$$

$$\psi_A + \frac{e_A}{\varepsilon_A} u_A = \psi_B + \frac{e_B}{\varepsilon_B} u_B, \tag{4.100}$$

$$D_A = D_B,$$

where $D = D_2$. Substituting Eqs. (4.90) and (4.93) into Eq. (4.100) and requiring the determinant of the coefficient matrix to vanish, we obtain

$$\bar{c}_A \left(1 - \frac{v^2}{v_A^2}\right)^{1/2} + \bar{c}_B \left(1 - \frac{v^2}{v_B^2}\right)^{1/2} = \frac{(e_A/\varepsilon_A - e_B/\varepsilon_B)^2}{1/\varepsilon_A + 1/\varepsilon_B}. \tag{4.101}$$

We note that the right-hand side of Eq. (4.101) depends on the difference between the two materials. As a special case, when mediums B is free space with Eq. (4.98), Eq. (4.101) reduces to

$$v^2 = v_A^2 \left[1 - \left(\frac{e_A^2}{\varepsilon_A \bar{c}_A (1 + \varepsilon_0 / \varepsilon_A)} \right)^2 \right], \tag{4.102}$$

which is the speed of the Bleustein–Gulyaev wave given in Eq. (4.20).

4.6. An Imperfectly Bonded Interface

Acoustic wave devices are usually nonhomogeneous structures with components of various dielectrics and conductors. In the analyses of these devices perfect bonding at the interfaces between two materials is routinely assumed, i.e., the displacement and traction are continuous across the interface. In fact imperfect bonding often exists in devices. Sometimes a very thin layer of glue is applied at an interface which has its own physical properties. To understand the basic effects of an imperfectly bonded interface, we analyze the propagation of piezoelectric waves near such an interface [17]. Consider two ceramic half-spaces with an interface (see Fig. 4.8). The x_1 and x_3 axes are in the interface where $x_2 = 0$. $x_2 > 0$ is occupied by ceramic A and $x_2 < 0$ by ceramic B. The ceramics are poled along the x_3 or $-x_3$ direction.

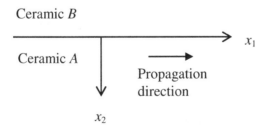

Fig. 4.8. Two ceramic half-spaces with an imperfect interface.

4.6.1. *Governing equations and fields*

The governing equations for u and ψ in the half-spaces are

$$\bar{c}_A \nabla^2 u_A = \rho_A \ddot{u}_A, \quad \nabla^2 \psi_A = 0, \quad \phi_A = \psi_A + \frac{e_A}{\varepsilon_A} u_A, \quad x_2 > 0,$$

$$\bar{c}_B \nabla^2 u_B = \rho_B \ddot{u}_B, \quad \nabla^2 \psi_B = 0, \quad \phi_B = \psi_B + \frac{e_B}{\varepsilon_B} u_B, \quad x_2 < 0. \tag{4.103}$$

The subscripts indicate quantities in ceramic A and ceramic B. For an interface wave, we require all fields to vanish when $x_2 \to \pm\infty$. For $x_2 > 0$, consider the possibility of the following waves propagating in the x_1 direction:

$$u_A = U_A \exp(-\eta_A x_2) \cos(\xi x_1 - \omega t),$$

$$\psi_A = \Psi_A \exp(-\xi x_2) \cos(\xi x_1 - \omega t), \tag{4.104}$$

where U_A and Ψ_A are undetermined constants, ω and ξ are the frequency and wave number, and

$$\eta_A^2 = \xi^2 - \frac{\rho_A \omega^2}{\bar{c}_A} = \xi^2 \left(1 - \frac{v^2}{v_A^2} \right) > 0, \tag{4.105}$$

where $v = \omega / \xi$ and $v_A^2 = \bar{c}_A / \rho_A$. For continuity conditions, we need T_{23} and D_2 in ceramic A, denoted by T_A and D_A:

$$T_A = -[\bar{c}_A \eta_A U_A \exp(-\eta_A x_2) + e_A \xi \Psi_A \exp(-\xi x_2)] \cos(\xi x_1 - \omega t),$$

$$D_A = \varepsilon_A \xi \Psi_A \exp(-\xi x_2) \cos(\xi x_1 - \omega t). \tag{4.106}$$

Similarly, for $x_2 < 0$, the solution is

$$u_B = U_B \exp(\eta_B x_2) \cos(\xi x_1 - \omega t),$$

$$\psi_B = \Psi_B \exp(\xi x_2) \cos(\xi x_1 - \omega t), \tag{4.107}$$

$$\eta_B^2 = \xi^2 - \frac{\rho_B \omega^2}{\bar{c}_B} = \xi^2 \left(1 - \frac{v^2}{v_B^2} \right) > 0, \tag{4.108}$$

$$v_B^2 = \frac{\bar{c}_B}{\rho_B},$$

$$T_B = [\bar{c}_B \eta_B U_B \exp(\eta_B x_2) + e_B \xi \Psi_B \exp(\xi x_2)] \cos(\xi x_1 - \omega t),$$

$$D_B = -\varepsilon_B \xi \Psi_B \exp(\xi x_2) \cos(\xi x_1 - \omega t). \tag{4.109}$$

For interface continuity conditions, we consider two cases of an electroded interface or an unelectroded interface separately below.

4.6.2. An electroded interface

First consider the case when the interface is a grounded electrode. We have, at $x_2 = 0$,

$$T_A = T_B = K(u_A - u_B),$$

$$\psi_A + \frac{e_A}{\varepsilon_A} u_A = 0, \quad \psi_B + \frac{e_B}{\varepsilon_B} u_B = 0. \tag{4.110}$$

Equation $(4.110)_1$ is the so-called shear-lag interface model. The displacement at the interface is allowed to be discontinuous. K is an interface elastic constant which describes how well the two half-spaces are bonded. $K = \infty$ is for perfect bonding. When $K = 0$, the two half-spaces lose their mechanical interaction at the interface. Physically, if we think the interface as a separate phase with a small thickness 2δ, then, the shear stress-strain relation of the interface can be written as $\tau = G[u(\delta) - u(-\delta)]/(2\delta)$, where τ is shear stress and G is the usual shear modulus of the interface material. Therefore the relation between the real interface shear modulus G and the effective shear modulus K is $G = 2\delta K$. Hence K in fact represents a combination of the interface elasticity and thickness. Substitution of Eqs. (4.104), (4.106), (4.107) and (4.109) into Eq. (4.110) results in a system of linear homogenous equations for U_A, Ψ_A, U_B and Ψ_B. For nontrivial solutions, the determinant of the coefficient matrix has to vanish, which yields the following frequency equation that determines the wave speed v:

$$\bar{c}_A \bar{c}_B \left(\eta_A - \bar{k}_A^2 \xi \right)\left(\eta_B - \bar{k}_B^2 \xi \right)$$

$$= -K\left[\bar{c}_A \left(\eta_A - \bar{k}_A^2 \xi \right) + \bar{c}_B \left(\eta_B - \bar{k}_B^2 \xi \right) \right], \tag{4.111}$$

where $\bar{k}_A^2 = e_A^2/(\bar{c}_A \varepsilon_A)$ and $\bar{k}_B^2 = e_B^2/(\bar{c}_B \varepsilon_B)$. Clearly, the waves are dispersive. This may be somewhat unexpected as there does not seem to exist a geometric characteristic length in the problem. Mathematically, the dispersion is caused by the shear-lag interface condition in Eq. $(4.110)_1$ which has both u and its first derivative. Note that $K/\bar{c}_A$ has the dimension of wave number. To gain some insight into Eq. (4.111), let us consider the following special cases.

(i) $K = 0$. There is no interface bonding. Mechanically the two half-spaces move independently. Equation (4.111) reduces to

$$v^2 = v_A^2(1 - \bar{k}_A^4), \quad \text{and} \quad v^2 = v_B^2(1 - \bar{k}_B^4), \tag{4.112}$$

which are the speeds of Bleustein–Gulyaev waves for A and B, respectively.

(ii) $K = \infty$. The interface is perfectly bonded. Equation (4.111) implies that

$$\bar{c}_A\left(1 - \frac{v^2}{v_A^2}\right)^{1/2} + \bar{c}_B\left(1 - \frac{v^2}{v_B^2}\right)^{1/2} = \frac{e_A^2}{\varepsilon_A} + \frac{e_B^2}{\varepsilon_B}, \tag{4.113}$$

which is the frequency equation for interface waves (see Eq. 4.97)).

(iii) When the two half-spaces are of the same ceramic but are poled in opposite directions, $\rho_B = \rho_A$, $c_B = c_A$, $e_B = -e_A$, $\varepsilon_B = \varepsilon_A$. In this case two roots can be found from Eq. (4.111). One is

$$v^2 = v_A^2(1 - \bar{k}_A^4), \tag{4.114}$$

which is a wave with u symmetric about the interface and does not feel the imperfect bonding, and ϕ is antisymmetric about the interface and vanishes there so that the wave has the speed of the electroded Bleustein–Gulyaev wave. The other root is given by

$$\frac{v^2}{v_A^2} = 1 - \bar{k}_A^4 + 4\frac{K}{\bar{c}_A\xi}\left(\bar{k}_A^2 - \frac{K}{\bar{c}_A\xi}\right). \tag{4.115}$$

This is a wave with u antisymmetric about the interface and can feel the imperfect bonding. The effect of the bonding is more pronounced for long waves with a small wave number ξ. Equation (4.115) provides a relation that can be used to measure K from v.

(iv) When the two half-spaces are of the same ceramic and are poled in the same direction, $\rho_B = \rho_A$, $c_B = c_A$, $e_B = e_A$, $\varepsilon_B = \varepsilon_A$. In this case two roots can be found from Eq. (4.111). They are the same as Eqs. (4.114) and (4.115) which are the same whether $e_B = \pm e_A$.

4.6.3. *An unelectroded interface*

Next consider the case when the interface is unelectroded. The continuity conditions at $x_2 = 0$ are

$$T_A = T_B = K(u_A - u_B),$$

$$\psi_A + \frac{e_A}{\varepsilon_A} u_A = \psi_B + \frac{e_B}{\varepsilon_B} u_B, \tag{4.116}$$

$$D_A = D_B.$$

Substitution of Eqs. (4.104), (4.106), (4.107) and (4.109) into Eq. (4.116) implies the following equation that determines the wave speed:

$$(P_A P_B - Q^2) + (P_A + P_B + 2Q)K = 0, \tag{4.117}$$

where

$$P_A = \overline{c}_A \left(\sqrt{1 - \frac{v^2}{v_A^2} - \frac{\overline{k}_A^2}{1 + \varepsilon_A / \varepsilon_B}} \right) \xi,$$

$$P_B = \overline{c}_B \left(\sqrt{1 - \frac{v^2}{v_B^2} - \frac{\overline{k}_B^2}{1 + \varepsilon_B / \varepsilon_A}} \right) \xi, \tag{4.118}$$

$$Q = \frac{e_A e_B}{\varepsilon_A + \varepsilon_B} \xi.$$

Special cases:

(i) When $K = 0$, the interface has no mechanical interaction. However, the two half-spaces can still interact electrically. Equation (4.117) reduces to

$$\sqrt{1 - \frac{v^2}{v_A^2}} \sqrt{1 - \frac{v^2}{v_B^2} - \frac{\overline{k}_A^2}{(1 + \varepsilon_A / \varepsilon_B)}} \sqrt{1 - \frac{v^2}{v_B^2} - \frac{\overline{k}_B^2}{(1 + \varepsilon_B / \varepsilon_A)}} \sqrt{1 - \frac{v^2}{v_A^2}} = 0. \tag{4.119}$$

(ii) $K = \infty$. The interface is perfectly bonded. Equation (4.117) gives the following frequency equation:

$$\overline{c}_A \left(1 - \frac{v^2}{v_A^2} \right)^{1/2} + \overline{c}_B \left(1 - \frac{v^2}{v_B^2} \right)^{1/2} = \frac{(e_A / \varepsilon_A - e_B / \varepsilon_B)^2}{1 / \varepsilon_A + 1 / \varepsilon_B}, \tag{4.120}$$

which is the frequency equation for interface waves when the bonding is perfect (see Eq. (4.101)).

(iii) When the two half-spaces are of the same ceramic but are poled in opposite directions, Eq. (4.117) can be written as:

$$\bar{c}_A^2 \xi^2 \left(\sqrt{1 - \frac{v^2}{v_A^2}} - \bar{k}_A^2 \right) \left(\sqrt{1 - \frac{v^2}{v_A^2}} - \frac{2K}{\bar{c}_A \xi} \right) = 0 . \tag{4.121}$$

It leads to two solutions, namely,

$$v^2 = v_A^2 (1 - \bar{k}_A^4) , \tag{4.122}$$

and

$$\frac{v^2}{v_A^2} = 1 - \left(\frac{2K}{\bar{c}_A \xi} \right)^2 . \tag{4.123}$$

(iv) When the two half-spaces are of the same ceramic and are poled in the same direction, the two roots of Eq. (4.117) are

$$v^2 = v_A^2 ,$$
$$\frac{v^2}{v_A^2} = 1 - \bar{k}_A^4 + 4 \frac{K}{\bar{c}_A \xi} \left(\bar{k}_A^2 - \frac{K}{\bar{c}_A \xi} \right) . \tag{4.124}$$

Equation (4.124)$_1$ is no longer an interface wave.

4.7. An Interface between Two FGM Half-spaces

Next we study interface waves between two FGM half-spaces (see Fig. 4.9) [18]. The interface may be electroded or unelectroded.

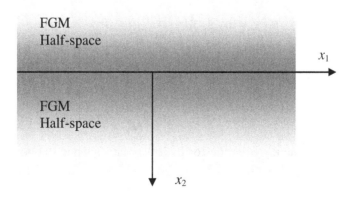

Fig. 4.9. An interface between two FGM half-spaces.

4.7.1. Fields in the lower half-space

We are interested in modes decaying from the interface only without oscillations. The fields in the lower half-space with $x_2 > 0$ are still the same as those in Sec. 4.3. However, the α in Eq. (4.44) will be replaced by 2α in this section. This will simplify the equations somewhat. We summarize the major expressions below.

$$u = A_1 \exp(rx_2) \exp\left[i\xi(x_1 - vt)\right],$$

$$\phi = \left(A_2 \exp(sx_2) + \frac{e^0}{\varepsilon^0} A_1 \exp(rx_2)\right) \exp\left[i\xi(x_1 - vt)\right], \quad (4.125)$$

where A_1 and A_2 are arbitrary constants,

$$r = -\alpha \pm \sqrt{\alpha^2 + (1 - v^2/v_T^2)\xi^2}, \quad (4.126)$$

$$s = -\alpha \pm \sqrt{\alpha^2 + \xi^2}, \quad (4.127)$$

$$v_T^2 = \frac{c^0 + (e^0)^2/\varepsilon^0}{\rho^0}. \quad (4.128)$$

r needs to be real and negative. We choose the negative s.

4.7.2. Fields in the upper half-space

For the upper half-space $x_2 < 0$, we use an over bar for the parameters and fields. The material properties are represented by

$$\bar{c} = \bar{c}^0 \exp(2\beta x_2), \quad \bar{e} = \bar{e}^0 \exp(2\beta x_2),$$

$$\bar{\varepsilon} = \bar{\varepsilon}^0 \exp(2\beta x_2), \quad \bar{\rho} = \bar{\rho}^0 \exp(2\beta x_2). \quad (4.129)$$

The fields are

$$\bar{u} = B_1 \exp(\bar{r}x_2) \exp[i\xi(x_1 - vt)],$$

$$\bar{\phi} = \left(B_2 \exp(\bar{s}x_2) + \frac{\bar{e}^0}{\bar{\varepsilon}^0} B_1 \exp(\bar{r}x_2)\right) \exp[i\xi(x_1 - vt)], \quad (4.130)$$

where B_1 and B_2 are arbitrary constants,

$$\bar{r} = -\beta \pm \sqrt{\beta^2 + \xi^2\left(1 - \frac{v^2}{\bar{v}_T^2}\right)}, \quad (4.131)$$

$$\overline{s} = -\beta \pm \sqrt{\beta^2 + \xi^2} \, , \tag{4.132}$$

$$\overline{v}_T^2 = \frac{\overline{c}^0 + (\overline{e}^0)^2 / \overline{\varepsilon}^0}{\overline{\rho}^0} \, . \tag{4.133}$$

$\overline{r}$ should be real and positive. We choose the positive $\overline{s}$.

4.7.3. An electroded interface

In this case, at $x_2 = 0$, the boundary and continuity conditions are

$$\begin{aligned}
&u(x_1, 0^+, t) = \overline{u}(x_1, 0^-, t), \\
&T_{23}(x_1, 0^+, t) = \overline{T}_{23}(x_1, 0^-, t), \\
&\phi(x_1, 0^+, t) = 0, \\
&\overline{\phi}(x_1, 0^-, t) = 0.
\end{aligned} \tag{4.134}$$

Substituting Eqs. (4.125) and (4.130) into Eq. (4.134) results in the following linear equations for the undetermined constants:

$$\begin{cases}
A_1 - B_1 = 0, \\
\left(c^0 + \dfrac{(e^0)^2}{\varepsilon^0} \right) r A_1 + e^0 s A_2 - \left(\overline{c}^0 + \dfrac{(\overline{e}^0)^2}{\overline{\varepsilon}^0} \right) \overline{r} B_1 - \overline{e}^0 \overline{s} B_2 = 0, \\
\dfrac{e^0}{\varepsilon^0} A_1 + A_2 = 0, \\
\dfrac{\overline{e}^0}{\overline{\varepsilon}^0} B_1 + B_2 = 0.
\end{cases} \tag{4.135}$$

For nontrivial solutions, the determinant of the coefficient matrix has to vanish, which yields the frequency equation below that determines the wave speed.

$$\begin{aligned}
&Q_0 \left(-\beta + \sqrt{\beta^2 + \xi^2} \right) + Q_1 \left[-\beta + \sqrt{\beta^2 + \xi^2 \left(1 - \dfrac{v^2}{\overline{v}_T^2} \right)} \right] \\
&+ Q_2 \left[-\alpha - \sqrt{\alpha^2 + \xi^2} \right] + Q_3 \left[-\alpha - \sqrt{\alpha^2 + \xi^2 \left(1 - \dfrac{v^2}{v_T^2} \right)} \right] = 0,
\end{aligned} \tag{4.136}$$

where

$$Q_0 = -\frac{(\bar{e}^0)^2}{\bar{\varepsilon}^0}, \quad Q_1 = \bar{c}^0 + \frac{(\bar{e}^0)^2}{\bar{\varepsilon}^0},$$

$$Q_2 = \frac{(e^0)^2}{\varepsilon^0}, \quad Q_3 = -c^0 - \frac{(e^0)^2}{\varepsilon^0}.$$

(4.137)

When $\alpha = 0$ and $\beta = 0$, Eq. (4.136) reduces to

$$Q_0 - Q_2 + \left(Q_1 \sqrt{1 - \frac{v^2}{\bar{v}_T^2}} - Q_3 \sqrt{1 - \frac{v^2}{v_T^2}} \right) = 0,$$

(4.138)

which can be shown to be the same as Eq. (4.97), the result of [16] for interface waves between two electroded, homogeneous half-spaces.

As an example, consider the case when the lower half-space material is PZT-5H and the upper half-space material is PZT-5A. Wave velocities for various values of α and β are shown in Figs. 4.10 and 4.11.

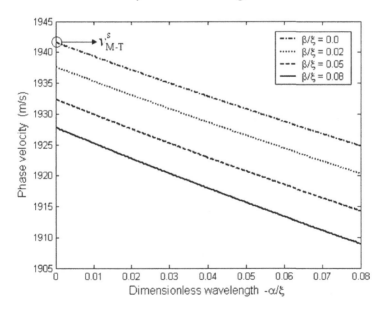

Fig. 4.10. Phase velocity versus dimensionless wavelength $-\alpha/\xi$ for selected values of β/ξ.

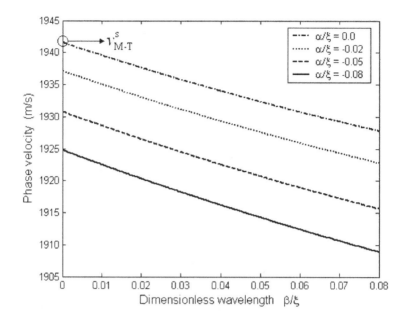

Fig. 4.11. Phase velocity versus dimensionless wavelength β/ξ for selected values of α/ξ.

4.7.4. An unelectroded interface

For an unelectroded interface, at $x_2 = 0$, we have:

$$u(x_1,0^+,t) = \bar{u}(x_1,0^-,t),$$
$$T_{23}(x_1,0^+,t) = \bar{T}_{23}(x_1,0^-,t),$$
(4.139)

$$\phi(x_1,0^+,t) = \bar{\phi}(x_1,0^-,t),$$
$$D_2(x_1,0^+,t) = \bar{D}_2(x_1,0^-,t).$$
(4.140)

The continuity conditions imply the following four linear equations for the undetermined constants:

$$A_1 - B_1 = 0,$$

$$\left(c^0 + \frac{(e^0)^2}{\varepsilon^0}\right)rA_1 + e^0 sA_2 - \left(\bar{c}^0 + \frac{(\bar{e}^0)^2}{\bar{\varepsilon}^0}\right)\bar{r}B_1 - \bar{e}^0\bar{s}B_2 = 0,$$

$$\frac{e^0}{\varepsilon^0}A_1 + A_2 - \frac{\bar{e}^0}{\bar{\varepsilon}^0}B_1 - B_2 = 0,$$ (4.141)

$$-\varepsilon^0 sA_2 + \bar{\varepsilon}^0\bar{s}B_2 = 0.$$

For nontrivial solutions, the determinant of the coefficient matrix has to vanish, which yields the following frequency equation:

$$P_0\left(\beta - b_1\right)\left(\alpha + a_1\right) - P_1\left(\beta - b_2\right)\left(\alpha + a_1\right) - P_2\left(\beta - b_1\right)\left(\alpha + a_2\right)$$
$$+ P_3\left(\beta - b_1\right)\left(\beta - b_2\right) + P_4\left(\alpha + a_1\right)\left(\alpha + a_2\right) = 0,$$ (4.142)

where

$$P_0 = \left(e^0\sqrt{\bar{\varepsilon}^0/\varepsilon^0} - \bar{e}^0\sqrt{\varepsilon^0/\bar{\varepsilon}^0}\right)^2,$$

$$P_1 = \varepsilon^0[\bar{c}^0 + (\bar{e}^0)^2/\bar{\varepsilon}^0], \quad P_2 = \bar{\varepsilon}^0[c^0 + (e^0)^2/\varepsilon^0],$$ (4.143)

$$P_3 = \bar{\varepsilon}^0[\bar{c}^0 + (\bar{e}^0)^2/\bar{\varepsilon}^0], \quad P_4 = \varepsilon^0[c^0 + (e^0)^2/\varepsilon^0],$$

$$a_1 = \sqrt{\alpha^2 + \xi^2}, \quad a_2 = \sqrt{\alpha^2 + \xi^2\left(1 - v^2/v_T^2\right)},$$

$$b_1 = \sqrt{\beta^2 + \xi^2}, \quad b_2 = \sqrt{\beta^2 + \xi^2\left(1 - v^2/\bar{v}_T^2\right)}.$$ (4.144)

When $\alpha = 0$ and $\beta = 0$, Eq. (4.142) reduces to

$$P_0 - \left(P_1 + P_3\right)\sqrt{1 - \frac{v^2}{\bar{v}_T^2}} - \left(P_2 + P_4\right)\sqrt{1 - \frac{v^2}{v_T^2}} = 0,$$ (4.145)

which can be shown to be the same as Eq. (4.101), the result of [16] for interface waves between two unelectroded, homogeneous half-spaces. Wave velocities for various values of α and β are shown in Figs. 4.12 and 4.13.

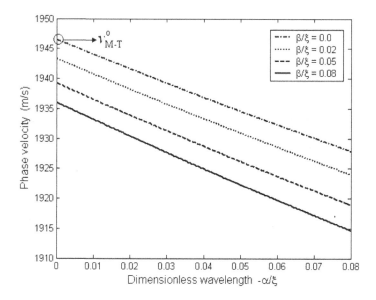

Fig. 4.12. Phase velocity versus dimensionless wavelength $-\alpha/\xi$ for selected values of β/ξ.

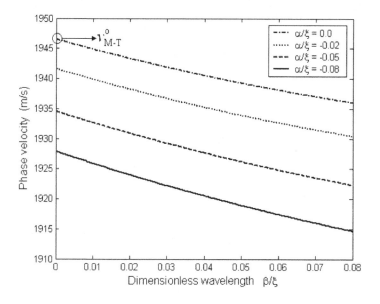

Fig. 4.13. Phase velocity versus dimensionless wavelength β/ξ for selected values of α/ξ.

4.8. Gap Waves between Two Half-spaces

Consider two piezoelectric half-spaces with a gap between them (see Fig. 4.14). The two surfaces of the half-spaces at $x_2 = \pm h$ are traction-free and unelectroded. Acoustic waves in the two half-spaces can be coupled by the electric field in the gap. We consider surface waves in the half-spaces near $x_2 = \pm h$ [19].

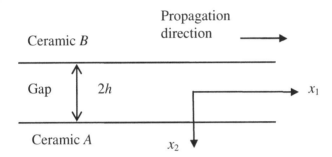

Fig. 4.14. Two half-spaces with a gap.

4.8.1. A gap between two half-spaces of different ceramics

First consider the general case when the two half-spaces are of different materials. The governing equations are

$$\bar{c}_A \nabla^2 u_A = \rho_A \ddot{u}_A, \quad \nabla^2 \psi_A = 0, \quad \phi_A = \psi_A + \frac{e_A}{\varepsilon_A} u_A, \quad x_2 > h, \quad (4.146)$$

$$\nabla^2 \phi = 0, \quad -h < x_2 < h, \quad (4.147)$$

$$\bar{c}_B \nabla^2 u_B = \rho_B \ddot{u}_B, \quad \nabla^2 \psi_B = 0, \quad \phi_B = \psi_B + \frac{e_B}{\varepsilon_B} u_B, \quad x_2 < -h, \quad (4.148)$$

where the subscripts A and B indicate fields in ceramic A and ceramic B. For waves near the gap, we require that

$$u_A, \psi_A \to 0, \quad x_2 \to +\infty, \quad (4.149)$$

$$u_B, \psi_B \to 0, \quad x_2 \to -\infty. \quad (4.150)$$

For $x_2 > h$, the solutions to Eq. (4.146) that satisfy Eq. (4.149) can be written as

$$u_A = U_A \exp(-\eta_A x_2)\cos(\xi x_1 - \omega t),$$
$$\psi_A = \Psi_A \exp(-\xi x_2)\cos(\xi x_1 - \omega t), \quad (4.151)$$

where U_A and Ψ_A are undetermined constants,

$$\eta_A^2 = \xi^2 - \frac{\rho_A \omega^2}{\overline{c}_A} = \xi^2 \left(1 - \frac{v^2}{v_A^2}\right) > 0, \qquad (4.152)$$

and

$$v = \frac{\omega}{\xi}, \quad v_A^2 = \frac{\overline{c}_A}{\rho_A}. \qquad (4.153)$$

For continuity conditions, we need T_{23} and D_2 in ceramic A, denoted by T_A and D_A:

$$\begin{aligned}
T_A &= \overline{c}_A u_{A,2} + e_A \psi_{A,2} \\
&= -[\overline{c}_A \eta_A U_A \exp(-\eta_A x_2) + e_A \xi \Psi_A \exp(-\xi x_2)]\cos(\xi x_1 - \omega t), \\
D_A &= -\varepsilon_A \psi_{A,2} \\
&= \varepsilon_A \xi \Psi_A \exp(-\xi x_2)\cos(\xi x_1 - \omega t).
\end{aligned} \qquad (4.154)$$

Similarly, for $x_2 < -h$, the solutions to Eq. (4.148) that satisfy Eq. (4.150) are

$$\begin{aligned}
u_B &= U_B \exp(\eta_B x_2)\cos(\xi x_1 - \omega t), \\
\psi_B &= \Psi_B \exp(\xi x_2)\cos(\xi x_1 - \omega t),
\end{aligned} \qquad (4.155)$$

where U_B and Ψ_B are undetermined constants,

$$\eta_B^2 = \xi^2 - \frac{\rho_B \omega^2}{\overline{c}_B} = \xi^2 \left(1 - \frac{v^2}{v_B^2}\right) > 0, \qquad (4.156)$$

and

$$v_B^2 = \frac{\overline{c}_B}{\rho_B}. \qquad (4.157)$$

For continuity conditions, we need T_{23} and D_2 in ceramic B, denoted by T_B and D_B:

$$\begin{aligned}
T_B &= \overline{c}_B u_{B,2} + e_B \psi_{B,2} \\
&= [\overline{c}_B \eta_B U_B \exp(\eta_B x_2) + e_B \xi \Psi_B \exp(\xi x_2)]\cos(\xi x_1 - \omega t), \\
D_B &= -\varepsilon_B \psi_{B,2} \\
&= -\varepsilon_B \xi \Psi_B \exp(\xi x_2)\cos(\xi x_1 - \omega t).
\end{aligned} \qquad (4.158)$$

The fields in the gap can be written as

$$\phi = (\Phi_1 \cosh \xi x_2 + \Phi_2 \sinh \xi x_2) \cos(\xi x_1 - \omega t), \tag{4.159}$$

where Φ_1 and Φ_2 are undetermined constants and Eq. (4.147) is already satisfied. For continuity conditions, we need D_2 in the gap:

$$\begin{aligned}
D_2 &= -\varepsilon_0 \phi_{,2} \\
&= -\varepsilon_0 (\xi \Phi_1 \sinh \xi x_2 + \xi \Phi_2 \cosh \xi x_2) \cos(\xi x_1 - \omega t).
\end{aligned} \tag{4.160}$$

For interface continuity conditions, we impose

$$T_A = 0, \quad \psi_A + \frac{e_A}{\varepsilon_A} u_A = \phi, \quad D_A = D_2, \quad x_2 = h,$$

$$T_B = 0, \quad \psi_B + \frac{e_B}{\varepsilon_B} u_B = \phi, \quad D_B = D_2, \quad x_2 = -h, \tag{4.161}$$

which implies that

$$\bar{c}_A \eta_A U_A \exp(-\eta_A h) + e_A \xi \Psi_A \exp(-\xi h) = 0,$$

$$\Psi_A \exp(-\xi h) + \frac{e_A}{\varepsilon_A} U_A \exp(-\eta_A h) = \Phi_1 \cosh \xi h + \Phi_2 \sinh \xi h,$$

$$\varepsilon_A \xi \Psi_A \exp(-\xi h) = -\varepsilon_0 (\xi \Phi_1 \sinh \xi h + \xi \Phi_2 \cosh \xi h),$$

$$\bar{c}_B \eta_B U_B \exp(-\eta_B h) + e_B \xi \Psi_B \exp(-\xi h) = 0, \tag{4.162}$$

$$\Psi_B \exp(-\xi h) + \frac{e_B}{\varepsilon_B} U_B \exp(-\eta_B h) = \Phi_1 \cosh \xi h - \Phi_2 \sinh \xi h,$$

$$\varepsilon_B \xi \Psi_B \exp(-\xi h) = \varepsilon_0 (-\xi \Phi_1 \sinh \xi h + \xi \Phi_2 \cosh \xi h).$$

Equation (4.162) can be rearranged into

$$\begin{pmatrix}
\bar{c}_A \eta_A e^{-\eta_A h} & e_A \xi e^{-\xi h} & 0 & 0 & 0 & 0 \\
\dfrac{e_A}{\varepsilon_A} e^{-\eta_A h} & e^{-\xi h} & -\cosh \xi h & -\sinh \xi h & 0 & 0 \\
0 & \varepsilon_A \xi e^{-\xi h} & \varepsilon_0 \xi \sinh \xi h & \varepsilon_0 \xi \cosh \xi h & 0 & 0 \\
0 & 0 & \varepsilon_0 \xi \sinh \xi h & -\varepsilon_0 \xi \cosh \xi h & 0 & \varepsilon_B \xi e^{-\xi h} \\
0 & 0 & -\cosh \xi h & \sinh \xi h & \dfrac{e_B}{\varepsilon_B} e^{-\eta_B h} & e^{-\xi h} \\
0 & 0 & 0 & 0 & \bar{c}_B \eta_B e^{-\eta_B h} & e_B \xi e^{-\xi h}
\end{pmatrix}
\begin{pmatrix}
U_A \\
\Psi_A \\
\Phi_1 \\
\Phi_2 \\
U_B \\
\Psi_B
\end{pmatrix} = 0.$$

$$\tag{4.163}$$

For nontrivial solutions the determinant of the coefficient matrix has to vanish, which yields the dispersion relations of the waves. One dispersion relation obtained by numerical solution is shown in Fig. 4.15.

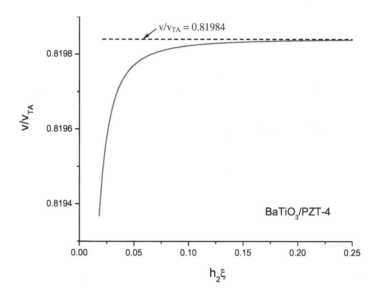

Fig. 4.15. v/v_A versus $2\xi h$ for two half-spaces of BaTiO₃/PZT-4 ($v_{TA} = v_A$, $h_2 = 2h$).

4.8.2. A gap between two half-spaces of the same material

We examine the special case when the two ceramic half-spaces are of the same material:

$$\bar{c}_A = \bar{c}_B = \bar{c}, \quad e_A = e_B = e,$$
$$\varepsilon_A = \varepsilon_B = \varepsilon, \quad \rho_A = \rho_B = \rho, \quad (4.164)$$
$$v_A = v_B = v_T.$$

Then the waves can be separated into symmetric and antisymmetric ones.

For symmetric waves we consider

$$U_A = U_B, \quad \Psi_A = \Psi_B, \quad \Phi_2 = 0, \tag{4.165}$$

for which the last three of Eq. (4.162) become identical to the first three, and the dispersion relation assumes the following simple form:

$$\xi \begin{vmatrix} \overline{c}\eta & e\xi & 0 \\ \dfrac{e}{\varepsilon} & 1 & -\cosh\xi h \\ 0 & \varepsilon & \varepsilon_0\sinh\xi h \end{vmatrix} = 0, \tag{4.166}$$

or

$$\cosh\xi h \begin{vmatrix} \overline{c}\eta & e\xi \\ 0 & \varepsilon \end{vmatrix} + \varepsilon_0\sinh\xi h \begin{vmatrix} \overline{c}\eta & e\xi \\ \dfrac{e}{\varepsilon} & 1 \end{vmatrix} \tag{4.167}$$

$$= \varepsilon\overline{c}\eta\cosh\xi h + \varepsilon_0(\overline{c}\eta - \xi\dfrac{e^2}{\varepsilon})\sinh\xi h = 0.$$

Equation (4.167) can be further written as

$$\tanh\xi h = \frac{\varepsilon\overline{c}\eta}{\varepsilon_0(\xi\dfrac{e^2}{\varepsilon} - \overline{c}\eta)} = \frac{n^2\eta}{\xi\overline{k}^2 - \eta}, \tag{4.168}$$

where we have denoted

$$n^2 = \varepsilon/\varepsilon_0. \tag{4.169}$$

With Eq. (4.152), Eq. (4.168) takes the following form:

$$\tanh\xi h = \frac{n^2\sqrt{1 - \dfrac{v^2}{v_T^2}}}{\overline{k}^2 - \sqrt{1 - \dfrac{v^2}{v_T^2}}}. \tag{4.170}$$

Equation (4.170) shows that the wave is dispersive.
For antisymmetric waves we consider

$$U_A = -U_B, \quad \Psi_A = -\Psi_B, \quad \Phi_1 = 0. \tag{4.171}$$

Then the last three of Eq. (4.162) become identical to the first three, and the dispersion relation assumes the following simple form:

$$\xi \begin{vmatrix} \bar{c}\eta & e\xi & 0 \\ \dfrac{e}{\varepsilon} & 1 & -\sinh\xi h \\ 0 & \varepsilon & \varepsilon_0 \cosh\xi h \end{vmatrix} = 0, \tag{4.172}$$

or

$$\sinh\xi h \begin{vmatrix} \bar{c}\eta & e\xi \\ 0 & \varepsilon \end{vmatrix} + \varepsilon_0 \cosh\xi h \begin{vmatrix} \bar{c}\eta & e\xi \\ \dfrac{e}{\varepsilon} & 1 \end{vmatrix}$$

$$= \varepsilon\bar{c}\eta\sinh\xi h + \varepsilon_0(\bar{c}\eta - \xi\dfrac{e^2}{\varepsilon})\cosh\xi h = 0. \tag{4.173}$$

Equation (4.173) can be further written as

$$\tanh\xi h = \frac{\varepsilon_0(\xi\dfrac{e^2}{\varepsilon} - \bar{c}\eta)}{\varepsilon\bar{c}\eta} = \frac{\xi\bar{k}^2 - \eta}{n^2\eta}. \tag{4.174}$$

With Eq. (4.152), Eq. (4.174) takes the following form:

$$\tanh\xi h = \left(\frac{n^2\sqrt{1 - \dfrac{v^2}{v_T^2}}}{\bar{k}^2 - \sqrt{1 - \dfrac{v^2}{v_T^2}}} \right)^{-1}. \tag{4.175}$$

We note that the right-hand side of Eq. (4.175) is the inverse of that of Eq. (4.170). Therefore for any $v < v_T$, one of the right-hand sides of Eqs. (4.175) and (4.170) is smaller than 1. Since the range of a hyperbolic tangent function is between -1 and 1, a root for ξ can be found from either Eqs. (4.175) or (4.170). Solutions to Eqs. (4.170) or (4.175) are shown in Fig. 4.16 for different materials.

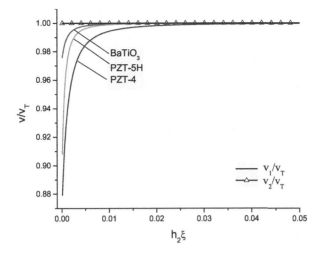

Fig. 4.16. v/v_T versus $h_2 = 2h$. v_1: antisymmetric wave. v_2: symmetric wave.

4.9. Waves over a Circular Cylindrical Surface

Consider waves propagating on the surface of a circular cylinder [20] as shown in Fig. 4.17. We choose (r, θ, z) to correspond to (1, 2, 3) so that the poling direction corresponds to 3.

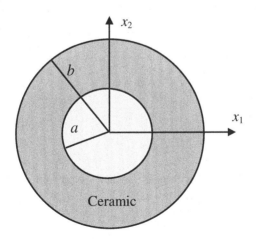

Fig. 4.17. A circular cylinder.

The governing equations are

$$\bar{c}\nabla^2 u = \rho\ddot{u}, \quad a < r < b,$$
$$\nabla^2 \psi = 0, \quad a < r < b, \tag{4.176}$$
$$\phi = \psi + \frac{e}{\varepsilon}u, \quad a < r < b.$$

The stress and electric displacement components relevant to boundary conditions are

$$T_{rz} = \bar{c}u_{,r} + e\psi_{,r},$$
$$D_r = -\varepsilon\psi_{,r}. \tag{4.177}$$

In polar coordinates we have

$$v_T^2\left(\frac{\partial^2 u}{\partial r^2} + \frac{1}{r}\frac{\partial u}{\partial r} + \frac{1}{r^2}\frac{\partial^2 u}{\partial\theta^2}\right) = \ddot{u},$$
$$\frac{\partial^2 \psi}{\partial r^2} + \frac{1}{r}\frac{\partial \psi}{\partial r} + \frac{1}{r^2}\frac{\partial^2 \psi}{\partial\theta^2} = 0, \tag{4.178}$$

where $v_T^2 = \bar{c}/\rho$. For waves propagating in the θ direction we consider

$$u_z(r,\theta,t) = u(r)\cos(v\theta - \omega t),$$
$$\psi(r,\theta,t) = \psi(r)\cos(v\theta - \omega t), \tag{4.179}$$

where v is allowed to assume any real value. Substitution of Eq. (4.179) into Eq. (4.178) results in

$$\frac{\partial^2 u}{\partial r^2} + \frac{1}{r}\frac{\partial u}{\partial r} + (\xi^2 - \frac{v^2}{r^2})u = 0,$$
$$\frac{\partial^2 \psi}{\partial r^2} + \frac{1}{r}\frac{\partial \psi}{\partial r} - \frac{v^2}{r^2}\psi = 0, \tag{4.180}$$

where we have denoted

$$\xi = \frac{\omega}{v_T}. \tag{4.181}$$

Equation (4.180)₁ can be written as Bessel's equation of order v. Then the general solution can be written as

$$u_z = [C_1 J_v(\xi r) + C_2 Y_v(\xi r)]\cos(v\theta - \omega t),$$
$$\psi = [C_3 r^v + C_4 r^{-v}]\cos(v\theta - \omega t), \tag{4.182}$$

where J_v and Y_v are the v-th order Bessel functions of the first and second kind, and C_1–C_4 are undetermined constants. The following expressions are needed for boundary conditions:

$$\phi = \{C_3 r^v + C_4 r^{-v} + \frac{e}{\varepsilon}[C_1 J_v(\xi r) + C_2 Y_v(\xi r)]\}\cos(v\theta - \omega t),$$

$$T_{rz} = \{\bar{c}[C_1 \xi J_v'(\xi r) + C_2 \xi Y_v'(\xi r)]$$

$$+ e[C_3 v r^{v-1} - C_4 v r^{-v-1}]]\}\cos(v\theta - \omega t), \tag{4.183}$$

$$D_r = -\varepsilon(C_3 v r^{v-1} - C_4 v r^{-v-1}))\}\cos(v\theta - \omega t),$$

where a superimposed prime indicates differentiation with respect to the whole argument of a function. Consider a solid cylinder ($a = 0$ in Fig. 4.17). The surface at $r = b$ is traction-free and carries a very thin electrode of a perfect conductor. Consider the case when v is greater than or equal to 1. Since Y_v and r^{-v-1} are singular at the origin, terms associated with C_2 and C_4 are dropped. T_{rz} and ϕ should both vanish at $r = b$. This leads to the following two homogeneous linear algebraic equations for C_1 and C_3:

$$\bar{c} C_1 \xi J_v'(\xi b) + e C_3 v b^{v-1} = 0,$$

$$C_3 b^v + \frac{e}{\varepsilon} C_1 J_v(\xi b) = 0. \tag{4.184}$$

For nontrivial solutions the determinant of the coefficient matrix has to vanish, which results in the following equation for ξ:

$$\begin{vmatrix} \bar{c}\xi J_v'(\xi b) & evb^{v-1} \\ \frac{e}{\varepsilon} J_v(\xi b) & b^v \end{vmatrix} \tag{4.185}$$

$$= \bar{c}\xi J_v'(\xi b) b^v - \frac{e^2}{\varepsilon} J_v(\xi b) v b^{v-1} = 0,$$

or

$$\frac{\xi b J_v'(\xi b)}{v J_v(\xi b)} = \bar{k}^2. \tag{4.186}$$

Consider the special case when the wavelength is much smaller than the cylinder radius. Then the cylinder is effectively like a half-space for these short waves. It is convenient to introduce a surface wave number η and a surface wave speed V by

$$\eta = \frac{v}{b}, \quad V = \frac{\omega}{\eta}. \tag{4.187}$$

Consider the limit when $v \to \infty$ and $b \to \infty$ but η remains finite. We have

$$\xi b = \frac{\omega}{v_T} b = \frac{\omega}{v_T} \frac{b}{v} v = \frac{\omega}{v_T} \frac{1}{\eta} v = \frac{V}{v_T} v. \tag{4.188}$$

With the following asymptotic expression due to Carlini for Bessel functions of large orders:

$$J_v(vx) \to \frac{x^v \exp(v\sqrt{1-x^2})}{\sqrt{2\pi v} \sqrt[4]{1-x^2} (1+\sqrt{1-x^2})^v}, \quad v \to \infty, \quad 0 < x < 1, \tag{4.189}$$

Eq. (4.186) reduces to

$$\sqrt{1 - \frac{V^2}{v_T^2}} = \bar{k}^2, \tag{4.190}$$

which gives the speed of Bleustein–Gulyaev surface waves (see Eq. (4.12)).

Chapter 5
Waves in Plates

.

This chapter is on waves in plates. These waves are widely used to make BAW devices. They are dispersive in general. In addition to basic propagation characteristics, the effects of electrodes, thin films and contact with fluids are examined. Important concepts like cutoff frequencies and energy trapping are discussed.

5.1. An Electroded Plate

First we consider an electroded plate with shorted electrodes (see Fig. 5.1) [21]. The surfaces of the plate are traction-free. The electrodes are grounded. We assume very thin electrodes. Their mechanical effects are negligible.

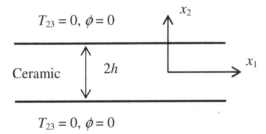

Fig. 5.1. A ceramic plate with shorted electrodes.

The governing equations are

$$\bar{c}\nabla^2 u = \rho \ddot{u}, \quad \nabla^2 \psi = 0, \quad \phi = \psi + \frac{e}{\varepsilon} u, \quad -h < x_2 < h. \tag{5.1}$$

The boundary conditions are

$$\begin{aligned} T_{23} &= 0, \quad x_2 = \pm h, \\ \phi &= 0, \quad x_2 = \pm h, \end{aligned} \tag{5.2}$$

or, in terms of u and ψ,

$$\bar{c}u_{,2} + e\psi_{,2} = 0, \quad x_2 = \pm h,$$

$$\psi + \frac{e}{\varepsilon}u = 0, \quad x_2 = \pm h. \tag{5.3}$$

There are two types of waves that can propagate in the plate. One may be called symmetric and the other antisymmetric. We discuss them separately below.

5.1.1. Antisymmetric waves

For antisymmetric waves we consider the possibility of

$$u = A\sin\xi_2 x_2 \cos(\xi_1 x_1 - \omega t),$$

$$\psi = B\sinh\xi_1 x_2 \cos(\xi_1 x_1 - \omega t), \tag{5.4}$$

where A and B are constants. For Eq. (5.4) to satisfy Eq. (5.1), we must have

$$\xi_2^2 = \frac{\rho\omega^2}{\bar{c}} - \xi_1^2 = \xi_1^2(\frac{v^2}{v_T^2} - 1),$$

$$v^2 = \frac{\omega^2}{\xi_1^2}, \quad v_T^2 = \frac{\bar{c}}{\rho}. \tag{5.5}$$

Substitution of Eq. (5.4) into Eq. (5.3) leads to

$$\bar{c}A\xi_2 \cos\xi_2 h + eB\xi_1 \cosh\xi_1 h = 0,$$

$$\frac{e}{\varepsilon}A\sin\xi_2 h + B\sinh\xi_1 h = 0, \tag{5.6}$$

which is a system of linear homogeneous equations for A and B. For nontrivial solutions we must have

$$\frac{\tan\xi_2 h}{\tanh\xi_1 h} = \frac{\xi_2}{\xi_1}\bar{k}^{-2}. \tag{5.7}$$

Equation (5.7) can be written as

$$\frac{\tan\frac{\pi}{2}(\Omega^2 - Z^2)^{1/2}}{\tanh\frac{\pi}{2}Z} = \frac{(\Omega^2 - Z^2)^{1/2}}{\bar{k}^2 Z}, \tag{5.8}$$

where the dimensionless frequency and the dimensionless wave number in the x_1 direction are defined by

$$\Omega^2 = \omega^2 \bigg/ \left(\frac{\pi^2 \bar{c}}{4\rho h^2} \right), \quad Z = \xi_1 \bigg/ \left(\frac{\pi}{2h} \right). \tag{5.9}$$

In the limit when $Z \to 0$, Eq. (5.8) reduces to

$$\tan \frac{\pi}{2} \Omega = \frac{\pi}{2} \frac{\Omega}{\bar{k}^2}, \tag{5.10}$$

which is the frequency equation for antisymmetric thickness-shear modes in a plate.

5.1.2. Symmetric waves

For symmetric waves we consider

$$u = A \cos \xi_2 x_2 \cos(\xi_1 x_1 - \omega t),$$
$$\psi = B \cosh \xi_1 x_2 \cos(\xi_1 x_1 - \omega t). \tag{5.11}$$

where A and B are constants. For Eq. (5.11) to satisfy Eq. (5.1), we still have Eq. (5.5). Substitution of Eq. (5.11) into Eq. (5.3) leads to

$$-\bar{c} A \xi_2 \sin \xi_2 h + eB \xi_1 \sinh \xi_1 h = 0,$$
$$\frac{e}{\varepsilon} A \cos \xi_2 h + B \cosh \xi_1 h = 0. \tag{5.12}$$

For nontrivial solutions of A and/or B, we must have

$$\frac{\tan \xi_2 h}{\tanh \xi_1 h} = -\bar{k}^2 \frac{\xi_1}{\xi_2}, \tag{5.13}$$

or

$$\frac{\tan \frac{\pi}{2}(\Omega^2 - Z^2)^{1/2}}{\tanh \frac{\pi}{2} Z} = -\frac{\bar{k}^2 Z}{(\Omega^2 - Z^2)^{1/2}}. \tag{5.14}$$

5.1.3. Numerical results

The first few branches of Eqs. (5.8) and (5.14) are plotted in Fig. 5.2. Waves in plates are dispersive in general. The lowest branch represents a wave that has no nodal points along the plate thickness and is called the face-shear wave (FS). Higher-order waves have nodal points along the plate thickness and are called thickness-twist waves (TT).

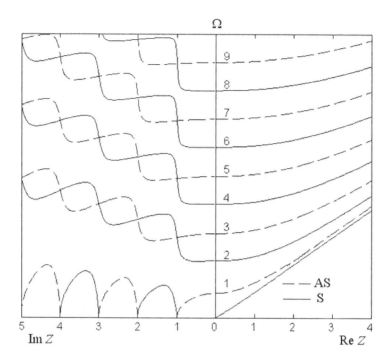

Fig. 5.2. Dispersion relations for waves in a plate.

5.2. An Unelectroded Plate

Next consider the case when the plate is unelectroded and there are electric fields in the adjacent half-spaces (see Fig. 5.3) [21]. We will discuss symmetric and antisymmetric waves separately.

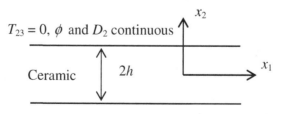

Fig. 5.3. An unelectroded plate.

5.2.1. Antisymmetric waves

For antisymmetric waves we have (see Eq. (5.4))

$$u = A \sin \xi_2 x_2 \cos(\xi_1 x_1 - \omega t),$$
$$\psi = B \sinh \xi_1 x_2 \cos(\xi_1 x_1 - \omega t), \tag{5.15}$$

where A and B are constants. In the free space the electric potential $\hat{\phi}$ is governed by

$$\nabla^2 \hat{\phi} = 0, \quad |x_2| > h,$$
$$\hat{\phi} \to 0, \quad x_2 \to \pm\infty. \tag{5.16}$$

The solution to Eq. (5.16) is

$$\hat{\phi} = \begin{cases} C \exp[\xi_1 (h - x_2)] \cos(\xi_1 x_1 - \omega t), & x_2 > h, \\ -C \exp[\xi_1 (h + x_2)] \cos(\xi_1 x_1 - \omega t), & x_2 < -h, \end{cases} \tag{5.17}$$

which C is an undetermined constant. The boundary and continuity conditions at the plate surfaces are

$$T_{23} = 0, \quad \phi = \hat{\phi}, \quad D_2 = \hat{D}_2, \quad x_2 = \pm h. \tag{5.18}$$

Substitution of Eqs. (5.15) and (5.17) into Eq. (5.18) yields three linear homogeneous equations for A, B and C:

$$\bar{c} A \xi_2 \cos \xi_2 h + e B \xi_1 \cosh \xi_1 h = 0,$$
$$\frac{e}{\varepsilon} A \sin \xi_2 h + B \sinh \xi_1 h = C, \tag{5.19}$$
$$- \varepsilon B \xi_1 \cosh \xi_1 h = -\varepsilon_0 C \xi_1.$$

For nontrivial solutions the determinant of the coefficient matrix of the linear equations has to vanish, which leads to the following frequency equation:

$$n^2 + \tanh \xi_1 h - \bar{k}^2 \frac{\xi_1}{\xi_2} \tan \xi_2 h = 0, \tag{5.20}$$

or

$$\frac{\tan \frac{\pi}{2} (\Omega^2 - Z^2)^{1/2}}{\tanh \frac{\pi}{2} Z + \frac{\varepsilon}{\varepsilon_0}} = \frac{(\Omega^2 - Z^2)^{1/2}}{\bar{k}^2 Z}. \tag{5.21}$$

5.2.2. Symmetric waves

For symmetric waves we consider (Eq. (5.11))

$$u = A\cos\xi_2 x_2 \cos(\xi_1 x_1 - \omega t),$$
$$\psi = B\cosh\xi_1 x_2 \cos(\xi_1 x_1 - \omega t). \tag{5.22}$$

The solution to Eq. (5.16) is

$$\hat{\phi} = \begin{cases} C\exp[\xi_1(h - x_2)]\cos(\xi_1 x_1 - \omega t), & x_2 > h, \\ C\exp[\xi_1(h + x_2)]\cos(\xi_1 x_1 - \omega t), & x_2 < -h. \end{cases} \tag{5.23}$$

Substitution of Eqs. (5.22) and (5.23) into Eq. (5.18) yields three linear homogeneous equations for A, B and C:

$$-\bar{c}A\xi_2\sin\xi_2 h + eB\xi_1\sinh\xi_1 h = 0,$$
$$\frac{e}{\varepsilon}A\cos\xi_2 h + B\cosh\xi_1 h = C, \tag{5.24}$$
$$-\varepsilon B\xi_1\sinh\xi_1 h = -\varepsilon_0 C\xi_1.$$

For nontrivial solutions the determinant of the coefficient matrix of the linear equations has to vanish, which leads to the following frequency equation:

$$n^2 + \mathrm{ctanh}\,\xi_1 h + \bar{k}^2\frac{\xi_1}{\xi_2}\mathrm{ctan}\,\xi_2 h = 0, \tag{5.25}$$

or

$$\frac{\left(1 + \dfrac{\varepsilon}{\varepsilon_0}\tanh\dfrac{\pi}{2}Z\right)\tan\dfrac{\pi}{2}(\Omega^2 - Z^2)^{1/2}}{\tanh\dfrac{\pi}{2}Z} = -\frac{\bar{k}^2 Z}{(\Omega^2 - Z^2)^{1/2}}. \tag{5.26}$$

Next we examine the effect of the electric field in the free space surrounding the plate [22]. For symmetric waves, if we neglect the free-space electric field as an approximation and impose the following boundary conditions at the plate surfaces:

$$D_2(x_3 = \pm h) = 0, \tag{5.27}$$

the dispersion relation becomes

$$\sinh(\frac{\pi}{2}Z)\sin[\frac{\pi}{2}(\Omega^2 - Z^2)^{1/2}] = 0. \tag{5.28}$$

Note that Eq. (5.28) can also be obtained from Eq. (5.26) by letting $\varepsilon \to \infty$. However, careful examination of this limit procedure shows that the procedure is questionable when Z is very small, i.e., when long waves are under consideration. In this case, $\varepsilon Z/\varepsilon_0$ may not be much larger than 1, which is needed in the reduction from Eqs. (5.26) to (5.28). For small Z, Ω may or may not be small depending on which branch of the dispersion curves is under consideration. We examine the face-shear wave which is the lowest order symmetric wave determined by Eqs. (5.26) or (5.28), and for which Ω is also small when Z is small. Then, from Eq. (5.26), for small Z and small Ω, we have

$$\Omega^2 \cong Z^2 - \bar{k}^2 \frac{Z^2}{1 + \dfrac{\varepsilon}{\varepsilon_0}\dfrac{\pi}{2}Z}, \tag{5.29}$$

which shows a dispersive wave. However, from Eq. (5.28) we obtain a nondispersive wave with

$$\Omega = Z . \tag{5.30}$$

Therefore there is a qualitative difference between the dispersion relation determined by Eq. (5.26) and the approximation in Eq. (5.28). The dispersion relation of face-shear waves from Eq. (5.26) is determined numerically and is plotted with Eq. (5.30) in Fig. 5.4. PZT-6B is used in the calculation, which has a fairly large dielectric constant of $\varepsilon = 407\,\varepsilon_0$. The quantitative difference between the two dispersion curves may or

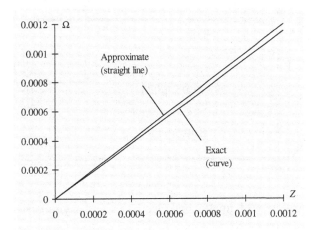

Fig. 5.4. Effect of the free-space electric field.

may not be considered as small in different applications. However, it should be emphasized that in Fig. 5.4 the straight line is not a tangent of the curve at the origin. Therefore prescribing $D_2 = 0$ at the plate surfaces does not yield an approximate dispersion relation for long waves in the sense that for the long wave limit the two dispersion relations do not share a common tangent.

5.3. A Plate with Unattached Electrodes

In device applications, unattached electrodes at some distance from the plate surfaces have been used for different purposes. Unattached or separated electrodes offer new design parameters and certain advantages over electrodes deposited on the plate surfaces. Various undesirable effects of deposited (or attached) electrodes disappear when the electrodes are unattached. This includes, e.g., the residual stress in the electrodes and the plates, the effect of electrode irregularity from manufacturing on frequency, and the field concentration at electrode edges. In this section we study thickness-twist waves propagating in a plate with unattached electrodes [23].

5.3.1. Governing equations and fields

Consider the piezoelectric plate shown in Fig. 5.5. The electrodes are shorted.

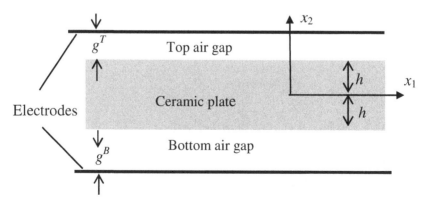

Fig. 5.5. A plate with unattached electrodes and asymmetric air gaps.

The governing equations are

$$\bar{c}\nabla^2 u = \rho\ddot{u}, \quad \nabla^2\psi = 0, \quad \phi = \psi + \frac{e}{\varepsilon}u, \quad |x_2| < h \tag{5.31}$$

$$\nabla^2\phi^T = 0, \quad h \le x_2 \le h + g^T, \tag{5.32}$$

$$\nabla^2\phi^B = 0, \quad -h - g^B \le x_2 \le -h, \tag{5.33}$$

where ϕ^T and ϕ^B are the electric potentials in the top and bottom gaps, respectively. The boundary and continuity conditions are

$$\begin{aligned}
\phi^T &= 0, \quad x_2 = h + g^T, \\
\phi &= \phi^T, \quad D_2 = D_2^T, \quad T_{23} = 0, \quad x_2 = h, \\
\phi &= \phi^B, \quad D_2 = D_2^B, \quad T_{23} = 0, \quad x_2 = -h, \\
\phi^B &= 0, \quad x_2 = -h - g^B.
\end{aligned} \tag{5.34}$$

Consider the following propagating waves in the plate:

$$\begin{aligned}
u &= \left(A_1 \sin\xi_2 x_2 + A_2 \cos\xi_2 x_2\right)\cos\left(\xi_1 x_1 - \omega t\right), \\
\psi &= \left(A_3 \sinh\xi_1 x_2 + A_4 \cosh\xi_1 x_2\right)\cos\left(\xi_1 x_1 - \omega t\right),
\end{aligned} \tag{5.35}$$

where A_1 through A_4 are undetermined constants, ω is the wave frequency, and ξ_1 and ξ_2 are waves numbers in the x_1 and x_2 directions. Equation (5.35) satisfies Eq. (5.31) when

$$\xi_2^2 = \xi_1^2\left(\frac{v^2}{v_T^2} - 1\right), \tag{5.36}$$

where the wave speed v is given by $\xi_1^2 = \omega^2/v^2$ and $v_T^2 = \bar{c}/\rho$. Similarly, the waves in the gaps are written as

$$\phi^T = A_5 \sinh\xi_1\left(x_2 - h - g^T\right)\cos\left(\xi_1 x_1 - \omega t\right), \tag{5.37}$$

$$\phi^B = A_6 \sinh\xi_1\left(x_2 + h + g^B\right)\cos\left(\xi_1 x_1 - \omega t\right), \tag{5.38}$$

where A_5 and A_6 are undetermined constants. Equations (5.37) and (5.38) satisfy Eqs. (5.32) and (5.33), respectively, as well as the boundary conditions at the top and bottom electrodes, i.e., the first and last equations in Eq. (5.34).

5.3.2. Dispersion relation

Substitution of Eqs. (5.35), (5.37) and (5.38) into the remaining boundary and continuity conditions at $x_2 = \pm h$ in Eq. (5.34) yields six linear homogeneous equations for A_1 through A_6. For nontrivial solutions of the undetermined constants, the determinant of the coefficient matrix of the linear equations has to vanish, which leads to the following dispersion relation of the waves:

$$\frac{n^2 \tanh \xi_1 g^T + \tanh \xi_1 h - \bar{k}^2 \frac{\xi_1}{\xi_2} \tan \xi_2 h}{n^2 \tanh \xi_1 g^B + \tanh \xi_1 h - \bar{k}^2 \frac{\xi_1}{\xi_2} \tan \xi_2 h}$$

$$= -\frac{n^2 \tanh \xi_1 g^T + \text{ctanh} \xi_1 h - \bar{k}^2 \frac{\xi_1}{\xi_2} \text{ctan} \xi_2 h}{n^2 \tanh \xi_1 g^B + \text{ctanh} \xi_1 h - \bar{k}^2 \frac{\xi_1}{\xi_2} \text{ctan} \xi_2 h},$$

(5.39)

where $n^2 = \varepsilon/\varepsilon_0$ is the refractive index. Equation (5.39) determines v versus ξ_1 or ω versus ξ_1. We examine some special cases below. When the gaps have the same thickness, i.e., $g^T = g^B = g$, Eq. (5.39) factors into two equations that determine two groups of waves. One may be called antisymmetric waves and the other symmetric. The displacement u of the antisymmetric or symmetric wave is an odd and or function of x_2, respectively. The corresponding dispersion relations are

$$n^2 \tanh \xi_1 g + \tanh \xi_1 h - \bar{k}^2 \frac{\xi_1}{\xi_2} \tan \xi_2 h = 0, \text{ (AS)}$$

(5.40)

$$n^2 \tanh \xi_1 g + \text{ctanh} \xi_1 h + \bar{k}^2 \frac{\xi_1}{\xi_2} \text{ctan} \xi_2 h = 0. \text{ (S)}$$

(5.41)

In particular, if the gaps are not present, i.e., when $g^T = g^B = 0$, Eqs. (5.40) and (5.41) reduce to (see Eqs. (5.7) and (5.13))

$$\frac{\tan \xi_2 h}{\tanh \xi_1 h} = \bar{k}^{-2} \frac{\xi_2}{\xi_1}, \text{ (AS)}$$

(5.42)

$$\frac{\tan \xi_2 h}{\tanh \xi_1 h} = -\bar{k}^2 \frac{\xi_1}{\xi_2}. \text{ (S)}$$

(5.43)

If the electrodes are very far away, i.e., when $g^T = g^B = \infty$, Eqs. (5.40) and (5.41) reduce to (see Eqs. (5.20) and (5.25))

$$n^2 + \tanh \xi_1 h - \overline{k}^2 \frac{\xi_1}{\xi_2} \tan \xi_2 h = 0 \text{, (AS)} \tag{5.44}$$

$$n^2 + \operatorname{ctanh} \xi_1 h + \overline{k}^2 \frac{\xi_1}{\xi_2} \operatorname{ctan} \xi_2 h = 0 \text{. (S)} \tag{5.45}$$

5.3.3. Numerical results

We introduce the following dimensionless frequency Ω and the dimensionless wave number Z by:

$$\Omega^2 = \omega^2 \Big/ \left(\frac{\pi^2 \overline{c}}{4\rho h^2} \right), \quad Z = \xi_1 \Big/ \left(\frac{\pi}{2h} \right). \tag{5.46}$$

We also introduce a thickness ration $m = g/h$. Numerical results for Ω versus Z are obtained by solving Eqs. (5.40) and (5.41) using PZT-5H, and are shown in Fig. 5.6 for $m = 0$ and $m = 0.0001$. The case of $m = 0$ is determined by Eqs. (5.42) and (5.43). The figure shows that the frequencies of short waves with a large wave number Z are more sensitive to m than long waves with a small Z. For a very small $m = 0.0001$, the effect on short waves is already significant. The gaps raise the frequencies of short waves with a large Z.

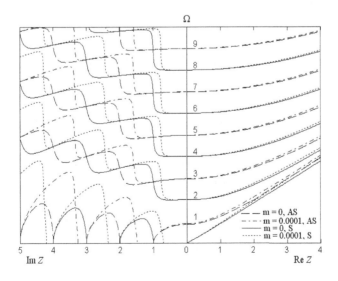

Fig. 5.6. Effect of air gaps on dispersion relations waves.

5.4. A Plate with Thin Films

In this section we consider the inertial effect of identical mass layers on waves in a plate (see Fig. 5.7) [13]. The mass layers are very thin. Their inertial effect is considered but the stiffness effect is neglected. The major surfaces of the plate are traction-free and electroded, and the electrodes are grounded.

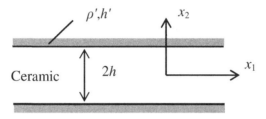

Fig. 5.7. A plate with identical mass layers.

5.4.1. Governing equations

The governing equations are

$$\bar{c}\nabla^2 u = \rho\ddot{u}, \quad \nabla^2\psi = 0, \quad \phi = \psi + \frac{e}{\varepsilon}u, \quad -h < x_2 < h. \quad (5.47)$$

The boundary conditions are

$$T_{23} = \mp\rho'h'\ddot{u}, \quad x_2 = \pm h,$$
$$\phi = 0, \quad x_2 = \pm h, \quad (5.48)$$

or, in terms of u and ψ,

$$\bar{c}u_{,2} + e\psi_{,2} = \mp\rho'h'\ddot{u}, \quad x_2 = \pm h,$$
$$\psi + \frac{e}{\varepsilon}u = 0, \quad x_2 = \pm h. \quad (5.49)$$

The waves that we are interested in may be classified as symmetric and antisymmetric. We discuss them separately below.

5.4.2. Antisymmetric waves

For antisymmetric waves we have (from Eqs. (5.4) and (5.5))

$$u = A\sin\xi_2 x_2 \cos(\xi_1 x_1 - \omega t),$$
$$\psi = B\sinh\xi_1 x_2 \cos(\xi_1 x_1 - \omega t), \quad (5.50)$$

$$\xi_2^2 = \frac{\rho\,\omega^2}{\overline{c}} - \xi_1^2 = \xi_1^2(\frac{v^2}{v_T^2} - 1),$$

$$v^2 = \frac{\omega^2}{\xi_1^2}, \quad v_T^2 = \frac{\overline{c}}{\rho}.$$

(5.51)

where A and B are undetermined constants. Substitution of Eq. (5.50) into Eq. (5.49) yields

$$\overline{c}A\xi_2\cos\xi_2 h + eB\xi_1\cosh\xi_1 h = \rho'h'\omega^2 A\sin\xi_2 h,$$

$$\frac{e}{\varepsilon}A\sin\xi_2 h + B\sinh\xi_1 h = 0.$$

(5.52)

For nontrivial solutions we must have

$$\begin{vmatrix} \overline{c}\,\xi_2\cos\xi_2 h - \rho'h'\omega^2\sin\xi_2 h & e\xi_1\cosh\xi_1 h \\ \dfrac{e}{\varepsilon}\sin\xi_2 h & \sinh\xi_1 h \end{vmatrix}$$

$$= \begin{vmatrix} \overline{c}\,\xi_2\cos\xi_2 h & e\xi_1\cosh\xi_1 h \\ \dfrac{e}{\varepsilon}\sin\xi_2 h & \sinh\xi_1 h \end{vmatrix} + \begin{vmatrix} -\rho'h'\omega^2\sin\xi_2 h & e\xi_1\cosh\xi_1 h \\ 0 & \sinh\xi_1 h \end{vmatrix}$$

(5.53)

$$= \overline{c}\,\xi_2\cos\xi_2 h\sinh\xi_1 h - \frac{e^2}{\varepsilon}\xi_1\sin\xi_2 h\cosh\xi_1 h$$

$$- \rho'h'\omega^2\sin\xi_2 h\sinh\xi_1 h = 0,$$

or

$$\tanh\xi_1 h - \overline{k}^2\frac{\xi_1}{\xi_2}\tan\xi_2 h = \frac{\rho'h'\omega^2}{\overline{c}\,\xi_2}\tan\xi_2 h\tanh\xi_1 h .$$

(5.54)

When $\rho'h' = 0$, Eq. (5.54) reduces to

$$\tanh\xi_1 h - \overline{k}^2\frac{\xi_1}{\xi_2}\tan\xi_2 h = 0 ,$$

(5.55)

which is Eq. (5.7). The difference in frequencies determined from Eqs. (5.54) and (5.55) can be used to design mass sensors for measuring $\rho'h'$.

5.4.3. Symmetric waves

For symmetric waves we consider (from Eq. (5.11))

$$u = A\cos\xi_2 x_2 \cos(\xi_1 x_1 - \omega t),$$
$$\psi = B\cosh\xi_1 x_2 \cos(\xi_1 x_1 - \omega t).$$

$$(5.56)$$

We still have Eq. (5.51). Substitution of Eq. (5.56) into Eq. (5.49) leads to

$$-\bar{c}A\xi_2 \sin\xi_2 h + eB\xi_1 \sinh\xi_1 h = \rho'h'\omega^2 A\cos\xi_2 h,$$
$$\frac{e}{\varepsilon}A\cos\xi_2 h + B\cosh\xi_1 h = 0.$$

$$(5.57)$$

For nontrivial solutions of A and/or B, we must have

$$\begin{vmatrix} -\bar{c}\xi_2 \sin\xi_2 h - \rho'h'\omega^2 \cos\xi_2 h & e\xi_1 \sinh\xi_1 h \\ \dfrac{e}{\varepsilon}\cos\xi_2 h & \cosh\xi_1 h \end{vmatrix}$$

$$= \begin{vmatrix} -\bar{c}\xi_2 \sin\xi_2 h & e\xi_1 \sinh\xi_1 h \\ \dfrac{e}{\varepsilon}\cos\xi_2 h & \cosh\xi_1 h \end{vmatrix} + \begin{vmatrix} -\rho'h'\omega^2 \cos\xi_2 h & e\xi_1 \sinh\xi_1 h \\ 0 & \cosh\xi_1 h \end{vmatrix}$$

$$(5.58)$$

$$= -\bar{c}\xi_2 \sin\xi_2 h\cosh\xi_1 h - \frac{e^2}{\varepsilon}\xi_1 \cos\xi_2 h\sinh\xi_1 h$$

$$- \rho'h'\omega^2 \cos\xi_2 h\cosh\xi_1 h = 0,$$

or

$$-\tan\xi_2 h - \bar{k}^2 \frac{\xi_1}{\xi_2}\tanh\xi_1 h = \frac{\rho'h'\omega^2}{\bar{c}\xi_2}.$$

$$(5.59)$$

When $\rho'h' = 0$, Eq. (5.59) reduces to

$$\tan\xi_2 h + \bar{k}^2 \frac{\xi_1}{\xi_2}\tanh\xi_1 h = 0,$$

$$(5.60)$$

which is Eq. (5.13).

5.5. Effect of Film Stiffness

In Secs. 4.2 and 5.4, the thin films on a half-space or a plate were assumed to be very thin. Only their inertial effect was considered. Their stiffness was neglected. In this section we examine the effect of film stiffness [24]. Consider the plate with asymmetric films in Fig. 5.8.

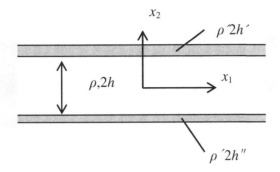

Fig. 5.8. A plate with asymmetric films.

5.5.1. An unelectroded plate

The electric field in the surrounding space is neglected as an approximation. The governing equations are

$$\bar{c}\nabla^2 u = \rho\ddot{u}, \quad \nabla^2 \psi = 0, \quad \phi = \psi + \frac{e}{\varepsilon}u, \quad -h < x_2 < h. \tag{5.61}$$

The boundary conditions are

$$2h'\mu u_{,11} - T_{23} = \rho'2h'\ddot{u}, \quad D_2 = 0, \quad x_2 = h,$$
$$2h''\mu u_{,11} + T_{23} = \rho'2h''\ddot{u}, \quad D_2 = 0, \quad x_2 = -h, \tag{5.62}$$

where μ is the shear elastic constant of the mass layers. Physically, the mechanical boundary conditions at $x_2 = \pm h$ in Eq. (5.62) state that the stress gradient in the mass layer and the stress between the mass layer and the plate are responsible for the layer acceleration according to Newton's law. Equation (5.62) is equivalent to

$$2h'\mu u_{,11} - \bar{c}u_{,2} - e\psi_{,2} = \rho'2h'\ddot{u}, \quad \psi_{,2} = 0, \quad x_2 = h,$$
$$2h''\mu u_{,11} + \bar{c}u_{,2} + e\psi_{,2} = \rho'2h''\ddot{u}, \quad \psi_{,2} = 0, \quad x_2 = -h. \tag{5.63}$$

Consider

$$u = (A_1 \cos\xi_2 x_2 + A_2 \sin\xi_2 x_2)\cos(\xi_1 x_1 - \omega t),$$
$$\psi = (B_1 \cosh\xi_1 x_2 + B_2 \sinh\xi_1 x_2)\cos(\xi_1 x_1 - \omega t), \tag{5.64}$$

where A_1, A_2, B_1 and B_2 are undetermined constants. Equation (5.64)$_2$ already satisfies Eq. (5.61)$_2$. For Eq. (5.64)$_1$ to satisfy Eq. (5.61)$_1$, we must have

$$\rho\omega^2 = \bar{c}(\xi_1^2 + \xi_2^2). \tag{5.65}$$

Substitution of Eq. (5.64) into Eq. (5.63) gives

$$2h'\mu\xi_1^2(A_1\cos\xi_2 h + A_2\sin\xi_2 h) + \bar{c}(-\xi_2 A_1\sin\xi_2 h + \xi_2 A_2\cos\xi_2 h)$$

$$= \rho'2h'\omega^2(A_1\cos\xi_2 h + A_2\sin\xi_2 h),$$

$$- 2h''\mu\xi_1^2(A_1\cos\xi_2 h - A_2\sin\xi_2 h) + \bar{c}(\xi_2 A_1\sin\xi_2 h + \xi_2 A_2\cos\xi_2 h)$$ (5.66)

$$= -\rho'2h''\omega^2(A_1\cos\xi_2 h - A_2\sin\xi_2 h),$$

$$B_1\xi_1\sinh\xi_1 h + B_2\xi_1\cosh\xi_1 h = 0,$$

$$- B_1\xi_1\sinh\xi_1 h + B_2\xi_1\cosh\xi_1 h = 0.$$

Equations (5.66)$_{3,4}$ imply that $B_1 = B_2 = 0$. Note that in this case $\phi = eu/\varepsilon$ so the electric field still exists although $\psi = 0$, but the electric displacement **D** vanishes. Equations (5.66)$_{1,2}$ then imply the following frequency equation of the modes:

$$\begin{vmatrix} -\bar{c}\xi_2\sin\xi_2 h + 2h'(\mu\xi_1^2 - \rho'\omega^2)\cos\xi_2 h & \bar{c}\xi_2\cos\xi_2 h + 2h'(\mu\xi_1^2 - \rho'\omega^2)\sin\xi_2 h \\ \bar{c}\xi_2\sin\xi_2 h + 2h''(-\mu\xi_1^2 + \rho'\omega^2)\cos\xi_2 h & \bar{c}\xi_2\cos\xi_2 h + 2h''(\mu\xi_1^2 - \rho'\omega^2)\sin\xi_2 h \end{vmatrix} = 0.$$ (5.67)

Equation (5.67) can be further written as

$$\tan 2\xi_2 h = \frac{\bar{c}\xi_2(\rho'\omega^2 - \mu\xi_1^2)2(h' + h'')}{4(\rho'\omega^2 - \mu\xi_1^2)^2 h'h'' - \bar{c}^2\xi_2^2}.$$ (5.68)

We have the following observations and discussion:

(i) For symmetric films with $h'' = h'$, Eq. (5.68) can be written as

$$\tan 2\xi_2 h = \frac{2\dfrac{\bar{c}\xi_2}{(\rho'\omega^2 - \mu\xi_1^2)2h'}}{1 - \left[\dfrac{\bar{c}\xi_2}{(\rho'\omega^2 - \mu\xi_1^2)2h'}\right]^2} = \frac{2\left[-\dfrac{(\rho'\omega^2 - \mu\xi_1^2)2h'}{\bar{c}\xi_2}\right]}{1 - \left[-\dfrac{(\rho'\omega^2 - \mu\xi_1^2)2h'}{\bar{c}\xi_2}\right]^2}.$$ (5.69)

Comparing Eq. (5.69) with the following trigonometric identity:

$$\tan 2\xi_2 h = \frac{2\tan\xi_2 h}{1 - (\tan\xi_2 h)^2},$$ (5.70)

we identify the following frequency equations for symmetric films:

$$\tan \xi_2 h = \frac{\bar{c}\xi_2}{(\rho'\omega^2 - \mu\xi_1^2)2h'},$$

$$\text{or} \quad \tan \xi_2 h = -\frac{(\rho'\omega^2 - \mu\xi_1^2)2h'}{\bar{c}\xi_2}. \tag{5.71}$$

For symmetric films, the modes can be classified as antisymmetric and symmetric about x_2, with separate frequency equations as given in Eq. (5.71).

(ii) When the films are not present, i.e., $h' = h'' = 0$, or $\rho' = 0$ and $\mu = 0$, we denote the solution to Eq. (5.68) by $\xi_2^{(0)}$ and $\omega^{(0)}$. Then we have $\sin 2\xi_2^{(0)} h = 0$, or $2\xi_2^{(0)} h = n\pi$, $n = 0, 1, 2, \cdots$, and

$$\rho(\omega^{(0)})^2 = \bar{c}[\xi_1^2 + (\xi_2^{(0)})^2] = \bar{c}\left[\xi_1^2 + \left(\frac{n\pi}{2h}\right)^2\right], \quad n = 0, 1, 2, \cdots. \tag{5.72}$$

(iii) For thin films, the inertial and stiffness effects of the films in the denominator of Eq. (5.68) (which are of second- or higher-order) can be neglected. Given ξ_1, denote the changes of ξ_2 and ω by $\Delta\xi_2$ and $\Delta\omega$. We have

$$\xi_2 = \xi_2^{(0)} + \Delta\xi_2, \quad \omega = \omega^{(0)} + \Delta\omega. \tag{5.73}$$

Substituting Eqs. (5.72) and (5.73) into Eq. (5.68), to the lowest-order effect, we obtain

$$2(\Delta\xi_2)h \cong \frac{2\left[\rho'(\omega^{(0)})^2 - \mu\xi_1^2\right](h' + h'')}{-\bar{c}\xi_2^{(0)}}, \tag{5.74}$$

which gives the perturbation of ξ_2 due to the films. From Eq. (5.65) we can obtain

$$2\rho\,\omega^{(0)}\Delta\omega = \bar{c}2\xi_2^{(0)}\Delta\xi_2, \tag{5.75}$$

which is a relation between the perturbations of ξ_2 and ω. Then, from Eqs. (5.74) and (5.75), the frequency perturbation is given by

$$\frac{\Delta\omega}{\omega^{(0)}} = -R^{(m)} + R^{(s)} \frac{X^2}{(\Omega^{(0)})^2}$$

$$R^{(m)} = \frac{\rho'(2h' + 2h'')}{\rho 2h}, \quad R^{(s)} = \frac{\mu}{\bar{c}} \frac{2h' + 2h''}{2h}, \tag{5.76}$$

$$X^2 = \xi_1^2 \bigg/ \frac{\pi^2}{4h^2}, \quad (\Omega^{(0)})^2 = (\omega^{(0)})^2 \bigg/ \frac{\pi^2 \bar{c}}{4h^2 \rho} = X^2 + n^2,$$

where $R^{(m)}$ is the mass ratio between the layers and the plate, $R^{(s)}$ is the stiffness ratio, X is the normalized wave number in the x_1 direction, and $\Omega^{(0)}$ is the normalized frequency when the films are not present. Equation (5.76) shows that the inertial effect lowers the frequencies and the stiffness effect raises them. A fundamental difference between the two effects is that the stiffness effect depends on the wave number and frequency but the inertial effect does not. In the special case when $n = 0$ and hence $\Omega^{(0)} = X$, we have face-shear modes. For these modes the inertial effect and the stiffness effect of the films may be comparable and are both important. In general, for thickness-twist modes with $n \geq 1$, we have $\Omega^{(0)} > X$. Usually lower-order thickness-twist modes with a small n and long waves with a small X are used for device applications. For these modes we have $\Omega^{(0)} \gg X$, which implies that the stiffness effect is much smaller than the inertial effect.

5.5.2. An electroded plate

Next consider the case when the plate boundaries at $x_2 = \pm h$ are electroded and the electrodes are shorted. The boundary conditions are

$$2h'\mu u_{,11} - T_{23} = \rho'2h'\ddot{u}, \quad \phi = 0, \quad x_2 = h,$$
$$2h'\mu u_{,11} + T_{23} = \rho''2h''\ddot{u}, \quad \phi = 0, \quad x_2 = -h, \tag{5.77}$$

which are equivalent to

$$2h'\mu u_{,11} - (\bar{c}u_{,2} + e\psi_{,2}) = \rho'2h'\ddot{u}, \quad \psi + \frac{e}{\varepsilon}u = 0, \quad x_2 = h,$$
$$2h''\mu u_{,11} + \bar{c}u_{,2} + e\psi_{,2} = \rho''2h''\ddot{u}, \quad \psi + \frac{e}{\varepsilon}u = 0, \quad x_2 = -h, \tag{5.78}$$

Substitute Eq. (5.64) into Eq. (5.78):

$$2h'\mu\xi_1^2(A_1\cos\xi_2h+A_2\sin\xi_2h)+\bar{c}\xi_2(-A_1\sin\xi_2h+A_2\cos\xi_2h)+$$
$$e\xi_1(B_1\sinh\xi_1h+B_2\cosh\xi_1h) \tag{5.79}$$
$$=\rho'2h'\omega^2(A_1\cos\xi_2h+A_2\sin\xi_2h),$$

$$-2h''\mu\xi_1^2(A_1\cos\xi_2h-A_2\sin\xi_2h)+\bar{c}\xi_2(A_1\sin\xi_2h+A_2\cos\xi_2h)+$$
$$e\xi_1(-B_1\sinh\xi_1h+B_2\cosh\xi_1h) \tag{5.80}$$
$$=-\rho'2h''\omega^2(A_1\cos\xi_2h-A_2\sin\xi_2h),$$

$$\frac{e}{\varepsilon}(A_1\cos\xi_2h+A_2\sin\xi_2h)+B_1\cosh\xi_1h+B_2\sinh\xi_1h=0,$$
$$\frac{e}{\varepsilon}(A_1\cos\xi_2h-A_2\sin\xi_2h)+B_1\cosh\xi_1h-B_2\sinh\xi_1h=0. \tag{5.81}$$

For nontrivial solutions, the determinant of the coefficient matrix of Eqs. (5.79), (5.80) and (5.81) must vanish, which yields

$$\begin{vmatrix} a_{11} & a_{12} & e\xi_1\sinh\xi_1h & e\xi_1\cosh\xi_1h \\ a_{21} & a_{22} & -e\xi_1\sinh\xi_1h & e\xi_1\cosh\xi_1h \\ \dfrac{e}{\varepsilon}\cos\xi_2h & \dfrac{e}{\varepsilon}\sin\xi_2h & \cosh\xi_1h & \sinh\xi_1h \\ \dfrac{e}{\varepsilon}\cos\xi_2h & -\dfrac{e}{\varepsilon}\sin\xi_2h & \cosh\xi_1h & -\sinh\xi_1h \end{vmatrix}=0, \tag{5.82}$$

where

$$a_{11}=-\bar{c}\xi_2\sin\xi_2h+2h'(\mu\xi_1^2-\rho'\omega^2)\cos\xi_2h,$$
$$a_{12}=\bar{c}\xi_2\cos\xi_2h+2h'(\mu\xi_1^2-\rho'\omega^2)\sin\xi_2h,$$
$$a_{21}=\bar{c}\xi_2\sin\xi_2h+2h''(-\mu\xi_1^2+\rho'\omega^2)\cos\xi_2h, \tag{5.83}$$
$$a_{22}=\bar{c}\xi_2\cos\xi_2h+2h''(\mu\xi_1^2-\rho'\omega^2)\sin\xi_2h.$$

The expansion of Eq. (5.82) is rather lengthy and therefore it is not provided. The case of symmetric films is much simpler, which will be examined further below. For symmetric films with $h''=h'$, Eq. (5.82) splits into two frequency equations for modes that are symmetric and antisymmetric in x_2:

$$\begin{vmatrix} \bar{c}\xi_2\sin\xi_2h+2h'(-\mu\xi_1^2+\rho'\omega^2)\cos\xi_2h & -e\xi_1\sinh\xi_1h \\ \dfrac{e}{\varepsilon}\cos\xi_2h & \cosh\xi_1h \end{vmatrix}=0, \tag{5.84}$$

$$\begin{vmatrix} \overline{c}\,\xi_2 \cos\xi_2 h + 2h'(\mu\xi_1^2 - \rho'\omega^2)\sin\xi_2 h & e\xi_1 \cosh\xi_1 h \\ \dfrac{e}{\varepsilon}\sin\xi_2 h & \sinh\xi_1 h \end{vmatrix} = 0 \, . \tag{5.85}$$

Equations (5.84) and (5.85) can be written as

$$\overline{c}\,\xi_2 \tan(\xi_2 h) + \frac{e^2}{\varepsilon}\xi_1 \tanh(\xi_1 h) = 2h'(\mu\xi_1^2 - \rho'\omega^2) \, , \tag{5.86}$$

$$\overline{c}\,\xi_2 \cot(\xi_2 h) - \frac{e^2}{\varepsilon}\xi_1 \coth(\xi_1 h) = -2h'(\mu\xi_1^2 - \rho'\omega^2) \, . \tag{5.87}$$

When the mass layers are not present, i.e., $\rho' = 0$ and $\mu = 0$, we denote the solution to Eqs. (5.86) and (5.87) by $\xi_2^{(0)}$ and $\omega^{(0)}$ such that

$$\overline{c}\,\xi_2^{(0)} \tan(\xi_2^{(0)}h) + \frac{e^2}{\varepsilon}\xi_1 \tanh(\xi_1 h) = 0 \, , \tag{5.88}$$

$$\overline{c}\,\xi_2^{(0)} \cot(\xi_2^{(0)}h) - \frac{e^2}{\varepsilon}\xi_1 \coth(\xi_1 h) = 0 \, . \tag{5.89}$$

Then, for thin mass layers, we look for a perturbation solution by substituting Eq. (5.73) into Eqs. (5.86) and (5.87). To the lowest-order effect, for a fixed ξ_1, we obtain

$$\overline{c}(\Delta\xi_2)\tan(\xi_2^{(0)}h) + \overline{c}\,\xi_2^{(0)}\frac{1}{\cos^2(\xi_2^{(0)}h)}(\Delta\xi_2 h) = 2h'[\mu\xi_1^2 - \rho'(\omega^{(0)})^2] \, , \tag{5.90}$$

$$\overline{c}(\Delta\xi_2)\cot(\xi_2^{(0)}h) + \overline{c}\,\xi_2^{(0)}\frac{-1}{\sin^2(\xi_2^{(0)}h)}(\Delta\xi_2 h) = -2h'[\mu\xi_1^2 - \rho'(\omega^{(0)})^2] \, . \tag{5.91}$$

From Eqs. (5.90) and (5.75), and Eqs. (5.91) and (5.75), we obtain

$$\frac{\Delta\omega}{\omega^{(0)}} = \frac{-R^{(m)} + R^{(s)}\dfrac{X^2}{(\Omega^{(0)})^2}}{1 + \tan^2(\xi_2^{(0)}h) + \tan(\xi_2^{(0)}h)\big/(\xi_2^{(0)}h)} \, , \tag{5.92}$$

$$\frac{\Delta\omega}{\omega^{(0)}} = \frac{-R^{(m)} + R^{(s)}\dfrac{X^2}{(\Omega^{(0)})^2}}{1 + \cot^2(\xi_2^{(0)}h) - \cot(\xi_2^{(0)}h)\big/(\xi_2^{(0)}h)} \, . \tag{5.93}$$

5.6. A Plate in Contact with Fluids

In this section we consider waves in a plate in contact with a fluid (see Fig. 5.9) [13]. The major surfaces of the plate are electroded, and the electrodes are grounded. The fluid is a viscous dielectric.

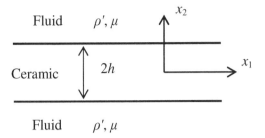

Fig. 5.9. A plate in a fluid.

5.6.1. *Governing equations and fields*

For the plate, the governing equations are

$$\bar{c}\nabla^2 u = \rho\ddot{u}, \quad \nabla^2\psi = 0, \quad \phi = \psi + \frac{e}{\varepsilon}u, \quad -h < x_2 < h. \tag{5.94}$$

The stress component needed for the interface continuity conditions is given by

$$T_{23} = \bar{c}u_{,2} + e\psi_{,2}. \tag{5.95}$$

For the fluid, Let $v_3(x_1, x_2, t)$ be the velocity field. The governing equation for v_3 and the relevant stress component for the interface continuity condition are

$$\mu(v_{3,11} + v_{3,22}) = \rho'\frac{\partial v_3}{\partial t}, \tag{5.96}$$

$$T_{23} = \mu v_{3,2},$$

where ρ' is the fluid mass density and μ is the fluid viscosity. Due to symmetry we only consider the upper fluid. For $x_2 > h$, the following fields are relevant:

$$v_3 = C\exp[-\lambda(x_2 - h)]\exp[i(\xi_1 x_1 - \omega t)], \tag{5.97}$$

$$T_{23} = -\mu\lambda C\exp[-\lambda(x_2 - h)]\exp[i(\xi_1 x_1 - \omega t)],$$

where C is an arbitrary constant. λ has a positive real part and is determined by

$$\mu(\lambda^2 - \xi_1^2) = \rho'i\omega. \tag{5.98}$$

5.6.2. Antisymmetric waves

For antisymmetric waves, in the plate we have, from Eq. (5.4),

$$u = A\sin\xi_2 x_2 \exp[i(\xi_1 x_1 - \omega t)],$$

$$\psi = B\sinh\xi_1 x_2 \exp[i(\xi_1 x_1 - \omega t)], \qquad (5.99)$$

$$\xi_2^2 = \frac{\rho\omega^2}{\bar{c}} - \xi_1^2,$$

where A and B are undetermined constants. At $x_2 = h$ we require that

$$i\omega u(x_2 = h^-) = v_3(x_2 = h^+),$$

$$T_{23}(x_2 = h^-) = T_{23}(x_2 = h^+), \qquad (5.100)$$

$$\frac{e}{\varepsilon}u + \psi = 0.$$

Substitution of Eqs. (5.97) and (5.99) into Eq. (5.100) gives

$$i\omega A\sin\xi_2 h = C,$$

$$\bar{c}A\xi_2\cos\xi_2 h + eB\xi_1\cosh\xi_1 h = -\mu\lambda C, \qquad (5.101)$$

$$\frac{e}{\varepsilon}A\sin\xi_2 h + B\sinh\xi_1 h = 0.$$

For nontrivial solutions we must have

$$
\begin{vmatrix}
i\omega\sin\xi_2 h & 0 & -1 \\
\bar{c}\xi_2\cos\xi_2 h & e\xi_1\cosh\xi_1 h & \mu\lambda \\
\dfrac{e}{\varepsilon}\sin\xi_2 h & \sinh\xi_1 h & 0
\end{vmatrix}
$$

$$
= i\omega\sin\xi_2 h
\begin{vmatrix}
e\xi_1\cosh\xi_1 h & \mu\lambda \\
\sinh\xi_1 h & 0
\end{vmatrix}
-
\begin{vmatrix}
\bar{c}\xi_2\cos\xi_2 h & e\xi_1\cosh\xi_1 h \\
\dfrac{e}{\varepsilon}\sin\xi_2 h & \sinh\xi_1 h
\end{vmatrix} \qquad (5.102)
$$

$$= -i\omega\mu\lambda\sin\xi_2 h\sinh\xi_1 h$$

$$-\bar{c}\xi_2\cos\xi_2 h\sinh\xi_1 h + \frac{e^2}{\varepsilon}\xi_1\sin\xi_2 h\cosh\xi_1 h = 0,$$

or

$$\tanh\xi_1 h - \bar{k}^2\frac{\xi_1}{\xi_2}\tan\xi_2 h = -\frac{i\omega\mu\lambda}{\bar{c}\xi_2}\tan\xi_2 h\tanh\xi_1 h. \qquad (5.103)$$

With Eqs. (5.98) and (5.99)$_3$, Eq. (5.103) is an equation for ω, which determines the wave frequency or speed. When $\mu = 0$, Eq. (5.103) reduces to

$$\tanh \xi_1 h - \bar{k}^2 \frac{\xi_1}{\xi_2} \tan \xi_2 h = 0, \tag{5.104}$$

which is Eq. (5.7). The difference in frequencies determined from Eqs. (5.103) and (5.104) can be used to measure μ or ρ'.

5.6.3. Symmetric waves

For symmetric waves we consider, from Eq. (5.11),

$$\begin{aligned}
u &= A \cos \xi_2 x_2 \cos(\xi_1 x_1 - \omega t), \\
\psi &= B \cosh \xi_1 x_2 \cos(\xi_1 x_1 - \omega t).
\end{aligned} \tag{5.105}$$

We still have Eq. (5.99)$_3$. Substitution of Eqs. (5.97) and (5.105) into Eq. (5.100) leads to

$$\begin{aligned}
i\omega A \cos \xi_2 h &= C, \\
-\bar{c} A \xi_2 \sin \xi_2 h + eB\xi_1 \sinh \xi_1 h &= -\mu\lambda C, \\
\frac{e}{\varepsilon} A \cos \xi_2 h + B \cosh \xi_1 h &= 0.
\end{aligned} \tag{5.106}$$

For nontrivial solutions we must have

$$\begin{vmatrix}
i\omega \cos \xi_2 h & 0 & -1 \\
-\bar{c}\xi_2 \sin \xi_2 h & e\xi_1 \sinh \xi_1 h & \mu\lambda \\
\dfrac{e}{\varepsilon} \cos \xi_2 h & \cosh \xi_1 h & 0
\end{vmatrix}$$

$$= i\omega \cos \xi_2 h \begin{vmatrix} e\xi_1 \sinh \xi_1 h & \mu\lambda \\ \cosh \xi_1 h & 0 \end{vmatrix} - \begin{vmatrix} -\bar{c}\xi_2 \sin \xi_2 h & e\xi_1 \sinh \xi_1 h \\ \dfrac{e}{\varepsilon} \cos \xi_2 h & \cosh \xi_1 h \end{vmatrix} \tag{5.107}$$

$$= -i\omega\mu\lambda \cos \xi_2 h \cosh \xi_1 h$$

$$+ \bar{c}\xi_2 \sin \xi_2 h \cosh \xi_1 h + \frac{e^2}{\varepsilon} \xi_1 \cos \xi_2 h \sinh \xi_1 h = 0,$$

or

$$\tan \xi_2 h + \bar{k}^2 \frac{\xi_1}{\xi_2} \tanh \xi_1 h = \frac{i\omega\mu\lambda}{\bar{c}\xi_2}. \tag{5.108}$$

When $\mu = 0$, Eq. (5.108) reduces to

$$\tan \xi_2 h + \overline{k}^2 \frac{\xi_1}{\xi_2} \tanh \xi_1 h = 0,$$ (5.109)

which is Eq. (5.13).

5.7. A Plate with Fluids under Unattached Electrodes

In this section we analyze the propagation of thickness-twist waves in an unbounded piezoelectric ceramic plate with unattached electrodes and viscous fluids between the plate surfaces and the electrodes [25]. Consider the plate shown in Fig. 5.10. The gaps are filled with two potentially different viscous fluids.

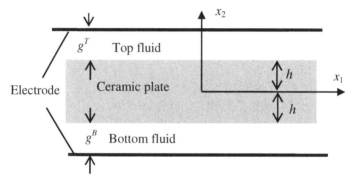

Fig. 5.10. A plate with unattached electrodes and gaps filled with fluids.

The governing equations are

$$\overline{c}\nabla^2 u = \rho \ddot{u},$$

$$\nabla^2 \psi = 0, \quad \phi = \psi + \frac{e}{\varepsilon}u, \qquad -h < x_2 < h,$$ (5.110)

$$\frac{\partial v^T}{\partial t} - \frac{\mu^T}{\rho^T}\nabla^2 v^T = 0,$$
$$\nabla^2 \phi^T = 0, \qquad h \le x_2 \le h + g^T,$$ (5.111)

$$\frac{\partial v^B}{\partial t} - \frac{\mu^B}{\rho^B}\nabla^2 v^B = 0,$$
$$\nabla^2 \phi^B = 0, \qquad -h - g^B \le x_2 \le -h,$$ (5.112)

where v^T and v^B are the fluid velocity fields, ρ^T and ρ^B are the fluid mass densities, and μ^T and μ^B are the fluid viscosities. The superscripts "T" and "B" are for the top and bottom fluids, respectively. In addition, we have the following boundary and continuity conditions:

$$
\begin{aligned}
&\phi^T = 0, \quad v^T = 0, \quad x_2 = h + g^T, \\
&\phi = \phi^T, \quad D_2 = D_2^T, \quad T_{23} = T_{23}^T, \quad v = v^T, \quad x_2 = h, \\
&\phi = \phi^B, \quad D_2 = D_2^B, \quad T_{23} = T_{23}^B, \quad v = v^B, \quad x_2 = -h, \\
&\phi^B = 0, \quad v^B = 0, \quad x_2 = -h - g^B.
\end{aligned}
\tag{5.113}
$$

We look for solutions representing waves propagating in the x_1 direction in the following form:

$$
\begin{aligned}
u &= \left(A_1 \sin\xi_2 x_2 + A_2 \cos\xi_2 x_2\right)\exp i\left(\xi_1 x_1 - \omega t\right), \\
\psi &= \left(A_3 \sinh\xi_1 x_2 + A_4 \cosh\xi_1 x_2\right)\exp i\left(\xi_1 x_1 - \omega t\right),
\end{aligned}
\quad -h < x_2 < h, \tag{5.114}
$$

$$
\begin{aligned}
v^T &= A_7 \sinh\xi_2^T \left(x_2 - h - g^T\right)\exp i\left(\xi_1 x_1 - \omega t\right), \\
\phi^T &= A_5 \sinh\xi_1 \left(x_2 - h - g^T\right)\exp i\left(\xi_1 x_1 - \omega t\right),
\end{aligned}
\quad h \le x_2 \le h + g^T, \tag{5.115}
$$

$$
\begin{aligned}
v^B &= A_8 \sinh\xi_2^B \left(x_2 + h + g^B\right)\exp i\left(\xi_1 x_1 - \omega t\right), \\
\phi^B &= A_6 \sinh\xi_1 \left(x_2 + h + g^B\right)\exp i\left(\xi_1 x_1 - \omega t\right),
\end{aligned}
\quad -h - g^B \le x_2 \le -h,
$$

$$\tag{5.116}$$

where A_1 through A_8 are undetermined constants. Equations (5.115) and (5.116) already satisfy the boundary conditions in Eq. (5.113) at $x_2 = h + g^T$ and $x_2 = -h - g^B$. Equations (5.114)–(5.116) satisfy Eqs. (5.110)–(5.112) provided that

$$
\xi_2^2 = \xi_1^2\left(\frac{v^2}{v_T^2} - 1\right),
\tag{5.117}
$$

$$
\left(\xi_2^T\right)^2 = \xi_1^2 - i\omega\frac{\rho^T}{\mu^T},
\tag{5.118}
$$

$$
\left(\xi_2^B\right)^2 = \xi_1^2 - i\omega\frac{\rho^B}{\mu^B},
\tag{5.119}
$$

where

$$v_T^2 = \frac{\bar{c}}{\rho}, \quad \xi_1^2 = \frac{\omega^2}{v^2}. \tag{5.120}$$

Substitution of Eqs. (5.114)–(5.116) into the remaining eight continuity conditions in Eq. (5.113) results in the following eight linear homogeneous equations for A_1 through A_8:

$$A_5 \sinh\left(-\xi_1 g^T\right)$$
$$= A_3 \sinh\xi_1 h + A_4 \cosh\xi_1 h + \frac{e}{\varepsilon}\left(A_1 \sin\xi_2 h + A_2 \cos\xi_2 h\right), \tag{5.121}$$

$$-\varepsilon^T A_5 \xi_1 \cosh\xi_1 g^T = -\varepsilon\left(A_3\xi_1 \cosh\xi_1 h + A_4\xi_1 \sinh\xi_1 h\right),$$

$$\bar{c}\left(A_1\xi_2 \cos\xi_2 h - A_2\xi_2 \sin\xi_2 h\right) + e\left(A_3\xi_1 \cosh\xi_1 h + A_4\xi_1 \sinh\xi_1 h\right)$$
$$= \mu^T \xi_2^T A_7 \cosh\xi_2^T g^T, \tag{5.122}$$

$$-\left(A_1 \sin\xi_2 h + A_2 \cos\xi_2 h\right)i\omega = A_7 \sinh\left(-\xi_2^T g^T\right),$$

$$\left(A_1 \sin\xi_2 h - A_2 \cos\xi_2 h\right)i\omega = A_8 \sinh\xi_2^B g^B,$$

$$\bar{c}\left(A_1\xi_2 \cos\xi_2 h + A_2\xi_2 \sin\xi_2 h\right) + e\left(A_3\xi_1 \cosh\xi_1 h - A_4\xi_1 \sinh\xi_1 h\right) \tag{5.123}$$
$$= \mu^B \xi_2^B A_8 \cosh\xi_2^B g^B,$$

$$-\varepsilon^B A_6 \xi_1 \cosh\xi_1 g^B = -\varepsilon\left(A_3\xi_1 \cosh\xi_1 h - A_4\xi_1 \sinh\xi_1 h\right),$$

$$A_6 \sinh\xi_1 g^B \tag{5.124}$$

$$= -A_3 \sinh\xi_1 h + A_4 \cosh\xi_1 h + \frac{e}{\varepsilon}\left(-A_1 \sin\xi_2 h + A_2 \cos\xi_2 h\right).$$

The dispersion relation of the waves, i.e., ω versus ξ_1, is determined from that the determinant of the coefficient matrix of Eqs. (5.121)–(5.124) vanishes. Below we consider the special case when the structure is symmetric about $x_2 = 0$, i.e.,

$$g^T = g^B = g, \quad \mu^T = \mu^B = \mu^L, \quad \rho^T = \rho^B = \rho^L, \quad \varepsilon^T = \varepsilon^B = \bar{\varepsilon}. \tag{5.125}$$

In this case,

$$\xi_2^T = \xi_2^B = \bar{\xi}. \tag{5.126}$$

The waves separate into antisymmetric and symmetric ones about $x_2 = 0$. They are odd or even functions of x_2, respectively. The determinant of the coefficient matrix of Eqs. (5.121)–(5.124) factors into two equations determining the dispersion relations of the antisymmetric (AS) and symmetric (S) waves:

$$\frac{\tanh \xi_1 h}{\tan \xi_2 h} + \bar{n}^2 \frac{\tanh \xi_1 g}{\tan \xi_2 h} - \bar{k}^2 \frac{\xi_1}{\xi_2}$$

$$-\frac{i\omega\mu\bar{\xi}}{\bar{c}\xi_2 \tanh \bar{\xi} g}\left(\bar{n}^2 \tanh \xi_1 g + \tanh \xi_1 h\right) = 0,$$

(AS)   (5.127)

$$\frac{\tan \xi_2 h}{\tanh \xi_1 h} + \bar{n}^2 \frac{\tanh \xi_1 g}{\text{ctan} \xi_2 h} + \bar{k}^2 \frac{\xi_1}{\xi_2}$$

$$+\frac{i\omega\mu\bar{\xi}}{\bar{c}\xi_2 \tanh \bar{\xi} g}\left(\bar{n}^2 \tanh \xi_1 g + \text{ctanh} \xi_1 h\right) = 0,$$

(S)   (5.128)

where $\bar{n}^2 = \varepsilon / \bar{\varepsilon}$.

If the fluids are not present, we have air gaps. Denoting the electric permittivity of free space by $\varepsilon^T = \varepsilon^B = \varepsilon_0$ and $n^2 = \varepsilon / \varepsilon_0$, we can reduce Eqs. (5.127) and (5.128) into

$$n^2 \tanh \xi_1 g + \tanh \xi_1 h - \bar{k}^2 \frac{\xi_1}{\xi_2} \tan \xi_2 h = 0, \text{ (AS)}$$

(5.129)

$$n^2 \tanh \xi_1 g + \text{ctanh} \xi_1 h + \bar{k}^2 \frac{\xi_1}{\xi_2} \text{ctan} \xi_2 h = 0, \text{ (S)}$$

(5.130)

which are Eqs. (5.40) and (5.41).

5.8. Waves through a Joint between Two Semi-infinite Plates

In this section we analyze the propagation of thickness-twist waves through a joint between two semi-infinite piezoelectric plates of different materials (see Fig. 5.11) [26]. The plate is unelectroded. The electric field in the free space is neglected. We need to analyze the two halves separately and then apply interface continuity conditions.

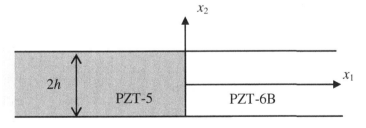

Fig. 5.11. A joint between two semi-infinite plates.

The equations for u, ψ and ϕ are:

$$\bar{c}\nabla^2 u = \rho\ddot{u}, \quad \nabla^2\psi = 0, \quad \phi = \psi + \frac{e}{\varepsilon}u . \tag{5.131}$$

At the plate major surfaces we consider traction-free boundary conditions with

$$\begin{aligned} T_{23} &= 0, \quad x_2 = \pm h, \\ D_2 &= 0, \quad x_2 = \pm h, \end{aligned} \tag{5.132}$$

or, equivalently, in terms of u and ψ,

$$u_{,2} = 0, \quad \psi_{,2} = 0, \quad x_2 = \pm h . \tag{5.133}$$

5.8.1. The left half

Consider incident waves coming from $x_1 = -\infty$. The waves may propagate through the joint and may also get reflected at the joint. Expressions of these waves can be obtained by separation of variables. It can be verified by direct substitution that the solutions to Eqs. (5.131) and (5.133) can be classified into waves symmetric or antisymmetric in x_2, and they are given by:

$$\begin{aligned} u &= \cos\xi_2 x_2 [A_1 \exp i(\xi_1 x_1) + A_2 \exp i(-\xi_1 x_1)]\exp(-i\omega t), \\ \psi &= \cos\xi_2 x_2 B \exp(\xi_2 x_1)\exp(-i\omega t), \\ \xi_2 &= \frac{m\pi}{2h}, \quad m = 0, 2, 4, \cdots, \end{aligned} \tag{5.134}$$

and

$$\begin{aligned} u &= \sin\xi_2 x_2 [A_1 \exp i(\xi_1 x_1) + A_2 \exp i(-\xi_1 x_1)]\exp(-i\omega t), \\ \psi &= \sin\xi_2 x_2 B \exp(\xi_2 x_1)\exp(-i\omega t), \\ \xi_2 &= \frac{m\pi}{2h}, \quad m = 1, 3, 5, \cdots, \end{aligned} \tag{5.135}$$

respectively, where

$$\begin{aligned} \xi_1 &= \sqrt{\frac{\rho\,\omega^2}{\bar{c}} - \xi_2^2} = \sqrt{\frac{\rho}{\bar{c}}}\sqrt{\omega^2 - \left(\frac{m\pi}{2h}\right)^2\frac{\bar{c}}{\rho}} = \frac{1}{v_T}\sqrt{\omega^2 - \omega_m^2}, \\ v_T &= \sqrt{\frac{\bar{c}}{\rho}}, \quad \omega_m^2 = \left(\frac{m\pi}{2h}\right)^2\frac{\bar{c}}{\rho}. \end{aligned} \tag{5.136}$$

A_2 and B are undetermined constants. A_1 is for the incident wave (right-traveling) and is considered known. A_2 is for the reflected wave (left-traveling). B represents a nonpropagating field localized near the joint. The localized behavior of B or ψ is dictated by Eq. (5.131)$_2$. ω_m is the cutoff frequency of thickness-twist waves obtained from the dispersion relations for these waves in an unbounded plate.

5.8.2. *The right half*

We use a prime to indicate the material constants for $x_1 > 0$. Similar to Eqs. (5.134)–(5.136), we have the following symmetric and antisymmetric fields and waves (right-traveling) satisfying the governing equations and the boundary conditions at $x_2 = \pm h$:

$$u = A' \cos \xi_2 x_2 \exp[i(\xi_1' x_1 - \omega t)],$$
$$\psi = B' \cos \xi_2 x_2 \exp(-\xi_2 x_1) \exp(-i\omega t), \tag{5.137}$$
$$\xi_2 = \frac{m\pi}{2h}, \quad m = 0, 2, 4, \cdots,$$

and

$$u = A' \sin \xi_2 x_2 \exp[i(\xi_1' x_1 - \omega t)],$$
$$\psi = B' \sin \xi_2 x_2 \exp(-\xi_2 x_1) \exp(-i\omega t), \tag{5.138}$$
$$\xi_2 = \frac{m\pi}{2h}, \quad m = 1, 3, 5, \cdots,$$

where

$$\xi_1' = \sqrt{\frac{\rho' \omega^2}{\overline{c}'} - \xi_2^2} = \frac{1}{v_T'} \sqrt{\omega^2 - (\omega_m')^2},$$

$$v_T' = \sqrt{\frac{\overline{c}'}{\rho'}}, \quad (\omega_m') = \left(\frac{m\pi}{2h}\right)^2 \frac{\overline{c}'}{\rho'}. \tag{5.139}$$

5.8.3. *Continuity conditions*

At the joint, the continuity of u, T_{13}, ϕ and D_1 need to be imposed. Consider the symmetric waves given by Eqs. (5.134) and (5.137). We have

$$A_1 + A_2 = A',$$

$$\overline{c}(A_1 i \xi_1 - A_2 i \xi_1) + e B \xi_2 = \overline{c}' A' i \xi_1' + e' B'(-\xi_2),$$

$$B + \frac{e}{\varepsilon}(A_1 + A_2) = B' + \frac{e'}{\varepsilon'} A', \tag{5.140}$$

$$-\varepsilon B \xi_2 = -\varepsilon' B'(-\xi_2).$$

Equation (5.140) can be solved symbolically, which yields:

$$\frac{A_2}{A_1} = \left[-\overline{c}' \varepsilon' \varepsilon \xi_1' (\varepsilon' + \varepsilon) + \overline{c} \varepsilon' \varepsilon \xi_1 (\varepsilon' + \varepsilon) + i \xi_2 (e \varepsilon' - e' \varepsilon)^2 \right] \Delta^{-1},$$

$$\frac{A'}{A_1} = 2 \overline{c} \varepsilon' \varepsilon \xi_1 (\varepsilon' + \varepsilon) \Delta^{-1},$$

$$\frac{B}{A_1} = -2 \overline{c} \varepsilon' \xi_1 (e \varepsilon' - e' \varepsilon) \Delta^{-1}, \tag{5.141}$$

$$\frac{B'}{A_1} = -2 \overline{c} \varepsilon \xi_1 (-e \varepsilon' + e' \varepsilon) \Delta^{-1},$$

where

$$\Delta = \overline{c}' \varepsilon' \varepsilon \xi_1' (\varepsilon' + \varepsilon) + \overline{c} \varepsilon' \varepsilon \xi_1 (\varepsilon' + \varepsilon) - i \xi_2 (e \varepsilon' - e' \varepsilon)^2. \tag{5.142}$$

Given the incident wave A_1, Eq. (5.141) yields the reflected wave A_2 and the transmitted wave A' as what normally happens at an interface. However, in this case, there also exist B and B' which determine the ψ field. ψ is nonpropagating and is localized near the interface.

5.8.4. Numerical results

For numerical results we consider the case when the two semi-infinite plates are PZT-5 and PZT-6B, respectively. For this particular choice of materials, $\omega_m' > \omega_m$. The face-shear wave with $m = 0$ does not have a cutoff frequency. For thickness-twist waves with $m \geq 1$, we choose a special case of $m = 2$ to show some common behavior of these modes specifically.

When $\omega > \omega_m' > \omega_m$, the u, ϕ and ψ fields are shown in Figs. 5.12, 5.13 and 5.14, respectively. u is sinusoidal on both sides of the joint. ϕ has a sinusoidal part and a decaying part on both sides of the joint. Their combination $\psi = \phi - eu/\varepsilon$ is a field localized near the joint and decays exponentially from the joint. The propagating part of ϕ is from u through piezoelectric coupling. The wave numbers in the two halves are different, as expected.

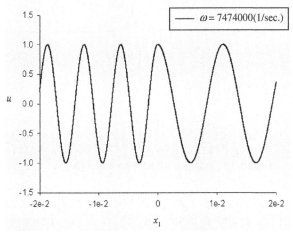

Fig. 5.12. u field when $\omega > \omega_2' > \omega_2$, $m = 2$.

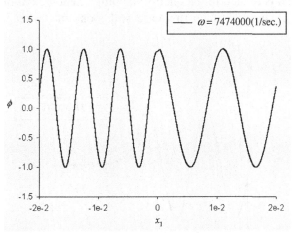

Fig. 5.13. ϕ field when $\omega > \omega_2' > \omega_2$, $m = 2$.

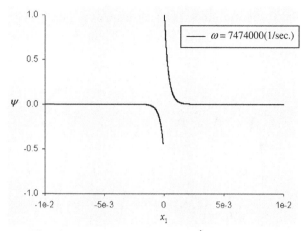

Fig. 5.14. ψ field when $\omega > \omega_2' > \omega_2$, $m = 2$.

The more interesting and probably more useful case is when $\omega_2 < \omega < \omega_2'$. The u and ϕ fields for this case are shown in Figs. 5.15 and 5.16, respectively. In the left half we still have sinusoidal waves, but in the right half u and ϕ are both exponentially decaying from the joint (cutoff). This is related to the energy trapping phenomenon of thickness-twist modes. The ψ field in this case still looks like what is shown in Fig. 5.14.

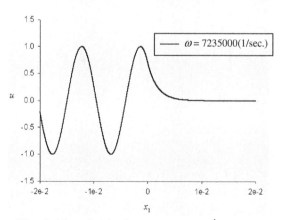

Fig. 5.15. u field when $\omega_2 < \omega < \omega_2'$, $m = 2$.

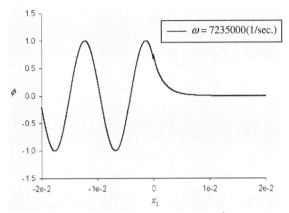

Fig. 5.16. ϕ field when $\omega_2 < \omega < \omega_2'$, $m = 2$.

For antisymmetric waves the results are similar except that m assumes odd numbers.

5.9. Trapped Modes in an Inhomogeneous Plate

In this section we show the existence of certain thickness-twist modes in an unbounded, inhomogeneous plate (see Fig. 5.17) [27]. The plate is unelectroded and the electric field in the surrounding free space is neglected. The major surfaces are traction-free. The modes found are trapped, i.e., with the vibration confined in a portion of the plate.

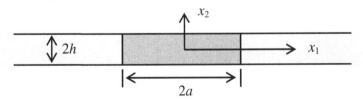

Fig. 5.17. An inhomogeneous plate.

5.9.1. Governing equations and fields

Due to the material inhomogeneity, we need to obtain solutions in different regions and apply interface conditions. First consider the central portion $(|x_1| < a)$. The governing equations for u and ψ are:

$$\overline{c}\nabla^2 u = \rho \ddot{u}, \quad \nabla^2 \psi = 0, \quad \phi = \psi + \frac{e}{\varepsilon}u. \tag{5.143}$$

For the unelectroded and traction-free surfaces at the top and bottom, we have

$$T_{23} = 0, \quad D_2 = 0, \quad x_2 = \pm h, \tag{5.144}$$

or, equivalently, in terms of u and ψ,

$$u_{,2} = 0, \quad \psi_{,2} = 0, \quad x_2 = \pm h. \tag{5.145}$$

The solutions to Eqs. (5.143) and (5.145) can be classified into waves symmetric or antisymmetric in x_2, and they are given by

$$u = (A_1 \cos \xi_1 x_1 + A_2 \sin \xi_1 x_1) \cos \xi_2 x_2 \exp(i\omega t),$$
$$\psi = (B_1 \cosh \xi_2 x_1 + B_2 \sinh \xi_2 x_1) \cos \xi_2 x_2 \exp(i\omega t), \tag{5.146}$$
$$\xi_2 = \frac{m\pi}{2h}, \quad m = 0, 2, 4, \cdots,$$

and

$$u = (A_1 \cos \xi_1 x_1 + A_2 \sin \xi_1 x_1) \sin \xi_2 x_2 \exp(i\omega t),$$
$$\psi = (B_1 \cosh \xi_2 x_1 + B_2 \sinh \xi_2 x_1) \sin \xi_2 x_2 \exp(i\omega t), \tag{5.147}$$
$$\xi_2 = \frac{m\pi}{2h}, \quad m = 1, 3, 5, \cdots,$$

respectively, where

$$\xi_1 = \sqrt{\frac{\rho \omega^2}{\overline{c}} - \xi_2^2} = \sqrt{\frac{\rho}{\overline{c}}} \sqrt{\omega^2 - \left(\frac{m\pi}{2h}\right)^2 \frac{\overline{c}}{\rho}} = \frac{1}{v_T}\sqrt{\omega^2 - \omega_m^2},$$

$$v_T = \sqrt{\frac{\overline{c}}{\rho}}, \quad \omega_m^2 = \left(\frac{m\pi}{2h}\right)^2 \frac{\overline{c}}{\rho}. \tag{5.148}$$

A_1, A_2, B_1 and B_2 are undetermined constants. ω_m is the cutoff frequency of thickness-twist waves.

For the outer regions of the plate where $|x_1| > a$, we use a prime to indicate the relevant material constants. Similar to Eqs. (5.146)–(5.148), we have the following fields satisfying the governing equations and the boundary conditions at $x_2 = \pm h$:

$$u = \begin{cases} A' \exp[-\xi_1'(x_1 - a)]\cos\xi_2 x_2, & x_1 > a, \\ A'' \exp[\xi_1'(x_1 + a)]\cos\xi_2 x_2, & x_1 < -a, \end{cases}$$

$$\psi = \begin{cases} B' \exp[-\xi_2(x_1 - a)]\cos\xi_2 x_2, & x_1 > a, \\ B'' \exp[\xi_2(x_1 + a)]\cos\xi_2 x_2, & x_1 < -a, \end{cases} \tag{5.149}$$

$$\xi_2 = \frac{m\pi}{2h}, \quad m = 0, 2, 4, \cdots,$$

$$u = \begin{cases} A' \exp[-\xi_1'(x_1 - a)]\sin\xi_2 x_2, & x_1 > a, \\ A'' \exp[\xi_1'(x_1 + a)]\sin\xi_2 x_2, & x_1 < -a, \end{cases}$$

$$\psi = \begin{cases} B' \exp[-\xi_2(x_1 - a)]\sin\xi_2 x_2, & x_1 > a, \\ B'' \exp[\xi_2(x_1 + a)]\sin\xi_2 x_2, & x_1 < -a, \end{cases} \tag{5.150}$$

$$\xi_2 = \frac{m\pi}{2h}, \quad m = 1, 3, 5, \cdots,$$

$$\xi_1' = \sqrt{\xi_2^2 - \frac{\rho' \omega^2}{\overline{c}'}} = \frac{1}{v_T'}\sqrt{(\omega_m')^2 - \omega^2},$$

$$\tag{5.151}$$

$$v_T' = \sqrt{\frac{\overline{c}'}{\rho'}}, \quad (\omega_m')^2 = \left(\frac{m\pi}{2h}\right)^2 \frac{\overline{c}'}{\rho'}.$$

In Eqs. (5.149) and (5.150), the fields in the outer regions are exponentially decaying from the interfaces at $x_1 = \pm a$. Modes having this behavior are called trapped modes and are particularly useful for certain device applications.

5.9.2. Continuity conditions and frequency equation

The solutions in the central and outer regions have to satisfy the continuity conditions of u, T_{13}, ϕ and D_1 at the interfaces where $x_1 = \pm a$. We have

$$A_1 \cos \xi_1 a + A_2 \sin \xi_1 a = A'$$

$$\bar{c}(-A_1\xi_1 \sin \xi_1 a + A_2\xi_1 \cos \xi_1 a) + e(B_1\xi_2 \sinh \xi_2 a + B_2\xi_2 \cosh \xi_2 a)$$
$$= \bar{c}'(-A'\xi_1') + e'(-B'\xi_2),$$

$$B_1 \cosh \xi_2 a + B_2 \sinh \xi_2 a + \frac{e}{\varepsilon}(A_1 \cos \xi_1 a + A_2 \sin \xi_1 a) = B' + \frac{e'}{\varepsilon'}A',$$

$$- \varepsilon(B_1\xi_2 \sinh \xi_2 a + B_2\xi_2 \cosh \xi_2 a) = -\varepsilon'(-B'\xi_2),$$

$$A_1 \cos \xi_1 a - A_2 \sin \xi_1 a = A'' \tag{5.152}$$

$$\bar{c}(A_1\xi_1 \sin \xi_1 a + A_2\xi_1 \cos \xi_1 a) + e(-B_1\xi_2 \sinh \xi_2 a + B_2\xi_2 \cosh \xi_2 a)$$
$$= \bar{c}'(A''\xi_1') + e'(B''\xi_2),$$

$$B_1 \cosh \xi_2 a - B_2 \sinh \xi_2 a + \frac{e}{\varepsilon}(A_1 \cos \xi_1 a - A_2 \sin \xi_1 a) = B'' + \frac{e'}{\varepsilon'}A'',$$

$$- \varepsilon(-B_1\xi_2 \sinh \xi_2 a + B_2\xi_2 \cosh \xi_2 a) = -\varepsilon'(B''\xi_2),$$

which is true for waves symmetric or antisymmetric about x_2. Equation (5.152) are eight linear homogeneous equations for A_1, A_2, B_1, B_2, A', B', A'' and B''. For nontrivial solutions, the determinant of the coefficient matrix has to vanish. This yields the frequency equation for possible modes. These modes can be separated into modes symmetric and antisymmetric in x_1. We discuss them separately below.

For modes that are symmetric in x_1, we have

$$A_2 = 0, \quad B_2 = 0, \quad A'' = A', \quad B'' = B'. \tag{5.153}$$

Then the last four equations in Eq. (5.152) become the same as the first four for A_1, B_1, A' and B':

$$\begin{bmatrix} \cos \xi_1 a & 0 & -1 & 0 \\ -\bar{c}\xi_1 \sin \xi_1 a & e\xi_2 \sinh \xi_2 a & \bar{c}'\xi_1' & e'\xi_2 \\ e\varepsilon^{-1} \cos \xi_1 a & \cosh \xi_2 a & -e'/\varepsilon' & -1 \\ 0 & \varepsilon\xi_2 \sinh \xi_2 a & 0 & \varepsilon'\xi_2 \end{bmatrix} \begin{Bmatrix} A_1 \\ B_1 \\ A' \\ B' \end{Bmatrix} = 0. \tag{5.154}$$

The frequency equation is obtained by setting the determinant of the coefficient matrix of Eq. (5.154) to zero:

$$(\bar{c}'\xi_1' - \bar{c}\xi_1 \tan \xi_1 a)(\varepsilon' + \varepsilon \tanh \xi_2 a) = \xi_2 \varepsilon \varepsilon' \left(\frac{e}{\varepsilon} - \frac{e'}{\varepsilon'} \right)^2 \tanh \xi_2 a. \tag{5.155}$$

For modes that are antisymmetric in x_1, we have

$$A_1 = 0, \quad B_1 = 0, \quad A'' = -A', \quad B'' = -B'. \tag{5.156}$$

Again the last four equations in Eq. (5.152) become the same as the first four for A_2, B_2, A' and B':

$$\begin{bmatrix} \sin\xi_1 a & 0 & -1 & 0 \\ \bar{c}\xi_1\cos\xi_1 a & e\xi_2\cosh\xi_2 a & \bar{c}'\xi_1' & e'\xi_2 \\ e\varepsilon^{-1}\sin\xi_1 a & \sinh\xi_2 a & -e'/\varepsilon' & -1 \\ 0 & \varepsilon\xi_2\cosh\xi_2 a & 0 & \varepsilon'\xi_2 \end{bmatrix} \begin{bmatrix} A_2 \\ B_2 \\ A' \\ B' \end{bmatrix} = 0. \tag{5.157}$$

The frequency equation is given by

$$(\bar{c}'\xi_1' + \bar{c}\xi_1\cot\xi_1 a)(\varepsilon' + \varepsilon\coth\xi_2 a) = \xi_2\varepsilon\varepsilon'\left(\frac{e}{\varepsilon} - \frac{e'}{\varepsilon'}\right)^2\coth\xi_2 a. \tag{5.158}$$

With the expressions for ξ_1 and ξ_1' in Eqs. (5.148) and (5.151), Eqs. (5.155) and (5.158) can be written as equations for ω. For trapped modes to exist, we must have

$$\left(\frac{m\pi}{2h}\right)^2\frac{\bar{c}}{\rho} = \omega_m^2 < \omega^2 < (\omega_m')^2 = \left(\frac{m\pi}{2h}\right)^2\frac{\bar{c}'}{\rho'}, \tag{5.159}$$

or $\bar{c}/\rho < \bar{c}'/\rho'$, which is equivalent to $v_T < v_T'$, i.e., the shear wave speed in the central region must be smaller that in the outer regions. The face-shear mode corresponding to $m = 0$ is not a trapped mode.

5.9.3. Numerical results

As a numerical example consider the case when $m = 2$. To show the existence of trapped thickness-twist modes, we need to find roots of the frequency equations. We solve Eq. (5.155) numerically which is for modes symmetric in x_1. We choose PZT-5 for the central region and PZT-6B for the outer regions, respectively, so that Eq. (5.159) is satisfied. The plate thickness is chosen to be $h = 1$mm.

Figures 5.18 and 5.19 show the u and ϕ fields for the first two symmetric modes when $a/h = 10$. Clearly, these modes are trapped modes decaying rapidly away from the interfaces.

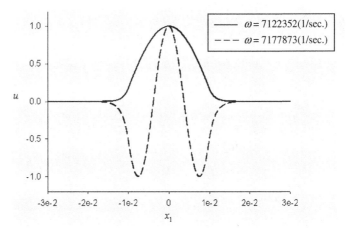

Fig. 5.18. Displacement field of the first two trapped modes.

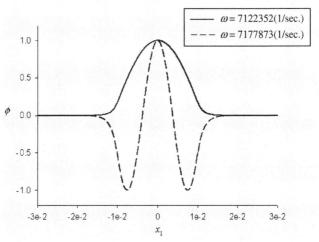

Fig. 5.19. Potential field of the first two trapped modes.

Numerical calculations show that the number of trapped modes depends strongly on a/h, with more modes for lager a/h. In device applications it is important to know for what a/h below which there exists only one mode symmetric in x_1. This particular value of a/h is called the Bechmann's number. For the numerical example we are considering, Fig. 5.20 shows that when a/h approaches approximately 6 from below the second mode will appear. Therefore Bechmann's number is no more than 6.

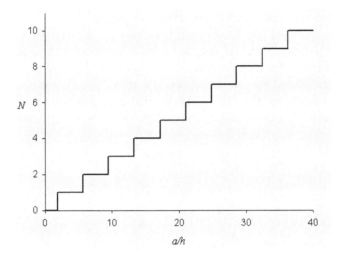

Fig. 5.20. Number of trapped modes versus a/h.

5.10. A Partially Electroded Plate

In this section we study thickness-twist modes in a partially electroded plate, or a plate with partial mass layers [28]. The Bechmann's number in the previous section is also relevant for the structure of this section. Consider a partially electroded plate as shown in Fig. 5.21. Bechmann's number can be obtained by studying two extreme cases of an unelectroded plate ($a = 0$) and a fully electroded plate $a = \infty$.

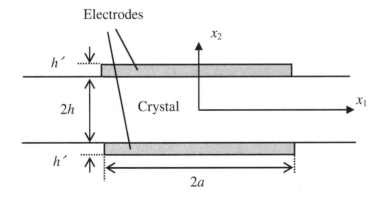

Fig. 5.21. A partially electroded plate.

5.10.1. Waves in an unbounded, unelectroded plate

This case was analyzed in Sec. 5.2 where the free-space electric field was considered. For the purpose of this section the free-space electric field is neglected. For waves symmetric in x_2 the dispersion relations were determined by Eq. (5.28). There also exist similar waves that are antisymmetric in x_2. The dispersions relations for both the symmetric and antisymmetric waves together are given by:

$$\Omega^2 = Z^2 + n^2, \quad n = 0, 1, 2, 3, 4, \cdots, \tag{5.160}$$

where the dimensionless frequency and wave number are

$$\Omega^2 = \omega^2 \bigg/ \left(\frac{\pi^2 \bar{c}}{4\rho h^2} \right), \quad Z = \xi_1 \bigg/ \left(\frac{\pi}{2h} \right). \tag{5.161}$$

When n is odd we have antisymmetric waves. When n is even we have symmetric waves. In particular, $n = 0$ is the face-shear wave. This wave has no cutoff frequency and will not be considered in the following. For later determination of Bechmann's number we need the cutoff frequencies corresponding to $Z = 0$. These frequencies are simply $\Omega = n$.

5.10.2. Waves in an unbounded, electroded plate

Electroded plates were treated in Sec. 5.4 where the inertial effect of the electrodes was considered. Let the mass density of the electrodes be ρ'. The dispersion relations for antisymmetric waves (from Eq. (5.54)) and symmetric waves (from Eq. (5.59)) are:

$$\tanh \xi_1 h - \bar{k}^2 \frac{\xi_1}{\xi_2} \tan \xi_2 h = \frac{\rho' h' \omega^2}{\bar{c} \xi_2} \tan \xi_2 h \tanh \xi_1 h , \tag{5.162}$$

$$-\tan \xi_2 h - \bar{k}^2 \frac{\xi_1}{\xi_2} \tanh \xi_1 h = \frac{\rho' h' \omega^2}{\bar{c} \xi_2} , \tag{5.163}$$

where

$$\xi_2^2 = \frac{\rho \omega^2}{\bar{c}} - \xi_1^2 . \tag{5.164}$$

With Eq. (5.164), Eqs (5.162) and (5.163) are equations for ω versus ξ_1. When both $\rho' h'$ and $\bar{k}^2$ are small, we write the solution to Eqs. (5.162) and (5.163) as

$$\xi_2 h = \frac{n\pi}{2} + \Delta, \quad n = 1, 2, 3, \ldots \tag{5.165}$$

Substituting Eq. (5.165) into Eqs. (5.162) and (5.163), for small Δ, we obtain the following approximate dispersion relations for long waves with a small Z:

$$\Omega^2 = n^2(1 - 2R) - \frac{8}{\pi^2}\overline{k}^2 + Z^2 - 2(R + \overline{k}^2)Z^2, \tag{5.166}$$

$$n = 1, 3, 5, \cdots,$$

$$\Omega^2 = n^2(1 - 2R) + Z^2 - 2(R + \overline{k}^2)Z^2, \tag{5.167}$$

$$n = 2, 4, 6, \cdots,$$

where the mass ratio $R = \rho'h'/(\rho h)$ is small. Equations (5.166) and (5.167) show that the electrode mass lowers the cutoff frequency as expected. It also affects the curvature near cutoff. In Eq. (5.166) the piezoelectric coupling lowers the cutoff frequency because shorted electrodes reduce the electric field in the plate and the related piezoelectric stiffening effect. The piezoelectric coupling also affects the curvature. In Eq. (5.167) the piezoelectric coupling does not lower the cutoff frequency because for symmetric waves values of the electric potential are the same at the two plate surfaces and shortening the electrodes does not affect the electric field.

5.10.3. Bechmann's number

Bechmann's number is defined as the ratio, to the plate thickness, of the wavelength in the electroded portion at the cutoff frequency of the unelectroded portion of the plate.

We begin with antisymmetric waves. Setting $\Omega^2 = n^2$ in Eq. (5.166), we obtain the corresponding wave number as

$$Z^2 = \frac{4}{\pi^2}(\xi_1 h)^2 \cong 2Rn^2 + \frac{8}{\pi^2}\overline{k}^2, \quad n = 1, 3, 5, \cdots. \tag{5.168}$$

Therefore, Bechmann's number for antisymmetric thickness-twist waves is given by

$$B_{TT} = \frac{\pi}{\xi_1 h} = \frac{\pi}{\sqrt{\dfrac{\pi^2}{2}Rn^2 + 2\overline{k}^2}}, \quad n = 1, 3, 5, \cdots. \tag{5.169}$$

We make the following observations from Eq. (5.169):

(i) If $\bar{k}^2$ is set to zero, Eq. (5.169) reduces to

$$B_{TT} = \frac{\sqrt{2}}{n\sqrt{R}}.$$ (5.170)

which has the same dependence on n and R as the result for thickness-twist waves in a quartz plate from an elastic analysis. Only the coefficient is different. For a larger R, Bechmann's number is smaller, as expected. Equation (5.170) also shows that higher-order modes imply a smaller Bechmann's number.

(ii) Both R and $\bar{k}^2$ contribute to Bechmann's number. R is usually of the order of 0.01 or less for resonator electrodes. For thin film resonators R can be an order of magnitude larger, of the order of 0.1. For materials like ZnO and AlN, $\bar{k}^2$ is of the order of 0.1. In applications usually lower order modes with a moderate n are used. In this case the contribution from $\bar{k}^2$ is comparable to the contribution from R. Therefore an accurate prediction of Bechmann's number requires a piezoelectric analysis. Even for quartz with $\bar{k}^2$ of the order of 0.01, a pure elastic analysis may be insufficient.

(iii) For symmetric waves, by setting $\Omega^2 = n^2$ in Eq. (5.167), we obtain Eq. (5.170).

5.11. Multi-sectioned Plates: Phononic Crystals

Motion of a mobile charge in a crystal is governed by Schroedinger's equation with a periodic potential and periodic boundary conditions, and the resulting eigenvalue spectrum has a banded structure. Mathematically this is a consequence of the variable coefficients in the differential equation and periodic boundary conditions. Therefore it allows other physical interpretations, e.g., the propagation of acoustic waves in nonhomogeneous media called phononic crystals. They can be either with gradually varying or piecewise constant material properties depending on the wavelength under consideration. In this section we study thickness-twist vibrations and waves in an unbounded, multi-sectioned plate [29–33]. Consider the plate in Fig. 5.22. It is unelectroded and we ignore the free-space electric field. The major surfaces are traction-free. The results below are from [33].

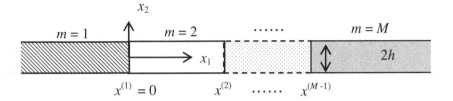

Fig. 5.22. A multi-sectioned plate.

5.11.1. Governing equations

For the mth section, the governing equations for u and ψ are:

$$\bar{c}^{(m)}\nabla^2 u = \rho^{(m)}\ddot{u}, \quad \nabla^2\psi = 0, \quad \phi = \psi + \frac{e^{(m)}}{\varepsilon^{(m)}}u, \tag{5.171}$$

where $\bar{c}^{(m)}$, $e^{(m)}$, $\varepsilon^{(m)}$, and $\rho^{(m)}$ are the elastic, piezoelectric, and dielectric constant as well as the mass density of the mth section. The nonzero stress and electric displacement components are given by

$$T_{23} = \bar{c}^{(m)}u_{,2} + e^{(m)}\psi_{,2}, \quad T_{31} = \bar{c}^{(m)}u_{,1} + e^{(m)}\psi_{,1},$$
$$D_1 = -\varepsilon^{(m)}\psi_{,1}, \quad D_2 = -\varepsilon^{(m)}\psi_{,2}. \tag{5.172}$$

For unelectroded and traction-free surfaces, $T_{23} = 0$ and $D_2 = 0$ at $x_2 = \pm h$. Or, equivalently, in terms of u and ψ,

$$u_{,2} = 0, \quad \psi_{,2} = 0, \quad x_2 = \pm h. \tag{5.173}$$

5.11.2. Solution

It can be obtained by separation of variables that the solutions to Eqs. (5.171) and (5.173) can be classified into waves antisymmetric or symmetric in x_2. For the mth section, they are given by

$$u^{(m,n)} = [A^{(m,n)}\exp(i\xi_1^{(m,n)}x_1) + B^{(m,n)}\exp(-i\xi_1^{(m,n)}x_1)]$$
$$\times \sin\xi_2^{(n)}x_2 \exp(i\omega t),$$
$$\psi^{(m,n)} = [C^{(m,n)}\exp(\xi_2^{(n)}x_1) + D^{(m,n)}\exp(-\xi_2^{(n)}x_1)] \tag{5.174}$$
$$\times \sin\xi_2^{(n)}x_2 \exp(i\omega t),$$
$$\xi_2^{(n)} = \frac{n\pi}{2h}, \quad n = 1, 3, 5, \cdots,$$

and

$$u^{(m,n)} = [A^{(m,n)} \exp(i\xi_1^{(m,n)} x_1) + B^{(m,n)} \exp(-i\xi_1^{(m,n)} x_1)]$$

$$\times \cos \xi_2^{(n)} x_2 \exp(i\omega t),$$

$$\psi^{(m,n)} = [C^{(m,n)} \exp(\xi_2^{(n)} x_1) + D^{(m,n)} \exp(-\xi_2^{(n)} x_1)] \qquad (5.175)$$

$$\pm \cos \xi_2^{(n)} x_2 \exp(i\omega t),$$

$$\xi_2^{(n)} = \frac{n\pi}{2h}, \quad n = 0, 2, 4, \cdots,$$

where

$$\xi_1^{(m,n)} = \sqrt{\frac{\rho^{(m)} \omega^2}{\overline{c}^{(m)}} - (\xi_2^{(n)})^2}$$

$$= \sqrt{\frac{\rho^{(m)}}{\overline{c}^{(m)}}} \sqrt{\omega^2 - \left(\frac{n\pi}{2h}\right)^2 \frac{\overline{c}^{(m)}}{\rho^{(m)}}} = \frac{1}{v_T^{(m)}} \sqrt{\omega^2 - (\omega^{(m,n)})^2}, \qquad (5.176)$$

$$v_T^{(m)} = \sqrt{\frac{\overline{c}^{(m)}}{\rho^{(m)}}}, \quad (\omega^{(m,n)})^2 = \left(\frac{n\pi}{2h}\right)^2 \frac{\overline{c}^{(m)}}{\rho^{(m)}}.$$

$A^{(m,n)}$, $B^{(m,n)}$, $C^{(m,n)}$ and $D^{(m,n)}$ are undetermined constants. The mode with $n = 0$ in Eq. (5.175) is independent of x_2 and is called a face-shear mode which will be excluded in the analysis below.

Consider the antisymmetric modes in Eq. (5.174) first. To apply the interface continuity conditions at the junctions between two neighboring sections, we calculate

$$\phi^{(m,n)} = [\frac{e^{(m)}}{\varepsilon^{(m)}} A^{(m,n)} \exp(i\xi_1^{(m,n)} x_1) + \frac{e^{(m)}}{\varepsilon^{(m)}} B^{(m,n)} \exp(-i\xi_1^{(m,n)} x_1)$$

$$+ C^{(m,n)} \exp(\xi_2^{(n)} x_1) + D^{(m,n)} \exp(-\xi_2^{(n)} x_1)] \sin \xi_2^{(n)} x_2 \exp(i\omega t),$$

$$(5.177)$$

$$D_1^{(m,n)} = [-\varepsilon^{(m)} C^{(m,n)} \xi_2^{(n)} \exp(\xi_2^{(n)} x_1) + \varepsilon^{(m)} D^{(m,n)} \xi_2^{(n)} \exp(-\xi_2^{(n)} x_1)]$$

$$\times \sin \xi_2^{(n)} x_2 \exp(i\omega t), \qquad (5.178)$$

$$T_{13}^{(mn)} = [\overline{c}^{(m)} A^{(m,n)} i\xi_1^{(m,n)} \exp(i\xi_1^{(m,n)} x_1) - \overline{c}^{(m)} B^{(m,n)} i\xi_1^{(m,n)} \exp(-i\xi_1^{(m,n)} x_1)$$

$$+ e^{(m)} C^{(m,n)} \xi_2^{(n)} \exp(\xi_2^{(n)} x_1) - e^{(m)} D^{(m,n)} \xi_2^{(n)} \exp(-\xi_2^{(n)} x_1)]$$

$$\times \sin \xi_2^{(n)} x_2 \exp(i\omega t).$$

$$(5.179)$$

At the mth interface with $x_1 = x^{(m)}$ between the mth section and the $(m + 1)$th section, we have the continuity of the displacement, potential, traction, and electric displacement at the interface:

$$u^{(m,n)}(x_1 = x^{(m)-}) = u^{(m+1,n)}(x_1 = x^{(m)+}),$$

$$\phi^{(m,n)}(x_1 = x^{(m)-}) = \phi^{(m+1,n)}(x_1 = x^{(m)+}),$$

$$T_{13}^{(m,n)}(x_1 = x^{(m)-}) = T_{13}^{(m+1,n)}(x_1 = x^{(m)+}),$$

$$D_1^{(m,n)}(x_1 = x^{(m)-}) = D_1^{(m+1,n)}(x_1 = x^{(m)+}).$$

(5.180)

Equations (5.180) can be written in the following matrix form:

$$\left[\mathbf{T}\right]_{x^{(m)}}^{(m,n)} \{\mathbf{F}\}^{(m,n)} = \left[\mathbf{T}\right]_{x^{(m)}}^{(m+1,n)} \{\mathbf{F}\}^{(m+1,n)},$$

(5.181)

where

$$\left[\mathbf{T}\right]_{x}^{(m,n)} = \begin{bmatrix} \exp(i\xi_1^{(m,n)}x) & \exp(-i\xi_1^{(m,n)}x) \\ \dfrac{e^{(m)}}{\varepsilon^{(m)}}\exp(i\xi_1^{(m,n)}x) & \dfrac{e^{(m)}}{\varepsilon^{(m)}}\exp(-i\xi_1^{(m,n)}x) \\ \overline{c}^{(m)}i\xi_1^{(m,n)}\exp(i\xi_1^{(m,n)}x) & -\overline{c}^{(m)}i\xi_1^{(m,n)}\exp(-i\xi_1^{(m,n)}x) \\ 0 & 0 \end{bmatrix}$$

(5.182)

$$\begin{bmatrix} 0 & 0 \\ \exp(\xi_2^{(n)}x) & \exp(-\xi_2^{(n)}x \\ e^{(m)}\xi_2^{(n)}\exp(\xi_2^{(n)}x) & -e^{(m)}\xi_2^{(n)}\exp(-\xi_2^{(n)}x) \\ -\varepsilon^{(m)}\xi_2^{(n)}\exp(\xi_2^{(n)}x) & \varepsilon^{(m)}\xi_2^{(n)}\exp(-\xi_2^{(n)}x) \end{bmatrix},$$

$$\{\mathbf{F}\}^{(m,n)} = \begin{Bmatrix} A^{(m,n)} \\ B^{(m,n)} \\ C^{(m,n)} \\ D^{(m,n)} \end{Bmatrix}.$$

(5.183)

Equation (5.181) can be further written as

$$\{\mathbf{F}\}^{(m+1,n)} = \left[\mathbf{K}\right]^{(m,n)}\{\mathbf{F}\}^{(m,n)},$$

(5.184)

where

$$\left[\mathbf{K}\right]^{(m,n)} = \left(\left[\mathbf{T}\right]_{x^{(m)}}^{(m+1,n)}\right)^{-1}\left[\mathbf{T}\right]_{x^{(m)}}^{(m,n)}.$$

(5.185)

Suppose that the plate has M sections. With Eq. (5.184) we can write

$$\{\mathbf{F}\}^{(M,n)} = [\mathbf{K}]^{(M-1,n)}[\mathbf{K}]^{(M-2,n)} \cdots [\mathbf{K}]^{(2,n)}[\mathbf{K}]^{(1,n)}\{\mathbf{F}\}^{(1,n)}. \qquad (5.186)$$

Equation (5.186) is a relation between the fields in the first and the last sections. It has four component equations for $A^{(1,n)}$, $B^{(1,n)}$, $C^{(1,n)}$, $D^{(1,n)}$, $A^{(M,n)}$, $B^{(M,n)}$, $C^{(M,n)}$ and $D^{(M,n)}$. The additional four equations needed are from the boundary conditions at the beginning of the first section and the end of the last section. For the symmetric modes in Eq. (5.175), the result is the same except that n assumes even numbers.

5.11.3. Numerical results

The above formulation is rather general and allows us to analyze various problems of practical importance. As an example consider waves incident upon a plate with four sections (see Fig. 5.23).

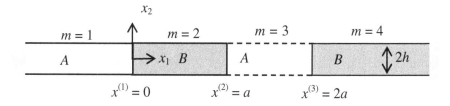

Fig. 5.23. A four-sectioned plate.

Two materials denoted by A and B are used alternatively. To be specific, we choose A and B to be PZT-7 and PZT-4, respectively. In this case

$$\omega_0^A = \frac{\pi}{2h}\sqrt{\frac{\overline{c}_A}{\rho_A}} < \frac{\pi}{2h}\sqrt{\frac{\overline{c}_B}{\rho_B}} = \omega_0^B, \qquad (5.187)$$

which is the fundamental thickness-twist frequency ($n = 1$) of an unbounded plate. $h = 1$ mm, $a = 5$ mm. Below a certain frequency called the cutoff frequency, thickness-twist waves cannot propagate and decay exponentially. Equation (5.187) in fact gives the cutoff frequencies of the fundamental ($n = 1$) thickness-twist waves. The behavior of an incident thickness-twist wave depends strongly on its frequency as compared to ω_0^A and ω_0^B. There are three possibilities which will be discussed separately below. When an incident wave is coming from $x_1 = -\infty$ in the first section, $B^{(1,n)}$ is known. Consider the case when $D^{(1,n)} = 0$. The fields in the other sections may vary according to ω, ω_0^A and ω_0^B.

When $\omega > \omega_0^B > \omega_0^A$, the wave can also propagate in material B. In the last section, there is a right-traveling wave for u only and a decaying field for ψ, with $A^{(M,n)} = 0$ and $C^{(M,n)} = 0$. Equation (5.186) can be used to solve for the reflected wave $A^{(1,n)}$ and $C^{(1,n)}$ in the first section, the transmitted wave $B^{(M,n)}$ and $D^{(M,n)}$ in the last section, and fields in every section of the structure. The displacement distribution along the plate is shown in Fig. 5.24. The four sections are separated by dotted lines. We have sinusoidal variations in all four sections, with different wavelengths in the two materials.

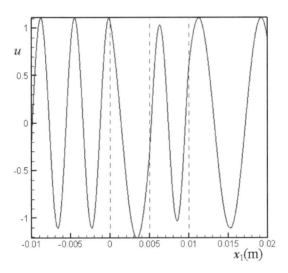

Fig. 5.24. Displacement distribution when $\omega = 1.12\omega_0^B$.

When $\omega_0^B > \omega > \omega_0^A$, the wave can propagate in material A but not in material B. In the last section a decaying field when $x_1 \to \infty$ is allowed but the growing field should be discarded. Therefore we require that $B^{(M,n)} = 0$ and $C^{(M,n)} = 0$. Equation (5.186) can be used to solve for the reflected wave $A^{(1,n)}$ and $C^{(1,n)}$ in the first section, the decaying fields

$A^{(M,n)}$ and $D^{(M,n)}$ in the last section, and fields in every section of the structure. The displacement distribution along the plate is shown in Fig. 5.25. We have sinusoidal variations in material A and exponential fields in material B.

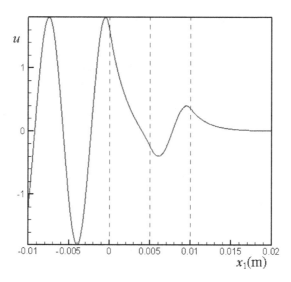

Fig. 5.25. Displacement distribution when $\omega = 0.95\omega_0^B$.

When $\omega_0^B > \omega_0^A > \omega$, the wave cannot propagate in both materials. We cannot specify an incident wave. Instead, we require the fields to be decaying when $x_1 \to \pm\infty$. We have an eigenvalue problem. The situation is similar to the trapped modes in Sec. 5.9.

5.11.4. Modes in a periodic plate

Next we examine the frequency spectrum of modes in a period plate. In this case, it is sufficient to consider a two-portion unit cell (see Fig. 5.26). The periodic plate is obtained by repeating the unit sell in the x_1 direction and is unbounded. Material A is PZT-5H and B is Pz-34. $h = 1$ mm, $a = b = 10$ mm.

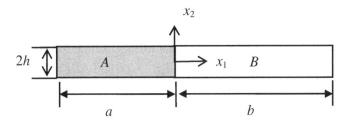

Fig. 5.26. Unit cell of a periodic plate.

The continuity conditions at the junction and periodic conditions at both ends lead to eight homogeneous equations representing an eigenvalue problem. For nontrivial solutions the determinant of the coefficient matrix, denoted by Δ, has to vanish, which yields the frequency equation whose roots are the eigenvalues. We plot in Fig. 5.27 the curve of Δ versus frequency. The intersections of the curve with the horizontal axis are the eigenvalues. To see the intersections better, only a small range of Δ is shown. The intersections with the horizontal axes are discrete. They are dense at certain places and sparse at other places, showing a banded

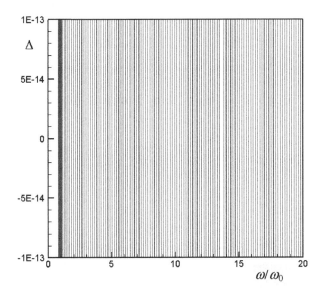

Fig. 5.27. Frequency spectrum of the periodic plate.

structure. Each thick line represents an eigenvalue band which in fact consists of many discrete roots. In particular, if we magnify the first wide band to the left of $\omega/\omega_0 = 1$, we see the further detailed structure of the band as shown in Fig. 5.28. In both Figs. 5.27 and 5.28, the horizontal axis is normalized by $\omega_0 = \omega_0^B$ from Pz-34.

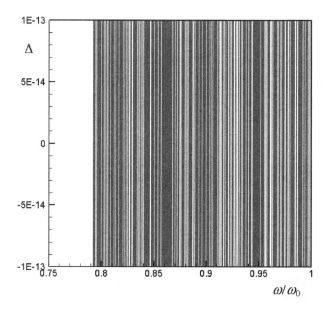

Fig. 5.28. Fine structure of the frequency spectrum.

Chapter 6
Waves in a Layer on a Substrate

This chapter is on propagating waves in a layer on a substrate. In elasticity, the corresponding waves in a plate on a half-space are called Love waves. Similar waves can also propagate near the surface of a circular cylinder with a layer of another material.

6.1. A Metal Plate on a Ceramic Half-space

Consider the case of a metal plate on a ceramic half-space (see Fig. 6.1) [34].

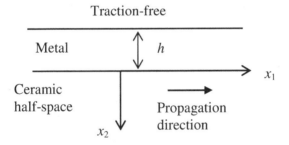

Fig. 6.1. A metal plate on a ceramic half-space.

The governing equations are

$$\bar{c}\,\nabla^2 u = \rho\ddot{u}, \quad \nabla^2\psi = 0, \quad \phi = \psi + \frac{e}{\varepsilon}u, \quad x_2 > 0,$$

$$\hat{c}\,\nabla^2 u = \hat{\rho}\ddot{u}, \quad -h < x_2 < 0,$$

(6.1)

where $\hat{\rho}$ and $\hat{c}$ are the mass density and shear constant of the metal plate. We look for solutions satisfying

$$u, \ \psi \to 0, \quad x_2 \to +\infty. \tag{6.2}$$

For $x_2 > 0$, the solutions to Eq. $(6.1)_1$ satisfying Eq. (6.2) can be written as

$$u = A\exp(-\xi_2 x_2)\cos(\xi x_1 - \omega t),$$
$$\psi = B\exp(-\xi_1 x_2)\cos(\xi x_1 - \omega t),$$
(6.3)

where A and B are constants. For Eq. $(6.3)_1$ to satisfy Eq. $(6.1)_1$, the following must be true

$$\xi_2^2 = \xi_1^2 - \frac{\rho\omega^2}{\overline{c}} = \xi_1^2\left(1 - \frac{v^2}{v_T^2}\right) > 0,$$
(6.4)

where

$$v^2 = \frac{\omega^2}{\xi_1^2}, \quad v_T^2 = \frac{\overline{c}}{\rho}.$$
(6.5)

The electric potential and the stress component needed for the boundary and continuity conditions are

$$T_{23} = \overline{c}u_{,2} + e\psi_{,2}$$
$$= -[\overline{c}A\xi_2\exp(-\xi_2 x_2) + eB\xi_1\exp(-\xi_1 x_2)]\cos(\xi x_1 - \omega t),$$

$$\phi = \psi + \frac{e}{\varepsilon}u$$
(6.6)

$$= [\frac{e}{\varepsilon}A\exp(-\xi_2 x_2) + B\exp(-\xi_1 x_2)]\cos(\xi x_1 - \omega t).$$

For $-h < x_2 < 0$, we write

$$u = (\hat{A}\cos\hat{\xi}_2 x_2 + \hat{B}\sin\hat{\xi}_2 x_2)\cos(\xi_1 x_1 - \omega t),$$
(6.7)

where

$$\hat{\xi}_2^2 = \frac{\hat{\rho}\omega^2}{\hat{c}} - \xi_1^2 = \xi_1^2\left(\frac{v^2}{\hat{v}_T^2} - 1\right),$$
(6.8)

and

$$\hat{v}_T^2 = \frac{\hat{c}}{\hat{\rho}}.$$
(6.9)

For boundary conditions, we need

$$T_{23} = \hat{c}u_{,2} = \hat{c}(-\hat{A}\hat{\xi}_2\sin\hat{\xi}_2 x_2 + \hat{B}\hat{\xi}_2\cos\hat{\xi}_2 x_2)\cos(\xi_1 x_1 - \omega t).$$
(6.10)

The continuity and boundary conditions are (except for a factor of $\cos(\xi_1 x_1 - \omega t)$)

$$\phi(0^+) = \frac{e}{\varepsilon} A + B = 0,$$

$$u(0^+) = A = \hat{A} = u(0^-),$$

$$T_{23}(0^+) = -\bar{c} A \xi_2 - eB\xi_1 = \hat{c}\hat{B}\hat{\xi}_2 = T_{23}(0^-),$$

$$T_{23}(-h) = \hat{c}(\hat{A}\hat{\xi}_2 \sin \hat{\xi}_2 h + \hat{B}\hat{\xi}_2 \cos \hat{\xi}_2 h) = 0.$$

(6.11)

Using Eqs. $(6.11)_{1,2}$ to eliminate A and B, we obtain

$$-\bar{c}\hat{A}\xi_2 + e\frac{e}{\varepsilon}\hat{A}\xi_1 - \hat{c}\hat{B}\hat{\xi}_2 = 0,$$

$$\hat{A}\hat{\xi}_2 \sin \hat{\xi}_2 h + \hat{B}\hat{\xi}_2 \cos \hat{\xi}_2 h = 0.$$

(6.12)

For nontrivial solutions,

$$\begin{vmatrix} -\bar{c}\xi_2 + e\dfrac{e}{\varepsilon}\xi_1 & -\hat{c}\hat{\xi}_2 \\ \hat{\xi}_2 \sin \hat{\xi}_2 h & \hat{\xi}_2 \cos \hat{\xi}_2 h \end{vmatrix} = 0,$$

(6.13)

or

$$\frac{\xi_2}{\xi_1} - \frac{\hat{c}}{\bar{c}} \frac{\hat{\xi}_2}{\xi_1} \tan \hat{\xi}_2 h = \bar{k}^2.$$

(6.14)

Substituting from Eqs. (6.4) and (6.8), we obtain

$$\sqrt{1 - \frac{v^2}{v_T^2}} - \frac{\hat{c}}{\bar{c}} \sqrt{\frac{v^2}{\hat{v}_T^2} - 1} \tan\left[\xi_1 h \sqrt{\frac{v^2}{\hat{v}_T^2} - 1}\right] = \bar{k}^2,$$

(6.15)

which determines the surface wave speed v as a function of the wave number ξ_1. The waves determined by Eq. (6.15) are dispersive.

When the electromechanical coupling factor $\bar{k} = 0$, Eq. (6.15) reduces to the equation that determines the dispersion relation for the well-known Love wave in an elastic layer on an elastic half-space. The dispersion relations for Love waves are real and multi-valued when $\hat{v}_T^2 < v^2 < v_T^2$, for which the shear wave speed of the layer has to be smaller than that of the half-space. In other words, the half-space is more shear-rigid than the layer.

When $\hat{c} = 0$ or $h = 0$, Eq. (6.15) reduces to Eq. (4.12) for the Bleustein–Gulyaev wave in a ceramic half-space with an electroded surface.

6.2. A Dielectric Plate on a Ceramic Half-space

For the case of a dielectric plate on a ceramic half-space, a few solutions can found in [35–38]. Below we study the relatively simple case when there is an electrode on the top surface of the plate so that the electric field in the free space does not need to be considered (see Fig. 6.2).

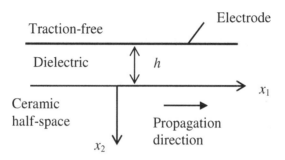

Fig. 6.2. A dielectric plate on a ceramic half-space.

In the substrate we still have Eqs. (6.3) and (6.6):

$$u = A\exp(-\xi_2 x_2)\cos(\xi x_1 - \omega t),$$
$$\psi = B\exp(-\xi_1 x_2)\cos(\xi x_1 - \omega t),$$
(6.16)

$$\phi = [\frac{e}{\varepsilon}A\exp(-\xi_2 x_2) + B\exp(-\xi_1 x_2)]\cos(\xi x_1 - \omega t),$$

$$T_{23} = -[\bar{c}A\xi_2\exp(-\xi_2 x_2) + eB\xi_1\exp(-\xi_1 x_2)]\cos(\xi x_1 - \omega t),$$
(6.17)

$$D_2 = \varepsilon\xi_1 B\exp(-\xi_1 x_2)\cos(\xi x_1 - \omega t).$$

For the dielectric layer we use a superimposed hat (caret) to represent the material parameters and undetermined constants. The governing equations are

$$\hat{c}(u_{,11} + u_{,22}) = \hat{\rho}\ddot{u},$$
(6.18)

$$\nabla^2\phi = 0.$$
(6.19)

The relevant fields are

$$u = (\hat{U}\cos\hat{\xi}_2 x_2 + \hat{V}\sin\hat{\xi}_2 x_2)\cos(\hat{\xi}_1 x_1 - \omega t),$$
(6.20)

$$T_4 = (-\hat{c}\hat{\xi}_2\hat{U}\sin\hat{\xi}_2 x_2 + \hat{c}\hat{\xi}_2\hat{V}\cos\hat{\xi}_2 x_2)\cos(\hat{\xi}_1 x_1 - \omega t),$$
(6.21)

$$\phi = (\hat{G}\cosh\hat{\eta}_2 x_2 + \hat{H}\sinh\hat{\eta}_2 x_2)\cos(\xi_1 x_1 - \omega t), \qquad (6.22)$$

$$D_2 = -\hat{\varepsilon}(\hat{G}\hat{\eta}_2 \sinh\hat{\eta}_2 x_2 + \hat{H}\hat{\eta}_2 \cosh\hat{\eta}_2 x_2)\cos(\xi_1 x_1 - \omega t), \qquad (6.23)$$

where

$$\hat{\xi}_2^2 = \hat{\rho}\omega^2 / \hat{c} - \xi_1^2 = \xi_1^2(\frac{v^2}{\hat{v}_T^2} - 1), \quad \hat{v}_T^2 = \frac{\hat{c}}{\hat{\rho}}. \qquad (6.24)$$

$\hat{U}$, $\hat{V}$, $\hat{G}$ and $\hat{H}$ are undetermined constants. At the interface $x_2 = 0$ and the boundary $x_2 = -h$, we have the following continuity and boundary conditions which represent six homogeneous linear algebraic equations for A, B, $\hat{U}$, $\hat{V}$, $\hat{G}$ and $\hat{H}$:

$$u(0^+) = A = \hat{U} = u(0^-),$$

$$T_4(0^+) = -(\bar{c}A\xi_2 + eB\xi_1) = \hat{c}\hat{\xi}_2\hat{V} = T_4(0^-),$$

$$\phi(0^+) = \frac{e}{\varepsilon}A + B = \hat{G} = \phi(0^-), \qquad (6.25)$$

$$D_2(0^+) = \varepsilon\xi_1 B = -\hat{\varepsilon}\hat{H}\hat{\eta}_2 = D_2(0^-),$$

$$T_4(-h) = -\hat{c}\hat{\xi}_2\hat{U}\sin\hat{\xi}_2 h + \hat{c}\hat{\xi}_2\hat{V}\cos\hat{\xi}_2 h = 0,$$

$$\phi(-h) = \hat{G}\cosh\hat{\eta}_2 h - \hat{H}\sinh\hat{\eta}_2 h = 0.$$

For nontrivial solutions the determinant of the coefficient matrix has to vanish, which yields the following equation that determines the dispersion relations of the waves:

$$\left(1 + \frac{\varepsilon}{\hat{\varepsilon}}\tanh\xi_1 h\right)\left(\sqrt{1 - \frac{v^2}{v_T^2}} - \frac{\hat{c}}{c}\sqrt{\frac{v^2}{\hat{v}_T^2} - 1}\ \tan\xi_1 h\sqrt{\frac{v^2}{\hat{v}_T^2} - 1}\right) = k^2. \qquad (6.26)$$

6.3. An FGM Ceramic Plate on an Elastic Half-space

Consider a semi-infinite elastic substrate covered by a functionally graded piezoelectric material layer as illustrated in Fig. 6.3 [39]. The material properties of the plate change gradually along the thickness of the plate. The elastic substrate is homogeneous. The case of a homogeneous plate on an FGM substrate is treated in [40].

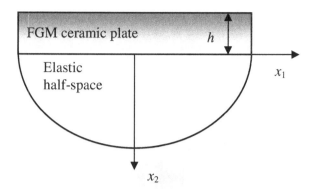

Fig. 6.3. An FGM ceramic plate on an elastic substrate.

6.3.1. Governing equations

For the FGM ceramic plate, when the material properties are varying along x_2, the relevant constitutive relations are

$$T_{13} = c(x_2)u_{,1} + e(x_2)\phi_{,1}, \quad T_{23} = c(x_2)u_{,2} + e(x_2)\phi_{,2}, \tag{6.27}$$

$$D_1 = e(x_2)u_{,1} - \varepsilon(x_2)\phi_{,1}, \quad D_2 = e(x_2)u_{,2} - \varepsilon(x_2)\phi_{,2}. \tag{6.28}$$

The relevant equation of motion and the charge equation of electrostatics are

$$T_{13,1} + T_{23,2} = \rho\ddot{u}, \tag{6.29}$$

$$D_{1,1} + D_{2,2} = 0. \tag{6.30}$$

Substituting Eqs. (6.27) and (6.28) into Eqs. (6.29) and (6.30), we obtain

$$c_{,2}u_{,2} + c\nabla^2 u + e_{,2}\phi_{,2} + e\nabla^2\phi = \rho\ddot{u}, \tag{6.31}$$

$$e_{,2}u_{,2} + e\nabla^2 u - \varepsilon_{,2}\phi_{,2} - \varepsilon\nabla^2\phi = 0. \tag{6.32}$$

We assume that all material properties of the piezoelectric layer have the same exponential variation along the x_2 direction, i.e.,

$$c = c^0 \exp(\beta x_2), \quad e = e^0 \exp(\beta x_2),$$
$$\varepsilon = \varepsilon^0 \exp(\beta x_2), \quad \rho = \rho^0 \exp(\beta x_2), \tag{6.33}$$

where β is a constant. Then Eqs. (6.31) and (6.32) become equations with constant coefficients:

$$c^0 \beta u_{,2} + c^0 \nabla^2 u + e^0 \beta \phi_{,2} + e^0 \nabla^2 \phi = \rho^0 \ddot{u} , \qquad (6.34)$$

$$e^0 \beta u_{,2} + e^0 \nabla^2 u - \varepsilon^0 \beta \phi_{,2} - \varepsilon^0 \nabla^2 \phi = 0 . \qquad (6.35)$$

The governing equations for the elastic substrate are as follows:

$$\hat{c} \nabla^2 u = \hat{\rho} \ddot{u} , \quad \nabla^2 \phi = 0 . \qquad (6.36)$$

where the caret indicates the parameters of the substrate. The shear stress components are given by

$$T_{31} = \hat{c} u_{,1} , \quad T_{32} = \hat{c} u_{,2} . \qquad (6.37)$$

In the free space above the plate, the electric potential is governed by

$$\nabla^2 \phi = 0 . \qquad (6.38)$$

For far fields we have the following limit conditions:

$$x \to +\infty , \quad u = 0, \quad \phi = 0 . \qquad (6.39)$$

$$x \to -\infty , \quad \phi = 0 . \qquad (6.40)$$

The continuity conditions at the interface between the plate and the substrate are

$$u\left(x_1, 0^+, t\right) = u\left(x_1, 0^-, t\right), \quad T_{23}\left(x_1, 0^+, t\right) = T_{23}\left(x_1, 0^-, t\right), \qquad (6.41)$$

$$\phi\left(x_1, 0^+, t\right) = \phi\left(x_1, 0^-, t\right), \quad D_2\left(x_1, 0^+, t\right) = D_2\left(x_1, 0^-, t\right). \qquad (6.42)$$

The traction-free boundary condition at the upper surface of the plate is

$$T_{23}\left(x_1, -h^+, t\right) = 0 . \qquad (6.43)$$

If the upper surface of the plate is unelectroded,

$$\phi\left(x_1, -h^+, t\right) = \phi\left(x_1, -h^-, t\right),$$
$$D_2\left(x_1 - h^+, t\right) = D_2\left(x_1, -h^-, t\right). \qquad (6.44)$$

If the upper surface of the plate is electroded,

$$\phi\left(x_1, -h^+, t\right) = 0 . \qquad (6.45)$$

6.3.2. Fields in different regions

The fields in the elastic half-space can be expressed as

$$u = \hat{A}\exp(-\xi\hat{b}x_2)\exp\left[i\xi(x_1 - vt)\right],\tag{6.46}$$

$$\phi = \hat{B}\exp(-\xi x_2)\exp\left[i\xi(x_1 - vt)\right],\tag{6.47}$$

where $\hat{A}$ and $\hat{B}$ are undetermined constants, and

$$\hat{b} = \sqrt{1 - v^2/\left(\hat{v}_T\right)^2}\,,\quad \hat{v}_T = \sqrt{\hat{c}/\hat{\rho}}\,.\tag{6.48}$$

We consider the case when $v < \hat{v}_T$. The shear stresses and the electric displacement of the substrate can be obtained as follows

$$T_{31} = i\hat{c}\xi\hat{A}\exp(-\xi\hat{b}x_2)\exp[i\xi(x_1 - vt)]\,,\tag{6.49}$$

$$T_{32} = -\hat{c}\xi\hat{b}\hat{A}\exp(-\xi\hat{b}x_2)\exp[i\xi(x_1 - vt)]\,,\tag{6.50}$$

$$D_2 = \hat{\varepsilon}\xi\hat{B}\exp(-\xi x_2)\exp[i\xi(x_1 - vt)]\,,\tag{6.51}$$

For the FGM piezoelectric plate, Eqs. (6.34) and (6.35) can be rewritten as follows:

$$\frac{1}{\rho^0}\left(c^0 + \frac{(e^0)^2}{\varepsilon^0}\right)\left(\beta u_{,2} + \nabla^2 u\right) = \ddot{u}\,,\tag{6.52}$$

$$\beta\psi_{,2} + \nabla^2\psi = 0\,,\tag{6.53}$$

where $\psi = \phi - (e/\varepsilon)u$. Letting

$$\psi = \psi(x_2)\exp\left[i\xi(x_1 - vt)\right],\tag{6.54}$$

$$u = u(x_2)\exp\left[i\xi(x_1 - vt)\right],\tag{6.55}$$

we obtain the following ordinary differential equations:

$$\psi'' + \beta\psi' - \xi^2\psi = 0\,,\tag{6.56}$$

$$u'' + \beta u' + \left(\frac{v^2}{v_T^2} - 1\right)\xi^2 u = 0\,,\tag{6.57}$$

where a prime indicates a derivative with respect to x_2, and

$$v_T^2 = \frac{1}{\rho^0}\left(c^0 + \frac{(e^0)^2}{\varepsilon^0}\right).$$ (6.58)

The solutions to Eqs. (6.56) and (6.57) are

$$\psi(x) = A_1 \exp(r_1 x_2) + A_2 \exp(r_2 x_2),$$ (6.59)

$$u(x) = C_1 \exp(s x_2)\cos\lambda x_2 + C_2 \exp(s x_2)\sin\lambda x_2,$$ (6.60)

where

$$r_{1,2} = \frac{-\beta \pm \sqrt{\beta^2 + 4\xi^2}}{2},$$ (6.61)

$$s = -\beta/2,$$ (6.62)

$$\lambda = \frac{1}{2}\sqrt{4\xi^2\left(\frac{v^2}{v_T^2} - 1\right) - \beta^2}.$$ (6.63)

We limit ourselves to the case when

$$v^2 > v_T^2\left(1 + \frac{\beta^2}{4\xi^2}\right).$$ (6.64)

The corresponding electric potential can be obtained as

$$\phi = A_1 \exp(r_1 x_2) + A_2 \exp(r_2 x_2)$$

$$+ \frac{e^0}{\varepsilon^0}\left[C_1 \exp(s x_2)\cos\lambda x_2 + C_2 \exp(s x_2)\sin\lambda x_2\right]$$ (6.65)

$$\times \exp\left[i\xi(x_1 - vt)\right].$$

The stress and the electric displacement components are

$$T_{31} = c^0 \exp(\beta x_2)i\xi[C_1 \exp(s x_2)\cos\lambda x_2 + C_2 \exp(s x_2)\sin\lambda x_2]$$
$$\times \exp[i\xi(x_1 - vt)]$$
$$+ e^0 \exp(\beta x_2)i\xi\{A_1 \exp(r_1 x_2) + A_2 \exp(r_2 x_2)$$
$$+ \frac{e^0}{\varepsilon^0}[C_1 \exp(s x_2)\cos\lambda x_2 + C_2 \exp(s x_2)\sin\lambda x_2]\}\exp[i\xi(x_1 - vt)],$$ (6.66)

$$T_{32} = c^0 \exp(\beta x_2)[C_1 s \exp(sx_2)\cos \lambda x_2 - C_1 \lambda \exp(sx_2)\sin \lambda x_2$$
$$+ C_2 s \exp(sx_2)\sin \lambda x_2 + C_2 \lambda \exp(sx_2)\cos \lambda x_2]\exp[i\xi(x_1 - vt)]$$
$$+ e^0 \exp(\beta x_2)\{A_1 r_1 \exp(r_1 x_2) + A_2 r_2 \exp(r_2 x_2) \tag{6.67}$$
$$+ \frac{e^0}{\varepsilon^0}[C_1 s \exp(sx_2)\cos \lambda x_2 - C_1 \lambda \exp(sx_2)\sin \lambda x_2$$
$$+ C_2 s \exp(sx_2)\sin \lambda x_2 + C_2 \lambda \exp(sx_2)\cos \lambda x_2]\}\exp[i\xi(x_1 - vt)],$$

$$D_1 = [-ik\varepsilon^0 \exp(\beta x_2 + r_1 x_2)$$
$$- ik\varepsilon^0 \exp(\beta x_2 + r_2 x_2)]\exp[i\xi(x_1 - vt)], \tag{6.68}$$

$$D_2 = [-A_1 r_1 \varepsilon^0 \exp(\beta x_2 + r_1 x_2)$$
$$- A_2 r_2 \varepsilon^0 \exp(\beta x_2 + r_2 x_2)]\exp[i\xi(x_1 - vt)]. \tag{6.69}$$

The electric potential in the vacuum is

$$\phi = A_0 \exp(\xi x_2)\exp\left[i\xi(x_1 - vt)\right]. \tag{6.70}$$

6.3.3. An unelectroded plate

Substituting the relevant fields into Eqs. (6.41)–(6.44), we obtain the following algebraic equations for the unknown constants C_1, C_2, A_1, A_2, A_0, $\hat{A}$ and $\hat{B}$:

$$(c^0 e^{-\beta h - sh} s \cos \lambda h + c^0 e^{-\beta h - sh} \lambda \sin \lambda h + \frac{(e^0)^2}{\varepsilon^0} se^{-\beta h - sh} \cos \lambda h$$

$$+ \frac{(e^0)^2}{\varepsilon^0} \lambda e^{-\beta h - sh} \sin \lambda h)C_1 + (c^0 e^{-\beta x - sh} \lambda \cos \lambda h - c^0 se^{-\beta h - sh} \sin \lambda h$$
$$\tag{6.71}$$
$$+ \frac{(e^0)^2}{\varepsilon^0} \lambda e^{-\beta h - sh} \cos \lambda h - \frac{(e^0)^2}{\varepsilon^0} se^{-\beta h - sh} \sin \lambda h)C_2$$

$$+ e^0 r_1 e^{-\beta h - r_1 h} A_1 + e^0 r_2 e^{-\beta h - r_2 h} A_2 = 0,$$

$$\hat{A} = C_1, \tag{6.72}$$

$$[c^0 s + \frac{(e^0)^2}{\varepsilon^0} s]C_1 + [c^0 \lambda + \frac{(e^0)^2}{\varepsilon^0} \lambda]C_2$$
$$+ A_1 r_1 e^0 + A_2 r_2 e^0 + \hat{A}\hat{c}\xi\hat{b} = 0,$$
(6.73)

$$A_1 + A_2 + \frac{e^0}{\varepsilon^0} C_1 - \hat{B} = 0,$$
(6.74)

$$A_1 r_1 \varepsilon^0 + A_2 r_2 \varepsilon^0 + \hat{B}\xi\hat{e} = 0.$$
(6.75)

$$\frac{e^0}{\varepsilon^0} e^{-sh} C_1 \cos\lambda h - \frac{e^0}{\varepsilon^0} e^{-sh} C_2 \sin\lambda h + A_1 e^{-r_1 h} + A_1 e^{-r_2 h} - A_0 e^{-\xi h} = 0,$$
(6.76)

$$A_1 r_1 \varepsilon^0 e^{-\beta h - r_1 h} + A_2 r_2 \varepsilon^0 e^{-\beta h - r_2 h} - A_0 k \varepsilon_0 e^{-\xi h} = 0,$$
(6.77)

For nontrivial solutions the determinant of the coefficient matrix has to vanish, which gives the frequency equation. The expression is long [39] and is not presented here.

When the piezoelectric plate is of PZT-5H and the elastic substrate is of SiO$_2$, the dispersion relation and the displacement distribution of the waves determined in the above are shown in Figs. 6.4 and 6.5, respectively. Similar to waves in plates, the dispersion curves have multiple braches. Only one is shown.

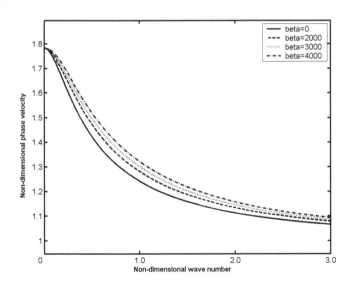

Fig. 6.4. v/v_T versus ξh of the first mode when $h = 0.0002$ m.

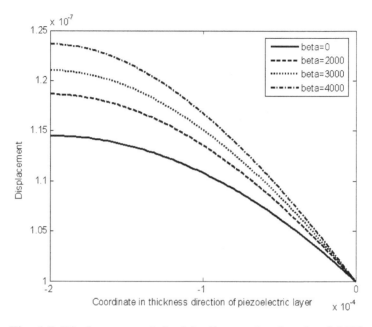

Fig. 6.5. Displacement u(m) of the first mode when $h = 0.0002$ m.

6.3.4. An electroded plate

For an electroded plate, when the relevant fields are substituted into Eqs. (6.41)–(6.43) and (6.45), we still have Eqs. (6.71)–(6.75). However, Eqs. (6.76) and (6.77) are replaced by

$$A_1 e^{-r_1 h} + A_2 e^{-r_2 h} + \frac{e^0}{\varepsilon^0} C_1 e^{-sh} \cos \lambda h - \frac{e^0}{\varepsilon^0} C_2 e^{-sh} \sin \lambda h = 0. \quad (6.78)$$

We have six equations for A_1, A_2, C_1, C_2, $\hat{A}$ and $\hat{B}$. The corresponding frequency equation is

$$-(r_2 - r_1)\left(e^{hs}\left(e^{hr_1} + e^{hr_2}\right)\cos(h\lambda) - \left(e^{2hs} + e^{h(r_2 + r_1)}\right)\right)(e^0)^2 \xi \hat{e} \lambda$$

$$+ (r_2 + r_1)e^{hs} \sin(h\lambda)\left(e^{hr_1} - e^{hr_2}\right)(e^0)^2 \xi s \hat{e}$$

$$+ (r_2 - r_1)e^{hs} \sin(h\lambda)e^{hr_1}[(e^0)^2 + c^0\varepsilon^0]\varepsilon^0\left(s^2 + \lambda^2\right)$$

$$+ r_1 e^{hs} \left(e^{hr_1} - e^{hr_2} \right) (e^0)^2$$

$$\times \left(r_2 \varepsilon^0 \left(\sin(h\lambda)s + \lambda \cos(h\lambda) \right) - \frac{r_2 (e^0)^2 \xi \hat{\varepsilon} \sin(h\lambda)}{(e^0)^2 + c^0 \varepsilon^0} \right) \qquad (6.79)$$

$$- e^{hs} \left(e^{hr_1} - e^{hr_2} \right) \xi [(e^0)^2 + c^0 \varepsilon^0] \hat{\varepsilon} \left(s^2 + \lambda^2 \right) \sin(h\lambda)$$

$$+ \hat{c} \xi \varepsilon^0 \hat{b} \left(e^{hs} \left(e^{hr_1} - e^{hr_2} \right) \xi \hat{\varepsilon} \left(-\sin(h\lambda)s + \lambda \cos(h\lambda) \right) \right.$$

$$+ e^{hs} \left(r_1 e^{hr_1} - r_2 e^{hr_2} \right) \varepsilon^0 \left(-\sin(h\lambda)s + \lambda \cos(h\lambda) \right)$$

$$+ e^{hs} \left(r_2 e^{hr_1} - r_1 e^{hr_2} \right) (e^0)^2 \xi \hat{\varepsilon} \sin(h\lambda) / [(e^0)^2 + c^0 \varepsilon^0]$$

$$+ e^{hs} r_1 r_2 \left(e^{hr_1} - e^{hr_2} \right) (e^0)^2 \varepsilon^0 \sin(h\lambda) / [(e^0)^2 + c^0 \varepsilon^0] \Big) = 0.$$

The corresponding dispersion relation and displacement distribution are shown in Figs. 6.6 and 6.7, respectively.

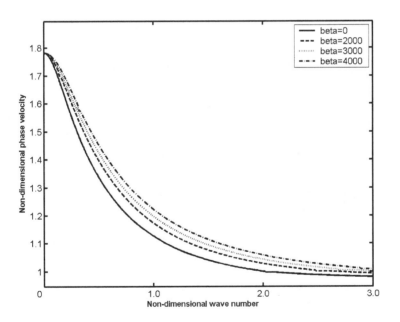

Fig. 6.6. v / v_T versus ξh of the first mode when $h = 0.0002$ m.

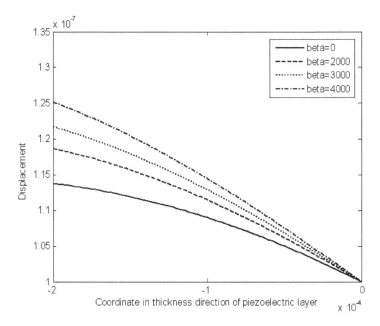

Fig. 6.7. Displacement u(m) of the first mode when $h = 0.0002$ m.

6.4. A Plate Imperfectly Bonded to a Half-space

Consider a ceramic half-space with a layer (see Fig. 6.8) [41,42]. The interface has imperfect bonding. There is an ideal electrode at the interface which is grounded. The layer is an isotropic elastic material.

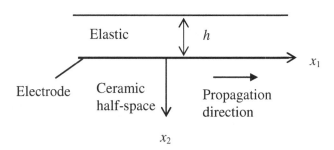

Fig. 6.8. An elastic plate on a ceramic half-space.

The governing equations for u, ψ and ϕ in the half-space and u in the layer are:

$$\bar{c}\nabla^2 u = \rho\ddot{u}, \quad x_2 > 0,$$
$$\nabla^2\psi = 0, \quad x_2 > 0,$$
$$\phi = \psi + \frac{e}{\varepsilon}u, \quad x_2 > 0, \quad (6.80)$$
$$\hat{c}\nabla^2 u = \hat{\rho}\ddot{u}, \quad -h < x_2 < 0,$$

where $\hat{\rho}$ and $\hat{c}$ are the mass density and shear modulus of the layer. We look for solutions satisfying

$$u, \psi \to 0, \quad x_2 \to +\infty. \quad (6.81)$$

The mechanical interaction at the interface is described by the shear-lag model. The continuity and boundary conditions are:

$$\phi(0^+) = 0,$$
$$T_{23}(0^+) = T_{23}(0^-),$$
$$T_{23} = K(u(0^+) - u(0^-)), \quad (6.82)$$
$$T_{23}(-h) = 0,$$

where Eq. (6.82)$_3$ represents the shear-lag model. The model describes an elastic interface with K as the effective interface elastic stiffness that describes how well the two materials are bonded. With this model the interface is allowed to deform and the displacement across the interface is no longer continuous. An infinite K represents a perfect interface with continuous displacement across the interface. A zero K means that the interface is without mechanical interaction (zero traction). For $x_2 > 0$, the solutions satisfying Eq. (6.81) can be written as

$$u = A\exp(-\xi_2 x_2)\cos(\xi_1 x_1 - \omega t),$$
$$\psi = B\exp(-\xi_1 x_2)\cos(\xi_1 x_1 - \omega t), \quad (6.83)$$

where A and B are undetermined constants. Equation (6.83)$_2$ already satisfies Eq. (6.80)$_2$. For Eq. (6.83)$_1$ to satisfy Eq. (6.80)$_1$, the following must be true:

$$\xi_2^2 = \xi_1^2 - \frac{\rho\omega^2}{\bar{c}} = \xi_1^2\left(1 - \frac{v^2}{v_T^2}\right) > 0, \quad v^2 = \frac{\omega^2}{\xi_1^2}, \quad v_T^2 = \frac{\bar{c}}{\rho}. \quad (6.84)$$

The electric potential and the stress component needed for the boundary and continuity conditions are

$$T_{23} = \bar{c}u_{,2} + e\psi_{,2}$$
$$= -[\bar{c}A\xi_2 \exp(-\xi_2 x_2) + eB\xi_1 \exp(-\xi_1 x_2)]\cos(\xi_1 x_1 - \omega t),$$

$$\phi = \psi + \frac{e}{\varepsilon}u \tag{6.85}$$

$$= [\frac{e}{\varepsilon}A\exp(-\xi_2 x_2) + B\exp(-\xi_1 x_2)]\cos(\xi_1 x_1 - \omega t).$$

For $-h < x_2 < 0$, we write

$$u = (\hat{A}\cos\hat{\xi}_2 x_2 + \hat{B}\sin\hat{\xi}_2 x_2)\cos(\xi_1 x_1 - \omega t) , \tag{6.86}$$

where

$$\hat{\xi}_2^2 = \frac{\hat{\rho}\omega^2}{\hat{c}} - \xi_1^2 = \xi_1^2\left(\frac{v^2}{\hat{v}_T^2} - 1\right), \quad \hat{v}_T^2 = \frac{\hat{c}}{\hat{\rho}}. \tag{6.87}$$

For boundary conditions, we need

$$T_{23} = \hat{c}u_{,2}$$
$$= \hat{c}(-\hat{A}\hat{\xi}_2 \sin\hat{\xi}_2 x_2 + \hat{B}\hat{\xi}_2 \cos\hat{\xi}_2 x_2)\cos(\xi_1 x_1 - \omega t). \tag{6.88}$$

The continuity and boundary conditions take the following form:

$$\phi(0^+) = \frac{e}{\varepsilon}A + B = 0,$$

$$T_{23}(0^+) = -\bar{c}A\xi_2 - eB\xi_1 = \hat{c}\hat{B}\hat{\xi}_2 = T_{23}(0^-), \tag{6.89}$$

$$T_{23} = K(u(0^+) - u(0^-)),$$

$$T_{23}(-h) = \hat{c}(\hat{A}\hat{\xi}_2 \sin\hat{\xi}_2 h + \hat{B}\hat{\xi}_2 \cos\hat{\xi}_2 h) = 0,$$

Using Eqs. $(6.89)_{1,3}$ to eliminate A and B, we obtain

$$\bar{c}(\frac{e^2}{\bar{c}\varepsilon}\xi_1 - \xi_2)\hat{A} - \hat{c}\left[(\frac{e^2}{\bar{c}\varepsilon}\xi_1 - \xi_2)\frac{\bar{c}\hat{\xi}_2}{K} - \hat{\xi}_2\right]\hat{B} = 0,$$
$$\hat{A}\sin\hat{\xi}_2 h + \hat{B}\cos\hat{\xi}_2 h = 0. \tag{6.90}$$

For nontrivial solutions, the determinant of the coefficient matrix of Eq. (6.90) has to vanish, i.e.,

$$\frac{\xi_2}{\xi_1} - \frac{\hat{c}}{\bar{c}}\frac{\hat{\xi}_2}{\xi_1}\left[1 - \frac{\bar{c}\hat{\xi}_2}{K}(\bar{k}^2 - \frac{\xi_2}{\xi_1})\right]\tan\hat{\xi}_2 h = \bar{k}^2. \tag{6.91}$$

Substituting from Eqs. (6.84) and (6.87), we obtain

$$\sqrt{1 - \frac{v^2}{v_T^2}} - \frac{\hat{c}}{\bar{c}}\sqrt{\frac{v^2}{\hat{v}_T^2} - 1}\left[1 - \frac{\bar{c}}{K}\xi_1\sqrt{\frac{v^2}{\hat{v}_T^2} - 1}\left(\bar{k}^2 - \sqrt{1 - \frac{v^2}{v_T^2}}\right)\right]$$

$$\times\tan\left[\xi_1 h\sqrt{\frac{v^2}{\hat{v}_T^2} - 1}\right] = \bar{k}^2, \tag{6.92}$$

which determines the wave speed v as a function of the wave number ξ_1. We make the following observations from Eq. (6.92):

(i) Bleustein–Gulyaev wave
When $h = 0$, Equation (6.92) reduces to

$$\sqrt{1 - \frac{v^2}{v_T^2}} = \bar{k}^2 \quad \text{or} \quad v^2 = v_T^2(1 - \bar{k}^4), \tag{6.93}$$

which is the well known Bleustein–Gulyaev (see Eq. (4.12)).

(ii) Perfect bonding
When $K \to \infty$, the layer is perfectly bonded to the half-space. Equation (6.92) reduces to

$$\sqrt{1 - \frac{v^2}{v_T^2}} - \frac{\hat{c}}{\bar{c}}\sqrt{\frac{v^2}{\hat{v}_T^2} - 1}\tan\left[\xi_1 h\sqrt{\frac{v^2}{\hat{v}_T^2} - 1}\right] = \bar{k}^2, \tag{6.94}$$

which is the result of [34] (see Eq. (6.15)). If we further set $\bar{k}^2 = 0$, Eq. (6.94) becomes the frequency equation for Love waves in an elastic half-space carrying an elastic layer.

(iii) Unbonded layer
When $K = 0$, the mechanical interaction at the interface disappears, Eq. (6.92) reduces to

$$-\frac{\hat{c}}{\bar{c}}\sqrt{\frac{v^2}{\hat{v}_T^2} - 1}\left[-\bar{c}\xi_1\sqrt{\frac{v^2}{\hat{v}_T^2} - 1}\left(\bar{k}^2 - \sqrt{1 - \frac{v^2}{v_T^2}}\right)\right]\tan\left[\xi_1 h\sqrt{\frac{v^2}{\hat{v}_T^2} - 1}\right] = 0. \tag{6.95}$$

Equation (6.95) has four factors. The first and the second are the same and they simply imply

$$v = \hat{v}_T , \qquad (6.96)$$

which is the face-shear wave in the elastic layer. The third factor yields Eqs. (6.93) for Bleustein–Gulyaev waves. The fourth factor gives

$$\tan\left[\xi_1 h \sqrt{\frac{v^2}{\hat{v}_T^2} - 1} \right] = 0 , \qquad (6.97)$$

which determines thickness-twist waves of the layer.

(iv) Nonpiezoelectric material

When the electromechanical coupling factor $\bar{k} = 0$, Eq. (6.92) reduces to

$$\sqrt{1 - \frac{v^2}{v_T^2}} - \frac{\hat{c}}{\bar{c}} \sqrt{\frac{v^2}{\hat{v}_T^2} - 1} \left[1 - \frac{\bar{c}}{K} \xi_1 \sqrt{\frac{v^2}{\hat{v}_T^2} - 1} \left(-\sqrt{1 - \frac{v^2}{v_T^2}} \right) \right]$$

$$\tan\left[\xi_1 h \sqrt{\frac{v^2}{\hat{v}_T^2} - 1} \right] = 0, \qquad (6.98)$$

which determines elastic surfaces waves in a half-space with an imperfectly bonded layer.

(v) Long waves

In applications we often encounter long waves whose wavelength is much larger than the layer thickness ($\xi_1 h \ll 1$). We examine the effects of K on long waves below. For small $\xi_1 h$, Eq. (92) can be approximated by

$$\sqrt{1 - \frac{v^2}{v_T^2}} - \alpha\left(\beta \frac{v^2}{v_T^2} - 1 \right)\left[1 - \gamma \xi_1 h \sqrt{\beta \frac{v^2}{v_T^2} - 1} \left(\bar{k}^2 - \sqrt{1 - \frac{v^2}{v_T^2}} \right) \right] \xi_1 h = \bar{k}^2 ,$$

$$(6.99)$$

where the dimensionless parameters, α, β and γ are defined as

$$\alpha = \frac{\hat{c}}{\bar{c}} , \quad \beta = \frac{v_T^2}{\hat{v}_T^2} = \frac{\bar{c}\hat{\rho}}{\rho\hat{c}} , \quad \gamma = \frac{\bar{c}}{Kh} . \qquad (6.100)$$

We perform some numerical calculations from Eq. (6.99). We use PZT-5H and gold for the substrate and the layer, respectively. For these materials we have $\beta > 1$.

In the long wave range shown in Fig. 6.9, only one root is present. The three curves are for a perfect interface ($\gamma = 0$) and $\gamma = 1$ and 10. The curves are close and therefore $\gamma = 1$ and 10 may be considered as cases of almost perfect bonding. The figure shows that the interface compliance reduces the wave speed. This is because that, when compared to the case of perfect interface bonding with $\gamma = 0$ which is shear-rigid, the cases for $\gamma = 1$ and 10 represent a shear-deformable interface with less stiffness and therefore lower frequency or wave speed.

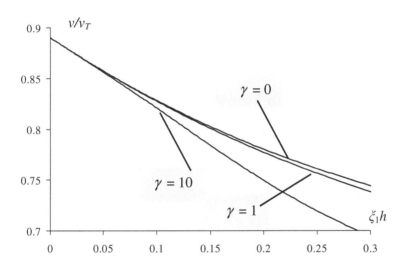

Fig. 6.9. Dispersion relations for a gold layer on PZT-5H (small γ, near perfect bonding).

Figure 6.10 is for larger values of $\gamma = 100$ and 1,000. They represent the cases of relatively loosely bonded interfaces. In this case two real roots of Eq. (6.99) can be found in the range shown. The two branches of the dispersion relations are bounded from above by $v/v_T < 1$. The wave speeds are reduced by the increase of the interface compliance.

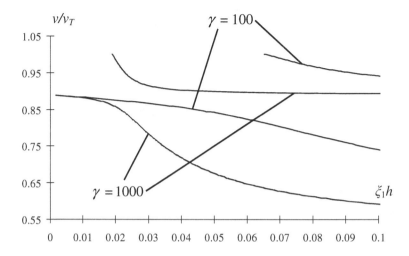

Fig. 6.10. Dispersion relations for a gold layer on PZT-5H (large γ, relatively loose bonding).

6.5. A Plate Imperfectly Bonded to Two Half-spaces

In this section we analyze wave propagation in a ceramic plate between two ceramic half-spaces (see Fig. 6.11)) [43]. The two surfaces of the plate can be electroded or not. The interfaces are modeled by the shear-lag model for imperfect bonding.

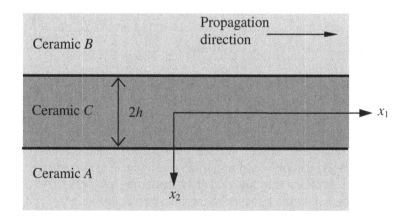

Fig. 6.11. A ceramic plate between two ceramic half-spaces.

6.5.1. Governing equations and fields in different regions

The governing equations for u, ψ and ϕ are:

$$\bar{c}_A \nabla^2 u_A = \rho_A \ddot{u}_A, \quad \nabla^2 \psi_A = 0, \quad \phi_A = \psi_A + \frac{e_A}{\varepsilon_A} u_A, \quad x_2 > h,$$

$$\bar{c}_C \nabla^2 u_C = \rho_C \ddot{u}_C, \quad \nabla^2 \psi_C = 0, \quad \phi_C = \psi_C + \frac{e_C}{\varepsilon_C} u_C, \quad -h < x_2 < h, \quad (6.101)$$

$$\bar{c}_B \nabla^2 u_B = \rho_B \ddot{u}_B, \quad \nabla^2 \psi_B = 0, \quad \phi_B = \psi_B + \frac{e_B}{\varepsilon_B} u_B, \quad x_2 < -h,$$

where the subscripts A, B and C are for ceramic A, ceramic B and ceramic C, respectively. For guided waves that decay from the plate surfaces, we require that

$$\begin{aligned} u_A, \ \psi_A &\to 0, \quad x_2 \to +\infty, \\ u_B, \ \psi_B &\to 0, \quad x_2 \to -\infty. \end{aligned} \quad (6.102)$$

Due to the material inhomogeneity in the structure in Fig. 6.11, we need to obtain solutions in different regions and apply interface conditions.

For $x_2 > h$, consider the possibility of the following waves propagating in the x_1 direction:

$$\begin{aligned} u_A &= U_A \exp\left[-\eta_A (x_2 - h)\right] \cos(\xi x_1 - \omega t), \\ \psi_A &= \Psi_A \exp\left[-\xi (x_2 - h)\right] \cos(\xi x_1 - \omega t), \end{aligned} \quad (6.103)$$

where U_A and Ψ_A are undetermined constants, and

$$\eta_A^2 = \xi^2 - \frac{\rho_A \omega^2}{\bar{c}_A} = \xi^2 \left(1 - \frac{v^2}{v_A^2}\right) > 0,$$

$$v = \frac{\omega}{\xi}, \quad v_A^2 = \frac{\bar{c}_A}{\rho_A}. \quad (6.104)$$

Equation (6.103) satisfies Eqs. (6.101)$_1$ and (6.102)$_1$. The inequality in Eq. (6.104)$_1$ is for guided waves that decay from the plate surfaces. For interface conditions, we need T_{23} and D_2 in ceramic A, denoted by T_A and D_A. They are given by

$$T_A = \bar{c}_A u_{A,2} + e_A \psi_{A,2}$$

$$= -\left\{\bar{c}_A \eta_A U_A \exp\left[-\eta_A (x_2 - h)\right] + e_A \xi \Psi_A \exp\left[-\xi (x_2 - h)\right]\right\}, \quad (6.105)$$

$$D_A = -\varepsilon_A \psi_{A,2} = \varepsilon_A \xi \Psi_A \exp\left[-\xi (x_2 - h)\right].$$

Similarly, for $x_2 < -h$, the solution to Eq. $(6.101)_3$ that satisfies Eq. $(6.102)_2$ are

$$u_B = U_B \exp\left[\eta_B\left(x_2 + h\right)\right]\cos\left(\xi x_1 - \omega t\right),$$
$$\psi_B = \Psi_B \exp\left[\xi\left(x_2 + h\right)\right]\cos\left(\xi x_1 - \omega t\right),$$

(6.106)

where U_B and Ψ_B are undetermined constants, and

$$\eta_B^2 = \xi^2 - \frac{\rho_B \omega^2}{\bar{c}_B} = \xi^2\left(1 - \frac{v^2}{v_B^2}\right) > 0,$$

$$v_B^2 = \frac{\bar{c}_B}{\rho_B}.$$

(6.107)

For interface conditions, we need:

$$T_B = \bar{c}_B u_{B,2} + e_B \psi_{B,2}$$
$$= \left\{\bar{c}_B \eta_B U_B \exp\left[\eta_B\left(x_2 + h\right)\right] + e_B \xi \Psi_B \exp\left[\xi\left(x_2 + h\right)\right]\right\},$$

(6.108)

$$D_B = -\varepsilon_B \psi_{B,2} = -\varepsilon_B \xi \Psi_B \exp\left[\xi\left(x_2 + h\right)\right].$$

For the piezoceramic plate (C), the fields can be represented by

$$u_C = \left(U_{C1}\cos\eta_C x_2 + U_{C2}\sin\eta_C x_2\right)\cos\left(\xi x_1 - \omega t\right),$$
$$\psi_C = \left(\Psi_{C1}\cosh\xi x_2 + \Psi_{C2}\sinh\xi x_2\right)\cos\left(\xi x_1 - \omega t\right),$$

(6.109)

where

$$\eta_C^2 = \frac{\rho_C \omega^2}{\bar{c}_C} - \xi^2 = \xi^2\left(\frac{v^2}{v_C^2} - 1\right),$$

$$v_C^2 = \frac{\bar{c}_C}{\rho_C}.$$

(6.110)

The field components relevant to interface conditions are found to be

$$T_C = \bar{c}_C u_{C,2} + e_C \psi_{C,2}$$
$$= \left[\bar{c}_C \eta_C\left(-U_{C1}\sin\eta_C x_2 + U_{C2}\cos\eta_C x_2\right) + \right.$$
$$\left. e_C \xi\left(\Psi_{C1}\sinh\xi x_2 + \Psi_{C2}\cosh\xi x_2\right)\right],$$

(6.111)

$$D_C = -\varepsilon_C \psi_{C,2} = -\varepsilon_C \xi\left(\Psi_{C1}\sinh\xi x_2 + \Psi_{C2}\cosh\xi x_2\right).$$

6.5.2. An electroded plate

First we consider the case of an electroded plate with the electrodes shorted and grounded. For interface conditions we have

$$T_A = T_C = K_1(u_A - u_C), \quad x_2 = h,$$

$$T_B = T_C = K_2(u_C - u_B), \quad x_2 = -h,$$

$$\psi_A + \frac{e_A}{\varepsilon_A} u_A = 0, \quad \psi_C + \frac{e_C}{\varepsilon_C} u_C = 0, \quad x_2 = h, \tag{6.112}$$

$$\psi_B + \frac{e_B}{\varepsilon_B} u_B = 0, \quad \psi_C + \frac{e_C}{\varepsilon_C} u_C = 0, \quad x_2 = -h.$$

Equations $(6.112)_{1,2}$ are according to the shear-lag interface model. The displacements at the interfaces are allowed to be discontinuous. K_1 and K_2 are the elastic constants of the interfaces. Substitution of Eqs. (6.103), (6.105), (6.106), (6.108), (6.109) and (6.111) into Eq. (6.112) yields eight linear homogeneous equations for U_A, U_B, U_{C1}, U_{C2}, Ψ_A, Ψ_B, Ψ_{C1}, and Ψ_{C2}. For nontrivial solutions of these constants, the determinant of the coefficient matrix has to vanish. This yields a rather long expression for the frequency equation and it is not provided here. We consider a few special cases below.

When the two half-spaces are of the same material, i.e.,

$$\rho_A = \rho_B, \quad \bar{c}_A = \bar{c}_B, \quad e_A = e_B, \quad \varepsilon_A = \varepsilon_B, \tag{6.113}$$

and $K_1 = K_2 = K$, the waves can be separated into symmetric and antisymmetric ones. For symmetric waves,

$$U_A = U_B, \quad \Psi_A = \Psi_B, \quad U_{C2} = 0, \quad \Psi_{C2} = 0. \tag{6.114}$$

In this case the frequency equation is

$$\begin{vmatrix} -\bar{c}_A \eta_A - K & K\cos\eta_C h & -e_A \xi & 0 \\ -K & -\bar{c}_C \eta_C \sin\eta_C h + K\cos\eta_C h & 0 & e_C \xi \sinh\xi h \\ \dfrac{e_A}{\varepsilon_A} & 0 & 1 & 0 \\ 0 & \dfrac{e_C}{\varepsilon_C}\cos\eta_C h & 0 & \cosh\xi h \end{vmatrix} = 0,$$

$$\tag{6.115}$$

which, when expanded, has the following form:

$$
\left(\frac{e_A^2}{\varepsilon_A} \xi - \overline{c}_A \eta_A \right) \left(\overline{c}_C \eta_C \tan \eta_C h + \frac{e_C^2}{\varepsilon_C} \xi \tanh \xi h \right)
$$
$$
= K \left[\left(\frac{e_A^2}{\varepsilon_A} \xi - \eta_A \overline{c}_A \right) + \left(\overline{c}_C \eta_C \tan \eta_C h + \frac{e_C^2}{\varepsilon_C} \xi \tanh \xi h \right) \right].
\tag{6.116}
$$

The following observations can be made from Eq. (6.116):

(i) $K = \infty$. The interfaces are perfectly bonded. Eq. (6.116) reduces to

$$
\left(\frac{e_A^2}{\varepsilon_A} \xi - \overline{c}_A \eta_A \right) + \left(\overline{c}_C \eta_C \tan \eta_C h + \frac{e_C^2}{\varepsilon_C} \xi \tanh \xi h \right) = 0 .
\tag{6.117}
$$

(ii) $K = 0$. There is no interface bonding. Mechanically the plate and the two half-spaces move independently. Since the plate is electroded and the electrodes are shorted, there is no electrical interaction between the plate and the half-spaces. Equation (6.116) implies that

$$
\frac{v^2}{v_A^2} = 1 - \overline{k}_A^4 ,
\tag{6.118}
$$

and

$$
\overline{c}_C \eta_C \tan \eta_C h + \frac{e_C^2}{\varepsilon_C} \xi \tanh \xi h = 0,
\tag{6.119}
$$

where $\overline{k}_A^2 = e_A^2 / (\overline{c}_A \varepsilon_A)$. Equation (6.118) is the speed of the well-known Blustein–Gulyaev surface wave over an electroded half-space (see Eq. (4.12)). Equation (6.119) is the dispersion relation for symmetric waves in an electroded plate (see Eq. (5.13)).

(iii) When the plate does not exist, i.e., $h = 0$, Eq. (6.116) reduces to

$$
\frac{v^2}{v_A^2} = 1 - \overline{k}_A^4 ,
\tag{6.120}
$$

which is the same as Eq. (6.118). This is because for symmetric waves there is no interface shear between the two half-spaces and therefore the waves do not feel the interface and we essentially have surface waves.

For numerical results, we write Eq. (6.116) as:

$$\gamma \left(\bar{k}_A^2 - \sqrt{1 - \frac{v^2}{v_A^2}} \right) \left(\sqrt{\frac{v^2}{v_A^2}\beta - 1} \tan \xi h \sqrt{\frac{v^2}{v_A^2}\beta - 1} + \bar{k}_C^2 \tanh \xi h \right)$$

$$= \alpha \left(\bar{k}_A^2 - \sqrt{1 - \frac{v^2}{v_A^2}} \right) + \left(\sqrt{\frac{v^2}{v_A^2}\beta - 1} \tan \xi h \sqrt{\frac{v^2}{v_A^2}\beta - 1} + \bar{k}_C^2 \tanh \xi h \right),$$

$$(6.121)$$

where $\bar{k}_C^2 = e_C^2 / (\bar{c}_C \varepsilon_C)$. The dimensionless parameters α, β, and γ are defined by:

$$\alpha = \frac{\bar{c}_A}{\bar{c}_C}, \quad \beta = \frac{v_A^2}{v_C^2} = \frac{\bar{c}_A \rho_C}{\bar{c}_C \rho_A}, \quad \gamma = \frac{\bar{c}_A}{Kh}. \quad (6.122)$$

For the half-spaces, we use PZT-4. For the plate, we consider two cases of PZT-5 and BaTiO$_3$. When the interfaces are perfect, the dispersion relations from Eq. (6.121) for different materials of the plate are shown in Fig. 6.12. For a BaTiO$_3$ plate, we have $\beta < 1$ and only one root is found in the range shown.

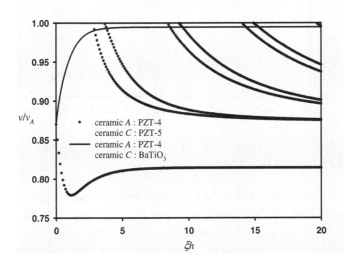

Fig. 6.12. Symmetric waves in an electroded plate with perfect bonding.

The effect of a large K or small γ on the waves determined by Eq. (6.121) is shown in Fig. 6.13. In the figure, for the plate we use PZT-5 with $\beta > 1$. When ξh is small (long waves), only one root is found in the range shown. The three curves correspond to different values of γ. The curves are close. Therefore, $\gamma = 1$ through 5 may be considered as cases of almost perfect bonding. The figure shows that imperfect bonding reduces the wave speed. This is because, when compared to the case of a perfect interface with $\gamma = 0$ which is shear-rigid, the cases for $\gamma = 1$ and 5 represent a shear-deformable interface with less stiffness and therefore lower frequency or wave speed.

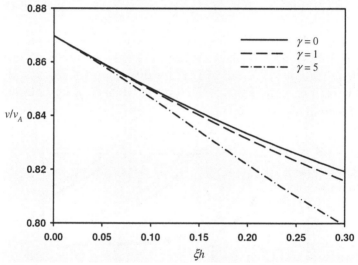

Fig. 6.13. Long symmetric waves in an electroded plate with a small γ (near perfect bonding).

Figure 6.14 is still for a PZT-5 plate, but it is for larger values of $\gamma = 50$ and 200. They represent cases of relatively loosely bonded interfaces. In this case, two real roots of Eq. (6.121) are found in the range shown. For an arbitrarily given γ, the equation may have more roots. In Figs. 6.13 and 6.14, we limit v/v_A to $(0,1)$ and ξh to $(0,0.1)$ or $(0,0.3)$.

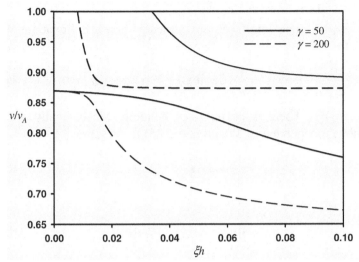

Fig. 6.14. Long symmetric waves in an electroded plate with a large γ
(loose bonding).

For antisymmetric waves, the solution satisfies

$$U_A = -U_B, \quad \Psi_A = -\Psi_B, \quad U_{C1} = 0, \quad \Psi_{C1} = 0. \tag{6.123}$$

In this case, the frequency equation is

$$\begin{vmatrix} -\bar{c}_A \eta_A - K & K \sin \eta_C h & -e_A \xi & 0 \\ -K & \bar{c}_C \eta_C \cos \eta_C h + K \sin \eta_C h & 0 & e_C \xi \cosh \xi h \\ \dfrac{e_A}{\varepsilon_A} & 0 & 1 & 0 \\ 0 & \dfrac{e_C}{\varepsilon_C} \sin \eta_C h & 0 & \sinh \xi h \end{vmatrix} = 0. \tag{6.124}$$

From Eq. (6.124), we obtain:

$$\begin{aligned}
&\left(\frac{e_A^2}{\varepsilon_A} \xi - \bar{c}_A \eta_A \right) \left(\frac{e_C^2}{\varepsilon_C} \xi \coth \xi h - \bar{c}_C \eta_C \cot \eta_C h \right) \\
&= K \left[\left(\frac{e_A^2}{\varepsilon_A} \xi - \eta_A \bar{c}_A \right) + \left(\frac{e_C^2}{\varepsilon_C} \xi \coth \xi h - \bar{c}_C \eta_C \cot \eta_C h \right) \right].
\end{aligned} \tag{6.125}$$

A few special cases are as follows.

(i) $K = \infty$. The interfaces are perfectly bonded. Equation (6.125) reduces to

$$\left(\frac{e_A^2}{\varepsilon_A} \xi - \eta_A \bar{c}_A \right) + \left(\frac{e_C^2}{\varepsilon_C} \xi \coth \xi h - \bar{c}_C \eta_C \cot \eta_C h \right) = 0. \qquad (6.126)$$

(ii) $K = 0$. There is no interface bonding. Mechanically the three parts move independently. Equation (6.125) factors into

$$\frac{v^2}{v_A^2} = 1 - \bar{k}_A^4, \qquad (6.127)$$

and

$$\frac{e_C^2}{\varepsilon_C} \xi \coth \xi h - \bar{c}_C \eta_C \cot \eta_C h = 0. \qquad (6.128)$$

Equation (6.128) is the dispersion relation for antisymmetric thickness-twist waves in an electroded plate (see Eq. (5.7)).

(iii) When the plate does not exist, i.e., $h = 0$, we have the interface waves treated in [17]. In this case Eq. (6.125) reduces to (compared to Eq. (4.115))

$$\frac{v^2}{v_A^2} = 1 - \left(\bar{k}_A^2 - \frac{K}{\bar{c}_A \xi} \right)^2. \qquad (6.129)$$

This antisymmetric wave has interface shear and therefore can feel the interface compliance. Equation (6.129) shows that the effect of K is more pronounced for long waves with a small wave number ξ.

6.5.3. An unelectroded plate

Next consider the case when the interfaces are unelectroded. The continuity conditions are

$$T_A = T_C = K_1(u_A - u_C), \quad x_2 = h,$$

$$T_B = T_C = K_2(u_C - u_B), \quad x_2 = -h,$$

$$\psi_A + \frac{e_A}{\varepsilon_A} u_A = \psi_C + \frac{e_C}{\varepsilon_C} u_C, \quad D_A = D_c, \quad x_2 = h, \qquad (6.130)$$

$$\psi_B + \frac{e_B}{\varepsilon_B} u_B = \psi_C + \frac{e_C}{\varepsilon_C} u_C, \quad D_B = D_c, \quad x_2 = -h,$$

Substitution of Eqs. (6.103), (6.105), (6.106), (6.108), (6.109) and (6.111) into Eq. (6.130) yields eight linear homogeneous equations for U_A, U_B, U_{C1}, U_{C2}, Ψ_A, Ψ_B, Ψ_{C1}, and Ψ_{C2}. For nontrivial solutions of these constants, the determinant of the coefficient matrix has to vanish. This yields a rather long expression for the frequency equation and it is not provided here. We consider a few special cases below.

Let Eq. (6.113) be true. For symmetric waves,

$$U_A = U_B, \quad \Psi_A = \Psi_B, \quad U_{C2} = 0, \quad \Psi_{C2} = 0. \tag{6.131}$$

The frequency equation is

$$\begin{vmatrix} -\bar{c}_A \eta_A - K & K \cos \eta_C h & -e_A \xi & 0 \\ -K & -\bar{c}_C \eta_C \sin \eta_C h + K \cos \eta_C h & 0 & e_C \xi \sinh \xi h \\ \dfrac{e_A}{\varepsilon_A} & -\dfrac{e_C}{\varepsilon_C} \cos \eta_C h & 1 & -\cosh \xi h \\ 0 & 0 & \varepsilon_A \xi & \varepsilon_C \xi \sinh \xi h \end{vmatrix} = 0, \tag{6.132}$$

or

$$\eta_A \bar{c}_A \varepsilon_A \left(\frac{e_C^2}{\varepsilon_C} \xi \tanh \xi h + \bar{c}_C \eta_C \tan \eta_C h \right)$$

$$- \bar{c}_C \eta_C \varepsilon_C \tan \eta_C h \tanh \xi h \left(\frac{e_A^2}{\varepsilon_A} \xi - \bar{c}_A \eta_A \right)$$

$$+ K \left[\varepsilon_C \tanh \xi h \left(\frac{e_A^2}{\varepsilon_A} \xi - \eta_A \bar{c}_A \right) - \eta_A \bar{c}_A \varepsilon_A - 2 e_A e_C \xi \tanh \xi h \right. \tag{6.133}$$

$$\left. + \varepsilon_A \left(\frac{e_C^2}{\varepsilon_C} \xi \tanh \xi h + \bar{c}_C \eta_C \tan \eta_C h \right) + \bar{c}_C \eta_C \varepsilon_C \tan \eta_C h \tanh \xi h \right] = 0.$$

The following observations can be made from Eq. (6.133):

(i) $K = \infty$. The interfaces are perfectly bonded. Equation (6.133) reduces to

$$\varepsilon_C \tanh \xi h \left(\frac{e_A^2}{\varepsilon_A} \xi - \eta_A \bar{c}_A \right) - \eta_A \bar{c}_A \varepsilon_A - 2 e_A e_C \xi \tanh \xi h$$

$$+ \varepsilon_A \left(\frac{e_C^2}{\varepsilon_C} \xi \tanh \xi h + \bar{c}_C \eta_C \tan \eta_C h \right) + \bar{c}_C \eta_C \varepsilon_C \tan \eta_C h \tanh \xi h = 0. \tag{6.134}$$

(ii) $K = 0$. The interfaces have no mechanical interaction. However, since the plate is unelectroded, the plate and the two half-spaces can still interact electrically. Equation (6.133) reduces to

$$\bar{c}_A \eta_A \varepsilon_A \left(\frac{e_C^2}{\varepsilon_C} \xi \tanh \xi h + \bar{c}_C \eta_C \tan \eta_C h \right)$$

$$- \bar{c}_C \eta_C \varepsilon_C \tan \eta_C h \tanh \xi h \left(\frac{e_A^2}{\varepsilon_A} \xi - \bar{c}_A \eta_A \right) = 0. \tag{6.135}$$

(iii) When the plate does not exist, i.e., $h = 0$, Eq. (6.133) reduces to

$$v^2 = v_A^2, \tag{6.136}$$

which is no longer a guided wave.

In terms of the wave speed v, Eq. (6.133) can be written as:

$$\gamma \left[\xi h \sqrt{1 - \frac{v^2}{v_A^2}} \varepsilon_A \left(\bar{k}_C^2 \tanh \xi h + \sqrt{\frac{v^2}{v_A^2} \beta - 1} \tan \xi h \sqrt{\frac{v^2}{v_A^2} \beta - 1} \right) \right.$$

$$\left. - \xi h \varepsilon_C \sqrt{\frac{v^2}{v_A^2} \beta - 1} \left(\bar{k}_A^2 - \sqrt{1 - \frac{v^2}{v_A^2}} \right) \tanh \xi h \tan \xi h \sqrt{\frac{v^2}{v_A^2} \beta - 1} \right]$$

$$+ \left[\varepsilon_C \alpha \tanh \xi h \left(\bar{k}_A^2 - \sqrt{1 - \frac{v^2}{v_A^2}} \right) - \varepsilon_A \alpha \sqrt{1 - \frac{v^2}{v_A^2}} - 2 \frac{e_A e_C}{\bar{c}_C} \tanh \xi h \tag{6.137} \right.$$

$$+ \varepsilon_A \left(\bar{k}_C^2 \tanh \xi h + \sqrt{\frac{v^2}{v_A^2} \beta - 1} \tan \xi h \sqrt{\frac{v^2}{v_A^2} \beta - 1} \right)$$

$$\left. + \varepsilon_C \sqrt{\frac{v^2}{v_A^2} \beta - 1} \tanh \xi h \tan \xi h \sqrt{\frac{v^2}{v_A^2} \beta - 1} \right] = 0.$$

Figures 6.15 and 6.16 show the results for the symmetric waves when the interfaces are unelectroded. The half-spaces are of PZT-4. PZT-5 is used for the plate with which $\beta > 1$.

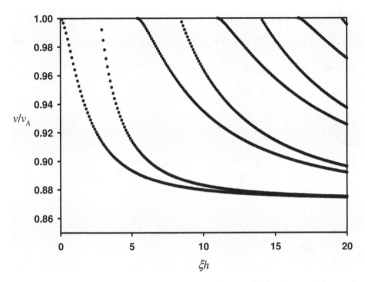

Fig. 6.15. Symmetric waves in an unelectroded plate with perfect bonding.

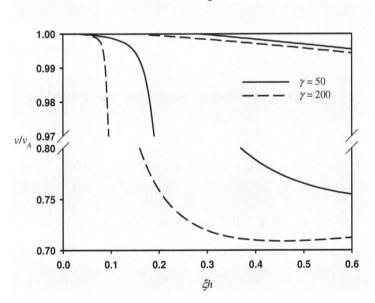

Fig. 6.16. Long symmetric waves in an unelectroded plate with a large γ.

For antisymmetric waves,

$$U_A = -U_B, \quad \Psi_A = -\Psi_B, \quad U_{C1} = 0, \quad \Psi_{C1} = 0. \tag{6.138}$$

The frequency equation is

$$\begin{vmatrix} -\overline{c}_A \eta_A - K & K \sin \eta_C h & -e_A \xi & 0 \\ -K & \overline{c}_C \eta_C \cos \eta_C h + K \sin \eta_C h & 0 & e_C \xi \cosh \xi h \\ \dfrac{e_A}{\varepsilon_A} & -\dfrac{e_C}{\varepsilon_C} \sin \eta_C h & 1 & -\sinh \xi h \\ 0 & 0 & \varepsilon_A \xi & \varepsilon_C \xi \cosh \xi h \end{vmatrix} = 0, \tag{6.139}$$

or

$$\eta_A \overline{c}_A \varepsilon_A \left(\frac{e_C^2}{\varepsilon_C} \xi \coth \xi h - \overline{c}_C \eta_C \cot \eta_C h \right)$$

$$+ \overline{c}_C \eta_C \varepsilon_C \cot \eta_C h \coth \xi h \left(\frac{e_A^2}{\varepsilon_A} \xi - \overline{c}_A \eta_A \right)$$

$$\tag{6.140}$$

$$+ K \left[\varepsilon_C \coth \xi h \left(\frac{e_A^2}{\varepsilon_A} \xi - \eta_A \overline{c}_A \right) - \eta_A \overline{c}_A \varepsilon_A - 2 e_A e_C \xi \coth \xi h \right.$$

$$\left. + \varepsilon_A \left(\frac{e_C^2}{\varepsilon_C} \xi \coth \xi h - \overline{c}_C \eta_C \cot \eta_C h \right) - \overline{c}_C \eta_C \varepsilon_C \cot \eta_C h \coth \xi h \right] = 0.$$

We have:

(i) When $K = \infty$ or the interfaces are perfectly bonded, Eq. (6.140) reduces to

$$\varepsilon_C \coth \xi h \left(\frac{e_A^2}{\varepsilon_A} \xi - \eta_A \overline{c}_A \right) - \eta_A \overline{c}_A \varepsilon_A - 2 e_A e_C \xi \tanh \xi h$$

$$\tag{6.141}$$

$$+ \varepsilon_A \left(\frac{e_C^2}{\varepsilon_C} \xi \coth \xi h - \overline{c}_C \eta_C \cot \eta_C h \right) - \overline{c}_C \eta_C \varepsilon_C \cot \eta_C h \coth \xi h = 0.$$

(ii) When $K = 0$, there is no interface bonding. Equation (6.140) reduces to

$$\eta_A \overline{c}_A \varepsilon_A \left(\frac{e_C^2}{\varepsilon_C} \xi \coth \xi h - \overline{c}_C \eta_C \cot \eta_C h \right)$$
$$+ \overline{c}_C \eta_C \varepsilon_C \cot \eta_C h \coth \xi h \left(\frac{e_A^2}{\varepsilon_A} \xi - \overline{c}_A \eta_A \right) = 0.$$
(6.142)

(iii) When the plate does not exist, i.e., $h = 0$, Eq. (6.140) reduces to

$$\frac{v^2}{v_A^2} = 1 - \left(\overline{k}_A^2 - \frac{K}{\overline{c}_A \xi} \right)^2 .$$
(6.143)

6.6. Gap Waves between a Plate and a Half-space

In this section we analyze wave propagation in a ceramic plate with an air gap from a ceramic half-space (see Fig. 6.17) [44]. The plate and the half-space are unelectroded and traction-free. The ceramic half-space, the air gap, the ceramic plate and the free half-space above the plate are denoted by *A, B, C* and *D*.

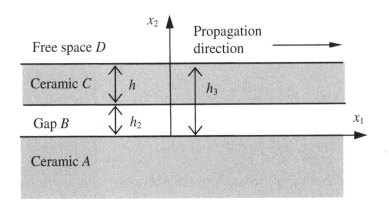

Fig. 6.17. A plate at a distance from a half-space.

6.6.1. Governing equations and fields

The governing equations for fields in different regions are:

$$\bar{c}_A \nabla^2 u_A = \rho_A \ddot{u}_A, \quad \nabla^2 \psi_A = 0, \quad \phi_A = \psi_A + \frac{e_A}{\varepsilon_A} u_A, \quad x_2 < 0, \qquad (6.144)$$

$$\nabla^2 \phi_B = 0, \quad 0 < x_2 < h_2, \qquad (6.145)$$

$$\bar{c}_C \nabla^2 u_C = \rho_C \ddot{u}_C, \quad \nabla^2 \psi_C = 0,$$

$$\phi_C = \psi_C + \frac{e_C}{\varepsilon_C} u_C, \quad h_2 < x_2 < h_3 = h_2 + h, \qquad (6.146)$$

$$\nabla^2 \phi_D = 0, \quad h_3 < x_2. \qquad (6.147)$$

For guided waves we require that

$$u_A, \psi_A \to 0, \quad x_2 \to -\infty,$$
$$\phi_D \to 0, \quad x_2 \to \infty. \qquad (6.148)$$

Due to the inhomogeneity in the structure, we need to obtain solutions in different regions and apply interface conditions. For the ceramic half-space at $x_2 < -h$, consider the possibility of the following waves propagating in the x_1 direction:

$$u_A = U_A \exp(\eta_A x_2) \cos(\xi x_1 - \omega t),$$
$$\psi_A = \Psi_A \exp(\xi x_2) \cos(\xi x_1 - \omega t), \qquad (6.149)$$

where U_A and Ψ_A are undetermined constants, and

$$\eta_A^2 = \xi^2 - \frac{\rho_A \omega^2}{\bar{c}_A} = \xi^2 \left(1 - \frac{v^2}{v_A^2}\right) > 0,$$

$$v = \frac{\omega}{\xi}, \quad v_A^2 = \frac{\bar{c}_A}{\rho_A}. \qquad (6.150)$$

The positive root for η_A should be taken. Equation (6.149) satisfies Eqs. (6.144) and (6.148)$_1$. For the air gap, we have

$$\phi_B = \left(\Phi_{B1} \sinh \xi x_2 + \Phi_{B2} \cosh \xi x_2\right) \cos(\xi x_1 - \omega t), \qquad (6.151)$$

where Φ_{B1} and Φ_{B2} are undetermined constants. Equation (6.151) satisfies Eq. (6.145). For the piezoceramic plate, the fields can be represented by

$$u_C = [U_{C1} \sin \eta_C (x_2 - h_2) + U_{C2} \cos \eta_C (x_2 - h_2)]$$
$$\times \cos(\xi x_1 - \omega t),$$
$$\psi_C = [\Psi_{C1} \sinh \xi (x_2 - h_2) + \Psi_{C2} \cosh \xi (x_2 - h_2)]$$
$$\times \cos(\xi x_1 - \omega t),$$

(6.152)

where U_{C1}, U_{C2}, Ψ_{C1} and Ψ_{C2} are undetermined constants, and

$$\eta_C^2 = \frac{\rho_C \omega^2}{\overline{c}_C} - \xi^2 = \xi^2 \left(\frac{v^2}{v_C^2} - 1 \right),$$
$$v_C^2 = \frac{\overline{c}_C}{\rho_C}.$$

(6.153)

Equation (6.152) satisfies Eq. (6.146). For the free half-space above the plate, we have

$$\phi_D = \Phi_D \exp[-\xi(x_2 - h_3)] \cos(\xi x_1 - \omega t),$$

(6.154)

where Φ_D is an undetermined constant. Equation (6.154) satisfies Eqs. (6.147) and (6.148)$_2$.

6.6.2. Dispersion relations

Substituting the above fields into the following boundary and continuity conditions:

$$T_{23}(0^-) = 0, \quad \phi(0^-) = \phi(0^+), \quad D_2(0^-) = D_2(0^+),$$
$$T_{23}(h_2^+) = 0, \quad \phi(h_2^-) = \phi(h_2^+), \quad D_2(h_2^-) = D_2(h_2^+),$$
$$T_{23}(h_3^-) = 0, \quad \phi(h_3^-) = \phi(h_3^+), \quad D_2(h_3^-) = D_2(h_3^+),$$

(6.155)

we obtain nine equations for the nine undetermined constants. For nontrivial solutions the determinant of the coefficient matrix has to vanish, which gives the frequency equation that determines the dispersion relations of the waves. The frequency equation is very long and is not presented here [44]. It can be reduced to the frequency equation for surfaces waves by setting $h = 0$ and $h_2 = 0$. If we set $h_2 = \infty$ only, we obtain separate surface and plate waves. Dispersion curves for different values of h/h_2 are shown in Figs. 6.18 and 6.19. For larger values of the ratio, the curves are lower.

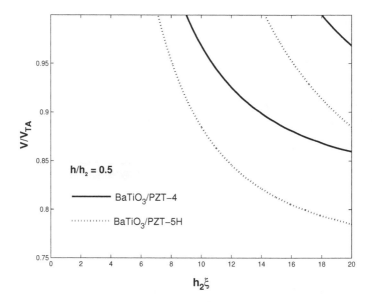

Fig. 6.18. Dispersion relations when $h/h_2 = 0.5$ ($v_{TA} = v_A$).

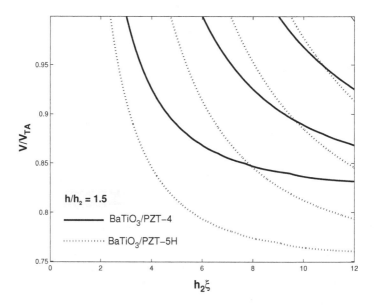

Fig. 6.19. Dispersion relations when $h/h_2 = 1.5$ ($v_{TA} = v_A$).

6.7. A Plate between a Half-space and a Fluid

Next we consider waves in a plate over a half-space and the plate is in contact with a viscous fluid (see Fig. 6.20) [45]. For simplicity we consider the case when the plate is a conductor and the half-space is ceramic.

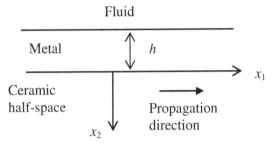

Fig. 6.20. A plate on a substrate and in contact with a fluid.

For the elastic plate,

$$\hat{c}\nabla^2 u = \hat{\rho}\ddot{u}, \quad -h < x_2 < 0, \tag{6.156}$$

where $\hat{\rho}$ and $\hat{c}$ are the mass density and shear elastic constant of the metal plate. From Eqs. (6.7)–(6.10), the fields in the plate are

$$u = (\hat{A}\cos\hat{\xi}_2 x_2 + \hat{B}\sin\hat{\xi}_2 x_2)\exp[i(\xi_1 x_1 - \omega t)], \tag{6.157}$$

$$T_{23} = \hat{c}u_{,2} = \hat{c}(-\hat{A}\hat{\xi}_2\sin\hat{\xi}_2 x_2 + \hat{B}\hat{\xi}_2\cos\hat{\xi}_2 x_2)\exp[i(\xi_1 x_1 - \omega t)], \tag{6.158}$$

where $\hat{A}$ and $\hat{B}$ are undetermined constants,

$$\hat{\xi}_2^2 = \frac{\hat{\rho}\omega^2}{\hat{c}} - \xi_1^2 = \xi_1^2\left(\frac{v^2}{\hat{v}_T^2} - 1\right), \tag{6.159}$$

and

$$\hat{v}_T^2 = \frac{\hat{c}}{\hat{\rho}}. \tag{6.160}$$

For the ceramic half-space, the governing equations are

$$\bar{c}\nabla^2 u = \rho\ddot{u}, \quad \nabla^2\psi = 0, \quad \phi = \psi + \frac{e}{\varepsilon}u, \quad x_2 > 0. \tag{6.161}$$

We look for solutions satisfying

$$u, \quad \psi \to 0, \quad x_2 \to +\infty. \tag{6.162}$$

From Eqs. (6.3)–(6.6), the fields satisfying Eqs. (6.161) and (6.162) can be written as

$$u = A\exp(-\xi_2 x_2)\exp[i(\xi_1 x_1 - \omega t)],$$
$$\psi = B\exp(-\xi_1 x_2)\exp[i(\xi_1 x_1 - \omega t)], \tag{6.163}$$

$$T_{23} = \bar{c}u_{,2} + e\psi_{,2}$$
$$= -[\bar{c}A\xi_2 \exp(-\xi_2 x_2) + eB\xi_1 \exp(-\xi_1 x_2)]\exp[i(\xi_1 x_1 - \omega t)],$$

$$\phi = \psi + \frac{e}{\varepsilon}u \tag{6.164}$$

$$= [\frac{e}{\varepsilon}A\exp(-\xi_2 x_2) + B\exp(-\xi_1 x_2)]\exp[i(\xi_1 x_1 - \omega t)],$$

where A and B are undetermined constants, and

$$\xi_2^2 = \xi_1^2 - \frac{\rho\omega^2}{\bar{c}} = \xi_1^2\left(1 - \frac{v^2}{v_T^2}\right) > 0, \tag{6.165}$$

$$v^2 = \frac{\omega^2}{\xi_1^2}, \quad v_T^2 = \frac{\bar{c}}{\rho}. \tag{6.166}$$

For the fluid region, let $v_3(x_1, x_2, t)$ be the velocity field. The governing equation for v_3 and the relevant stress component are (see Eq. (4.72))

$$\mu(v_{3,11} + v_{3,22}) = \rho'\frac{\partial v_3}{\partial t}, \tag{6.167}$$

$$T_{23} = \mu v_{3,2}.$$

We look for solutions satisfying

$$v_3 \to 0, \quad x_2 \to -\infty. \tag{6.168}$$

From Eq. (4.73), with slight modifications, we have the following fields:

$$v_3 = C\exp[\lambda(x_2 + h)]\exp[i(\xi_1 x_1 - \omega t)],$$
$$T_{23} = \mu\lambda C\exp[\lambda(x_2 + h)]\exp[i(\xi_1 x_1 - \omega t)], \tag{6.169}$$

where C is an arbitrary constant, λ has a positive real part and is determined from

$$\mu(\lambda^2 - \xi_1^2) = \rho' i \omega. \tag{6.170}$$

The continuity and boundary conditions are

$$\phi(0^+) = \frac{e}{\varepsilon} A + B = 0,$$

$$u(0^+) = A = \hat{A} = u(0^-),$$

$$T_{23}(0^+) = -\overline{c} A \xi_2 - e B \xi_1 = \hat{c} \hat{B} \hat{\xi}_2 = T_{23}(0^-), \tag{6.171}$$

$$T_{23}(-h^+) = \hat{c}(\hat{A} \hat{\xi}_2 \sin \hat{\xi}_2 h + \hat{B} \hat{\xi}_2 \cos \hat{\xi}_2 h) = \mu \lambda C = T_{23}(-h^-),$$

$$\dot{u}(-h^+) = -i\omega(\hat{A} \cos \hat{\xi}_2 h - \hat{B} \sin \hat{\xi}_2 h) = C = v_3(-h^-).$$

The dispersion relation is determined by

$$\begin{vmatrix} e/\varepsilon & 1 & 0 & 0 & 0 \\ 1 & 0 & -1 & 0 & 0 \\ -\overline{c}\xi_2 & -e\xi_1 & 0 & -\hat{c}\hat{\xi}_2 & 0 \\ 0 & 0 & \hat{c}\hat{\xi}_2 \sin \hat{\xi}_2 h & \hat{c}\hat{\xi}_2 \cos \hat{\xi}_2 h & -\mu\lambda \\ 0 & 0 & -i\omega\cos \hat{\xi}_2 h & i\omega\sin \hat{\xi}_2 h & -1 \end{vmatrix} = 0. \tag{6.172}$$

Chapter 7

Free Vibrations in Cartesian Coordinates

This chapter is on free vibrations in rectangular coordinates. Most problems are over finite domains. As a result, the frequency spectra of the eigenvalue problems are discrete. This is in contrast to the propagating waves in unbounded domains in previous chapters where the eigenvalues were continuously distributed. The first few problems in this chapter are special cases of the corresponding problems in previous chapters. Due to their importance in applications, some detailed discussions are given for these special cases in this chapter.

7.1. Thickness-shear in a Plate

Solutions to thickness vibrations of piezoelectric plates can be obtained in a general manner. We discuss a few special cases. Consider an unbounded plate (see Fig. 7.1). The plate surfaces at $x_2 = \pm h$ are traction-free and are electroded. A time-harmonic voltage may be applied across the plate thickness. For free vibrations the voltage will be set to zero.

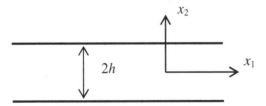

Fig. 7.1. An electroded plate.

From Eq. (1.43) the governing equations and boundary conditions are:

$$c\nabla^2 u + e\nabla^2 \phi = \rho \ddot{u},$$
$$e\nabla^2 u - \varepsilon\nabla^2 \phi = 0,$$
$$T_{23} = 0, \quad x_2 = \pm h, \tag{7.1}$$
$$\phi(x_2 = h) - \phi(x_2 = -h) = V\exp(i\omega t).$$

179

By thickness vibration we mean fields in the following form:

$$u = u(x_2), \quad \phi = \phi(x_2),$$ (7.2)

where the $\exp(i\omega t)$ factor has been dropped. The nontrivial components of the strain and electric fields are

$$S_4 = 2S_{23} = u_{,2}, \quad E_2 = -\phi_{,2}.$$ (7.3)

The nontrivial stress and electric displacement components are

$$T_4 = cu_{,2} + e\phi_{,2},$$
$$D_2 = eu_{,2} - \varepsilon\phi_{,2}.$$ (7.4)

The equations to be satisfied are

$$cu_{,22} + e\phi_{,22} = -\rho\omega^2 u,$$
$$eu_{,22} - \varepsilon\phi_{,22} = 0.$$ (7.5)

Equation (7.5)$_2$ can be integrated to yield

$$\phi = \frac{e}{\varepsilon}u + B_1 x_2 + B_2,$$ (7.6)

where B_1 and B_2 are integration constants, and B_2 is immaterial. Substitute Eq. (7.6) into the expressions for T_{23}, D_2, and Eq. (7.5)$_1$:

$$T_{23} = \bar{c}u_{,2} + eB_1, \quad D_2 = -\varepsilon B_1,$$ (7.7)

$$\bar{c}u_{,22} = -\rho\omega^2 u,$$ (7.8)

where

$$\bar{c} = c(1+k^2), \quad k^2 = \frac{e^2}{\varepsilon c}.$$ (7.9)

The general solution to Eq. (7.8) and the corresponding expression for the electric potential are

$$u = A_1 \sin\xi x_2 + A_2 \cos\xi x_2,$$
$$\phi = \frac{e}{\varepsilon}(A_1 \sin\xi x_2 + A_2 \cos\xi x_2) + B_1 x_2 + B_2,$$ (7.10)

where A_1 and A_2 are integration constants, and

$$\xi^2 = \frac{\rho}{\bar{c}}\omega^2.$$ (7.11)

The expression for stress is then

$$T_{23} = \bar{c}(A_1\xi\cos\xi x_2 - A_2\xi\sin\xi x_2) + eB_1.$$ (7.12)

The boundary conditions require that

$$\bar{c}A_1\xi\cos\xi h - \bar{c}A_2\xi\sin\xi h + eB_1 = 0,$$
$$\bar{c}A_1\xi\cos\xi h + \bar{c}A_2\xi\sin\xi h + eB_1 = 0, \tag{7.13}$$
$$2\frac{e}{\varepsilon}A_1\sin\xi h + 2B_1 h = V,$$

or, add the first two, and subtract the first two from each other:

$$\bar{c}A_1\xi\cos\xi h + eB_1 = 0,$$
$$\bar{c}A_2\xi\sin\xi h = 0, \tag{7.14}$$
$$2\frac{e}{\varepsilon}A_1\sin\xi h + 2B_1 h = V.$$

For free vibrations we have $V = 0$. Equation (7.14) decouples into two sets of equations. One set may be called symmetric modes for which

$$\bar{c}A_2\xi\sin\xi h = 0. \tag{7.15}$$

Nontrivial solutions may exist if

$$\sin\xi h = 0, \tag{7.16}$$

or

$$\xi^{(n)}h = \frac{n\pi}{2}, \quad n = 0, 2, 4, 6, \cdots, \tag{7.17}$$

which determines the resonant frequencies

$$\omega^{(n)} = \frac{n\pi}{2h}\sqrt{\frac{\bar{c}}{\rho}}, \quad n = 0, 2, 4, 6, \cdots. \tag{7.18}$$

Equation (7.16) implies that $B_1 = 0$ and $A_1 = 0$. The corresponding modes are

$$u^{(n)} = \cos\xi^{(n)}x_2, \quad \phi^{(n)} = \frac{e}{\varepsilon}\cos\xi^{(n)}x_2, \tag{7.19}$$

where $n = 0$ is a rigid-body mode.

For antisymmetric modes

$$\bar{c}A_1\xi\cos\xi h + eB_1 = 0,$$
$$2\frac{e}{\varepsilon}A_1\sin\xi h + 2B_1 h = 0. \tag{7.20}$$

The resonant frequencies are determined by

$$\begin{vmatrix} \bar{c}\xi\cos\xi h & e \\ \dfrac{e}{\varepsilon}\sin\xi h & h \end{vmatrix} = \bar{c}\xi h\cos\xi h - \dfrac{e^2}{\varepsilon}\sin\xi h = 0 , \qquad (7.21)$$

or

$$\tan\xi h = \dfrac{\xi h}{\bar{k}^2} , \qquad (7.22)$$

where

$$\bar{k}^2 = \dfrac{e^2}{\varepsilon\bar{c}} = \dfrac{e^2}{\varepsilon c(1+k^2)} = \dfrac{k^2}{1+k^2} . \qquad (7.23)$$

Equations (7.22) and (7.20) determine the resonant frequencies and modes. For antisymmetric modes, $A_2 = 0$. If the piezoelectric coupling is small and can be neglected in Eq. (7.22), a set of frequencies similar to Eq. (7.18) with n assuming odd numbers can be determined for a set of modes with sine dependence on the thickness coordinate. Static thickness-shear deformation and the first few thickness-shear modes in a plate are shown in Fig. 7.2.

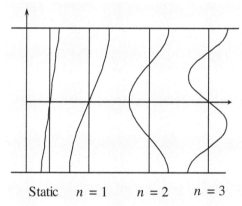

Static $n = 1$ $n = 2$ $n = 3$

Fig. 7.2. Thickness-shear deformation and modes in a plate.

When $\bar{k}^2$ is very small, Eq. (7.22) can be solved approximately. For example, for the most widely used fundamental mode of $n = 1$, we can write

$$\xi h = \dfrac{\pi}{2} - \Delta , \qquad (7.24)$$

where Δ is a small number. Substituting Eq. (7.24) into Eq. (7.22), for small $\bar{k}^2$ and small Δ, we obtain

$$\Delta \cong \frac{2}{\pi}\bar{k}^2, \qquad (7.25)$$

or

$$\omega^{(1)} \cong \frac{\pi}{2h}\sqrt{\frac{\bar{c}}{\rho}}(1-\frac{4}{\pi^2}\bar{k}^2) = \frac{\pi}{2h}\sqrt{\frac{c(1+k^2)}{\rho}}(1-\frac{4}{\pi^2}\bar{k}^2)$$

$$\cong \frac{\pi}{2h}\sqrt{\frac{c}{\rho}}(1+\frac{1}{2}k^2)(1-\frac{4}{\pi^2}\bar{k}^2) \qquad (7.26)$$

$$\cong \frac{\pi}{2h}\sqrt{\frac{c}{\rho}}(1+\frac{1}{2}k^2-\frac{4}{\pi^2}k^2) > \frac{\pi}{2h}\sqrt{\frac{c}{\rho}}.$$

Equation (7.26) shows that piezoelectric coupling raises the frequency. This effect is called piezoelectric stiffening. The approximation procedure in Eqs. (7.24) through (7.26) can be extended to higher-order modes.

7.2. Thickness-shear in a Plate with Unattached Electrodes

In this section we analyze thickness-shear vibration of a plate with unattached electrodes and asymmetric air gaps (see Fig. 7.3). It is under an applied voltage 2 *V*. For free vibrations the voltage will be set to zero.

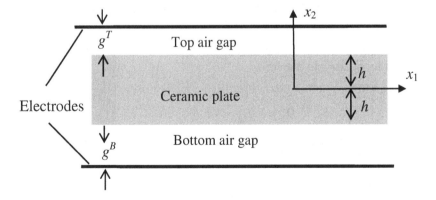

Fig. 7.3. A plate with unattached electrodes and asymmetric air gaps.

The governing equations and boundary conditions are:

$$c\nabla^2 u + e\nabla^2 \phi = \rho\ddot{u}, \quad |x_2| < h,$$
$$e\nabla^2 u - \varepsilon\nabla^2 \phi = 0, \quad |x_2| < h,$$
$$\nabla^2 \phi = 0, \quad h < x_2 < h + g^T,$$
$$\nabla^2 \phi = 0, \quad -h - g^B < x_2 < -h,$$

(7.27)

$$\phi = V \exp(i\omega t), \quad x_2 = h + g^T,$$
$$\phi = -V \exp(i\omega t), \quad x_2 = -h - g^B,$$
$$T_{23}(x_2 = h) = T_{23}(x_2 = -h) = 0,$$
$$\phi(x_2 = h^+) = \phi(x_2 = h^-),$$
$$\phi(x_2 = -h^+) = \phi(x_2 = -h^-),$$
$$D_2(x_2 = h^+) = D_2(x_2 = h^-),$$
$$D_2(x_2 = -h^+) = D_2(x_2 = -h^-).$$

(7.28)

From the previous section, the fields inside the plate are still given by

$$u = A_1 \sin\xi x_2 + A_2 \cos\xi x_2,$$
$$\phi = \frac{e}{\varepsilon}(A_1 \sin\xi x_2 + A_2 \cos\xi x_2) + B_1 x_2 + B_2,$$

(7.29)

$$T_{23} = \bar{c}(A_1 \xi \cos\xi x_2 - A_2 \xi \sin\xi x_2) + eB_1,$$

(7.30)

$$D_2 = -\varepsilon B_1,$$

(7.31)

where B_1, B_2, A_1 and A_2 are integration constants, and

$$\xi^2 = \frac{\rho}{\bar{c}}\omega^2.$$

(7.32)

In the top air gap, we have

$$\phi = F_1 x_2 + G_1,$$

(7.33)

$$D_2 = \varepsilon_0 E_2 = -\varepsilon_0 F_1,$$

(7.34)

where F_1 and G_1 are undetermined constants. Similarly, for the bottom air gap,

$$\phi = F_2 x_2 + G_2,$$

(7.35)

$$D_2 = -\varepsilon_0 F_2,$$

(7.36)

where F_2 and G_2 are undetermined constants.

Substituting the relevant fields into Eq. (7.28), we obtain

$$F_1(h + g^T) + G_1 = V,$$

$$F_2(-h - g^B) + G_2 = -V,$$

$$\bar{c}\xi(A_1 \cos \xi h - A_2 \sin \xi h) + eB_1 = 0,$$

$$\bar{c}\xi(A_1 \cos \xi h + A_2 \sin \xi h) + eB_1 = 0,$$

$$F_1 h + G_1 = \frac{e}{\varepsilon}(A_1 \sin \xi h + A_2 \cos \xi h) + B_1 h + B_2, \qquad (7.37)$$

$$\frac{e}{\varepsilon}(-A_1 \sin \xi h + A_2 \cos \xi h) - B_1 h + B_2 = -F_2 h + G_2,$$

$$-\varepsilon_0 F_1 = -\varepsilon B_1,$$

$$-\varepsilon B_1 = -\varepsilon_0 F_2.$$

For free vibrations $V = 0$. We require the determinant of the coefficient matrix of Eq. (7.37) to vanish. This yields the following equation that determines the resonant frequencies:

$$\bar{c}\xi \sin \xi h \left\{ \frac{2 \sin \xi h e^2}{\varepsilon} - \bar{c}\xi \cos \xi h \left[2h + \frac{\varepsilon(g^T + g^B)}{\varepsilon_0} \right] \right\} = 0. \qquad (7.38)$$

Equation (7.38) has two factors. $\sin \xi h = 0$ gives the mechanically symmetric modes. We are mainly interested in the second factor which can be written as

$$\cot \xi h = \frac{\bar{k}^2}{\xi[h + \dfrac{\varepsilon(g^T + g^B)}{2\varepsilon_0}]}. \qquad (7.39)$$

Equation (7.39) determines the frequencies of the mechanically antisymmetric modes that can be electrically excited by a thickness electric field. When there are no air gaps, Eq. (7.39) reduces to Eq. (7.22). From Eq. (7.39) it can be seen that it is the total thickness of the air gaps that matters when frequency is concerned. It can also be concluded from Eq. (7.39) that resonant frequencies increase when the air gaps become thicker. This is because when the electrodes are farther away the electrical shortening effect from the electrodes becomes less. Then the electric field is stronger and so is the related piezoelectric stiffening effect.

In resonator applications, the first root of Eq. (7.39) for the fundamental TSh frequency is most useful. For small $\bar{k}^2$ we write

$$\xi h = \frac{\pi}{2} - \Delta, \tag{7.40}$$

where Δ is an unknown small number. Substituting Eq. (7.40) into Eq. (7.39), for small $\bar{k}^2$ and small Δ, we obtain

$$\Delta \cong \frac{2\bar{k}^2}{\pi[1 + \dfrac{\varepsilon(g^T + g^B)}{2\varepsilon_0 h}]}. \tag{7.41}$$

Then, for the fundamental TSh frequency:

$$\omega^{(1)} \cong \frac{\pi}{2h}\sqrt{\frac{c}{\rho}}\left\{1 + \frac{1}{2}k^2 - \frac{4k^2}{\pi^2\left[1 + \dfrac{\varepsilon(g^T + g^B)}{2\varepsilon_0 h}\right]}\right\} > \frac{\pi}{2h}\sqrt{\frac{c}{\rho}} = \omega_0, \tag{7.42}$$

or

$$\frac{\omega^{(1)} - \omega_0}{\omega_0} \cong 1 + \frac{1}{2}k^2 - \frac{4k^2}{\pi^2\left[1 + \dfrac{\varepsilon(g^T + g^B)}{2\varepsilon_0 h}\right]}, \tag{7.43}$$

where ω_0 is the fundamental TSh frequency when piezoelectric coupling is neglected and is introduced as a reference. Equation (7.43) shows clearly that $(\omega^{(1)} - \omega_0)/\omega_0$ increases when the gaps become thicker. Due to the dependence of frequencies on air gaps, it is possible to use air gaps of varying thickness to create energy trapping in resonators. When the electrodes are attached to a crystal plate, energy trapping is due to the inertial effect of the electrodes and the effect of electrodes on the electric field in the plate and the related piezoelectric stiffening effect. When the electrodes are not attached to the crystal surface, the inertial effect disappears but the effect of electrodes on the electric field still exists.

7.3. Thickness-shear in a Plate with Thin Films

Consider a plate with electrodes or thin films of unequal thickness on its two surfaces as shown in Fig. 7.4. The two electrodes are of the same isotropic material. The outer surfaces of the electrodes are traction-free. The electrodes are shorted.

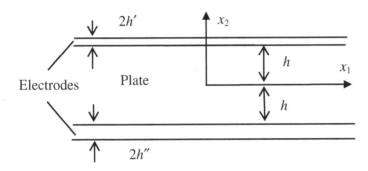

Fig. 7.4. A plate with electrodes of unequal thickness.

The governing equations and boundary conditions are:

$$c\nabla^2 u + e\nabla^2 \phi = \rho \ddot{u}, \quad |x_2| < h,$$
$$e\nabla^2 u - \varepsilon\nabla^2 \phi = 0, \quad |x_2| < h,$$
$$c'\nabla^2 u = \rho' \ddot{u}, \quad h < x_2 < h + 2h',$$
$$c'\nabla^2 u = \rho' \ddot{u}, \quad -h - 2h'' < x_2 < -h,$$
(7.44)
$$T_{23} = 0, \quad x_2 = h + 2h', \quad x_2 = -h - 2h'',$$
$$u(x_2 = h^-) = u(x_2 = h^+),$$
$$T_{23}(x_2 = h^-) = T_{23}(x_2 = h^+),$$
$$u(x_2 = -h^+) = u(x_2 = -h^-),$$
(7.45)
$$T_{23}(x_2 = -h^+) = T_{23}(x_2 = -h^-),$$
$$\phi(x_2 = h) = \phi(x_2 = -h),$$

where ρ' and c' are the mass density and the shear elastic constant of the electrodes. From Sec. 7.1, the trial fields inside the plate are still given by Eqs. (7.10)–(7.12)

$$u = A_1 \sin \xi x_2 + A_2 \cos \xi x_2,$$

$$\phi = \frac{e}{\varepsilon}(A_1 \sin \xi x_2 + A_2 \cos \xi x_2) + B_1 x_2 + B_2, \tag{7.46}$$

$$T_{23} = \overline{c}(A_1 \xi \cos \xi x_2 - A_2 \xi \sin \xi x_2) + eB_1, \tag{7.47}$$

$$D_2 = -\varepsilon B_1, \tag{7.48}$$

where A_1 and A_2 are integration constants, and

$$\xi^2 = \frac{\rho}{\overline{c}} \omega^2. \tag{7.49}$$

For the fields inside the electrodes, consider the upper electrode first:

$$u = A_1' \sin \xi'(x_2 - h) + A_2' \cos \xi'(x_2 - h), \tag{7.50}$$

$$T_{23} = c'[A_1' \xi' \cos \xi'(x_2 - h) - A_2' \xi' \sin \xi'(x_2 - h)], \tag{7.51}$$

where A'_1 and A'_2 are integration constants, and

$$(\xi')^2 = \frac{\rho'}{c'} \omega^2. \tag{7.52}$$

Similarly, for the lower electrode we have

$$u = A_1'' \sin \xi'(x_2 + h) + A_2'' \cos \xi'(x_2 + h), \tag{7.53}$$

$$T_{23} = c'[A_1'' \xi' \cos \xi'(x_2 + h) - A_2'' \xi' \sin \xi'(x_2 + h)], \tag{7.54}$$

where A_1'' and A_2'' are integration constants.

Substituting the relevant fields into Eq. (7.45), we obtain

$$\begin{aligned}
& A_1 \sin \xi h + A_2 \cos \xi h = A_2', \\
& -A_1 \sin \xi h + A_2 \cos \xi h = A_2'', \\
& \overline{c}\xi(A_1 \cos \xi h - A_2 \sin \xi h) + eB_1 = c'\xi' A_1', \\
& \overline{c}\xi(A_1 \cos \xi h + A_2 \sin \xi)h + eB_1 = c'\xi' A_1'', \\
& A_1' \cos \xi' 2h' - A_2' \sin \xi' 2h' = 0, \\
& A_1'' \cos \xi' 2h'' + A_2'' \sin \xi' 2h'' = 0, \\
& \frac{e}{\varepsilon} A_1 \sin \xi h + B_1 h = 0.
\end{aligned} \tag{7.55}$$

For nontrivial solutions of the undetermined constants, the determinant of the coefficient matrix of Eq. (7.55) has to vanish. This results in the following frequency equation:

$$\left(1-\bar{k}^{2}\frac{\tan\xi h}{\xi h}\right)\left[2\tan\xi h+\sqrt{\frac{\rho'c'}{\rho\bar{c}}}\left(\tan\xi'2h'+\tan\xi'2h''\right)\right]$$

$$=\sqrt{\frac{\rho'c'}{\rho\bar{c}}}\tan\xi h\left[\tan\xi h\left(\tan\xi'2h'+\tan\xi'2h''\right)\right. \tag{7.56}$$

$$\left.+2\sqrt{\frac{\rho'c'}{\rho\bar{c}}}\tan\xi'2h'\tan\xi'2h''\right].$$

We make the following observations from Eq. (7.56):

(i) In the limit of $h' \to 0$ and $h'' \to 0$, i.e., the mechanical effects of the electrodes are neglected, Eq. (7.56) reduces to

$$\left(1-\bar{k}^{2}\frac{\tan\xi h}{\xi h}\right)\tan\xi h=0, \tag{7.57}$$

which is the frequency equation for both symmetric and antisymmetric modes given in Eqs. (7.16) and (7.22).

(ii) When $h' = h''$, i.e., the electrodes are of the same thickness, Eq. (7.56) reduces to

$$\left(1-\bar{k}^{2}\frac{\tan\xi h}{\xi h}-\sqrt{\frac{\rho'c'}{\rho\bar{c}}}\tan\xi h\tan\xi'2h'\right)$$
$$\times\left(\tan\xi h+\sqrt{\frac{\rho'c'}{\rho\bar{c}}}\tan\xi'2h'\right)=0. \tag{7.58}$$

The first factor of Eq. (7.58) is the frequency equation for the antisymmetric modes. The second factor is for the symmetric modes. For small h', i.e., very thin electrodes, we approximately have

$$\tan\xi'2h' \cong \xi'2h', \quad \sqrt{\frac{\rho'c'}{\rho\bar{c}}}\tan\xi'2h' \cong R\xi h, \quad R=\frac{\rho'2h'}{\rho h}. \tag{7.59}$$

In this case the first and the second factors of Eq. (7.58) reduce to

$$\tan\xi h=\frac{\xi h}{\bar{k}^{2}+R(\xi h)^{2}}, \quad \tan\xi h+R\xi h=0. \tag{7.60}$$

Note that in Eq. (7.60) the shear stiffness of the electrodes (c') has disappeared. Only the mass effect of the electrodes is left and is represented by the mass ratio R.

(iii) For small h' and h'', i.e., thin and unequal electrodes, Eq. (7.56) reduces to

$$\left(1 - \bar{k}^2 \frac{\tan \xi h}{\xi h}\right)\left[2\tan \xi h + (R' + R'')\xi h\right]$$
$$= \xi h \tan \xi h\left[(R' + R'')\tan \xi h + 2R'R''\xi h\right],$$

(7.61)

where we have denoted

$$R' = \frac{\rho' 2h'}{\rho h}, \quad R'' = \frac{\rho' 2h''}{\rho h}.$$

(7.62)

To the lowest-order of the mass effect, the $R'R''$ term on the right-hand side of Eq. (7.62) can be dropped.

7.4. Thickness-shear in a Plate with Imperfectly Bonded Films

In this section we analyze thickness-shear vibrations of a plate with imperfectly bonded thin films or electrodes (see Fig. 7.5). The stiffness of the films is neglected. The interfaces between the plate and the thin films are described by the shear-lag model.

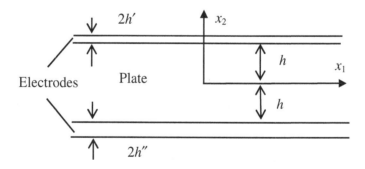

Fig. 7.5. A plate with imperfectly bonded electrodes of unequal thickness.

The governing equations and boundary conditions are

$$c\nabla^2 u + e\nabla^2 \phi = \rho \ddot{u}, \quad |x_2| < h,$$
$$e\nabla^2 u - \varepsilon\nabla^2 \phi = 0, \quad |x_2| < h, \tag{7.63}$$

$$-T_{23} = 2\rho' h' \ddot{u}', \quad x_2 = h,$$
$$T_{23} = 2\rho'' h'' \ddot{u}'', \quad x_2 = -h, \tag{7.64}$$
$$\phi(x_2 = h) = \phi(x_2 = -h).$$

ρ' and ρ'', and u' and u'' are the densities and displacements of the top and bottom mass layers, respectively. According to the shear-lag model, u' and u'' are allowed to be different from the plate surface displacements and the following constitutive relations describe the behavior of the interfaces:

$$T_{23} = k'(u' - u), \quad x_2 = h,$$
$$T_{23} = k''(u - u''), \quad x_2 = -h, \tag{7.65}$$

where k' and k'' are the elastic constants of the interfaces. With k' and k'', the interfaces are allowed to possess strain energies. Substituting Eq. (7.65) into Eq. (7.64) gives

$$-k'(u' - u) = 2\rho' h' \ddot{u}', \quad x_2 = h,$$
$$k''(u - u'') = 2\rho'' h'' \ddot{u}'', \quad x_2 = -h, \tag{7.66}$$
$$\phi(x_2 = h) = \phi(x_2 = -h).$$

The trial fields inside the plate are still the same as those in Sec. 7.1 and are given by Eqs. (7.10)–(7.12)

$$u = A_1 \sin \xi x_2 + A_2 \cos \xi x_2,$$
$$\varphi = \frac{e}{\varepsilon}(A_1 \sin \xi x_2 + A_2 \cos \xi x_2) + B_1 x_2 + B_2, \tag{7.67}$$

$$T_{23} = \overline{c}(A_1 \xi \cos \xi x_2 - A_2 \xi \sin \xi x_2) + eB_1, \tag{7.68}$$

$$D_2 = -\varepsilon B_1, \tag{7.69}$$

where A_1 and A_2 are integration constants, and

$$\xi^2 = \frac{\rho}{\overline{c}} \omega^2. \tag{7.70}$$

The displacements of the thin films are written as

$$u' = A' \exp(i\omega t), \quad u'' = A'' \exp(i\omega t), \tag{7.71}$$

where A' and A'' are undetermined constants. Substituting the relevant fields into Eqs. (7.65) and (7.66), we obtain

$$\bar{c}\xi(A_1\cos\xi h - A_2\sin\xi h) + eB_1$$
$$= k'[A' - (A_1\sin\xi h + A_2\cos\xi h)],$$
$$\bar{c}\xi(A_1\cos\xi h + A_2\sin\xi h) + eB_1$$
$$= k''[-A_1\sin\xi h + A_2\cos\xi h - A''], \qquad (7.72)$$
$$-k'[A' - (A_1\sin\xi h + A_2\cos\xi h)] = -\omega^2 2\rho'h'A',$$
$$k''[-A_1\sin\xi h + A_2\cos\xi h - A''] = -\omega^2 2\rho''h''A'',$$
$$\frac{e}{\varepsilon}A_1\sin\xi h + B_1 h = 0.$$

For nontrivial solutions of the undetermined constants, the determinant of the coefficient matrix of Eq. (7.72) has to vanish. This results in the following frequency equation:

$$\begin{vmatrix} a_{11} & a_{12} & e & -k' & 0 \\ a_{21} & a_{22} & e & 0 & k'' \\ k'\sin\xi h & k'\cos\xi h & 0 & \omega^2 2\rho'h'-k' & 0 \\ -k''\sin\xi h & k''\cos\xi h & 0 & 0 & \omega^2 2\rho''h''-k'' \\ \dfrac{e}{\varepsilon}\sin\xi h & 0 & h & 0 & 0 \end{vmatrix} = 0, \qquad (7.73)$$

where

$$a_{11} = \bar{c}\,\xi\cos\xi h + k'\sin\xi h, \quad a_{12} = -\bar{c}\,\xi\sin\xi h + k'\cos\xi h,$$
$$a_{21} = \bar{c}\,\xi\cos\xi h + k''\sin\xi h, \quad a_{22} = \bar{c}\,\xi\sin\xi h - k''\cos\xi h. \qquad (7.74)$$

In the special case of symmetric mass layers, $h'' = h'$, $\rho'' = \rho'$, and $k'' = k'$. The modes can be classified as symmetric and antisymmetric. Equation (7.73) has two factors. For symmetric modes $A_1 = B_1 = 0$ and $A'' = A'$. The frequency equation is

$$\bar{c}\,\xi h\tan\xi h = k'h + \frac{k'^2 h}{2\rho'h'\omega^2 - k'}. \qquad (7.75)$$

For antisymmetric modes $A_2 = 0$ and $A'' = -A'$. The frequency equation is

$$\bar{c}\,\xi h(\cot\xi h - \frac{\bar{k}_{26}^2}{\xi h}) = -k'h - \frac{k'^2 h}{2\rho'h'\omega^2 - k'}. \qquad (7.76)$$

Denoting

$$X = \xi h, \quad \alpha = \frac{\overline{c}}{k'h}, \qquad (7.77)$$

which may be interpreted as a normalized frequency and a relative compliance of the interface with respect to the plate, we can write Eqs. (7.75) and (7.76) as

$$\tan X = \frac{1}{\alpha X}\left(1 + \frac{1}{\alpha R X^2 - 1}\right), \qquad (7.78)$$

$$\cot X = \frac{\overline{k}^2}{X} - \frac{1}{\alpha X}\left(1 + \frac{1}{\alpha R X^2 - 1}\right), \qquad (7.79)$$

where

$$R = \frac{2\rho' h'}{\rho h} \qquad (7.80)$$

is the mass ratio between the layer and the plate. Next we examine two limit cases of Eqs. (7.78) and (7.79).

(i) Large k' Limit

For large k', $\alpha \to 0$. To the lowest order, Eqs. (7.78) and (7.79) are asymptotic to

$$\tan X = -RX, \qquad (7.81)$$

and

$$\cot X - \frac{\overline{k}^2}{X} = RX, \qquad (7.82)$$

which are the results for perfectly bonded mass layers (see Eq. (7.60)).

(ii) Small k' Limit

What is more interesting and less known is the case of relatively loosely bonded mass layers. In this case $k' \to 0$ or $\alpha \to \infty$. To the lowest order, Eqs. (7.78) and (7.79) are asymptotic to

$$\tan X = \frac{1}{\alpha X}, \qquad (7.83)$$

and

$$\cot X = \frac{\overline{k}^2 - \dfrac{1}{\alpha}}{X}, \qquad (7.84)$$

respectively. If α is set to infinity, Eqs. (7.83) and (7.84) reduce to the frequency equations for a plate without mass layers, as expected (see Eqs. (7.16) and (7.22)).

7.5. Thickness-shear in a Layered Plate with an Imperfect Interface

In this section we study thickness vibration of two elastic layers with an elastic or imperfectly bonded interface mounted on a plate piezoelectric resonator [46]. The effect of the interface elasticity on resonant frequencies is examined. The result obtained suggests an acoustic wave sensor for measuring the elastic property of an interface between two materials. Consider the structure in Fig. 7.6. The three layers are unbounded in the x_1 direction. Between the piezoelectric layer and the lower elastic layer perfect bonding is assumed with continuous displacement and traction. Between the two elastic layers an elastic interface with its own mechanical property is assumed which is what we want to measure. There are two electrodes at the surfaces of the piezoelectric layer as shown by the thick lines in the figure. A time-harmonic voltage $2\,V$ may be applied across the electrodes.

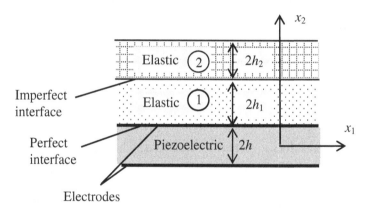

Fig. 7.6. Two elastic layers with an imperfect interface on a ceramic plate resonator.

The governing equations are

$$\bar{c}u_{,22} = \rho\ddot{u}, \quad \psi_{,22} = 0, \quad \phi = \psi + \frac{e}{\varepsilon}u, \tag{7.85}$$

$$c_1\nabla^2 u = \rho_1\ddot{u}, \tag{7.86}$$

where ρ_1 and c_1 are the mass density and the shear modulus of the lower elastic layer. For the upper elastic layer, the equations are similar to Eq. (7.86), with c_1 and ρ_1 replaced by c_2 and ρ_2. At $x_2 = h$ where perfect bonding is assumed, the conventional continuity conditions apply:

$$u(h)^+ = u(h)^-,$$
$$T_{23}(h)^+ = T_{23}(h)^-. \tag{7.87}$$

For the elastic interface at $x_2 = h + 2h_1$, we impose

$$T_{23}(h + 2h_1)^+ = T_{23}(h + 2h_1)^-$$
$$= K[u(h + 2h_1)^+ - u(h + 2h_1)^-]. \tag{7.88}$$

Equation (7.88) is according to the shear-lag interface model. At the bottom and top of the whole structure we have traction-free boundary conditions:

$$T_{23}(-h) = 0,$$
$$T_{23}(h + 2h_1 + 2h_2) = 0. \tag{7.89}$$

The applied voltage requires that

$$\phi(\pm h) = \pm V(t). \tag{7.90}$$

We look for solutions of each layer separately and then apply continuity and boundary conditions. All fields are with an $\exp(i\omega t)$ factor which will be dropped for convenience. The general solution for the piezoelectric layer can be directly taken from Sec. 7.1 as (see Eqs. (7.10)–(7.12))

$$u = A_1 \sin \xi x_2 + A_2 \cos \xi x_2,$$
$$\psi = B_1 x_2 + B_2, \tag{7.91}$$
$$\phi = \frac{e}{\varepsilon}(A_1 \sin \xi x_2 + A_2 \cos \xi x_2) + B_1 x_2 + B_2,$$

$$T_{23} = \bar{c}\xi(A_1 \cos \xi x_2 - A_2 \sin \xi x_2) + eB_1,$$
$$D_2 = -\varepsilon B_1. \tag{7.92}$$

where A_1, A_2, B_1 and B_2 are undetermined constants, and

$$\xi = \sqrt{\frac{\rho}{\bar{c}}}\,\omega. \tag{7.93}$$

For the lower elastic layer with $h < x_2 < h + 2h_1$, the solution can be written as

$$u = F_1 \sin \xi_1 x_2 + F_2 \cos \xi_1 x_2,$$

$$T_{23} = c_1 \xi_1 (F_1 \cos \xi_1 x_2 - F_2 \sin \xi_1 x_2), \qquad (7.94)$$

$$\xi_1 = \sqrt{\frac{\rho_1}{c_1}} \, \omega,$$

where F_1 and F_2 are undetermined constants. Similarly, for the upper elastic layer with $h + 2h_1 < x_2 < h + 2h_1 + 2h_2$, we have

$$u = G_1 \sin \xi_2 x_2 + G_2 \cos \xi_2 x_2,$$

$$T_{23} = c_2 \xi_2 (G_1 \cos \xi_2 x_2 - G_2 \sin \xi_2 x_2), \qquad (7.95)$$

$$\xi_2 = \sqrt{\frac{\rho_2}{c_2}} \, \omega,$$

where G_1 and G_2 are undetermined constants.

Substitution of the relevant fields into the boundary and continuity conditions in Eqs. (7.87)–(7.90) gives the following:

$$A_1 \sin \xi h + A_2 \cos \xi h = F_1 \sin \xi_1 h + F_2 \cos \xi_1 h,$$

$$\bar{c} \xi (A_1 \cos \xi h - A_2 \sin \xi h) + e B_1 = c_1 \xi_1 (F_1 \cos \xi_1 h - F_2 \sin \xi_1 h), \qquad (7.96)$$

$$c_2 \xi_2 [G_1 \cos \xi_2 (h + 2h_1) - G_2 \sin \xi_2 (h + 2h_1)]$$
$$= c_1 \xi_1 [F_1 \cos \xi_1 (h + 2h_1) - F_2 \sin \xi_1 (h + 2h_1)]$$

$$c_1 \xi_1 [F_1 \cos \xi_1 (h + 2h_1) - F_2 \sin \xi_1 (h + 2h_1)] \qquad (7.97)$$
$$= K[G_1 \sin \xi_2 (h + 2h_1) + G_2 \cos \xi_2 (h + 2h_1)$$
$$- F_1 \sin \xi_1 (h + 2h_1) - F_2 \cos \xi_1 (h + 2h_1)],$$

$$\bar{c} \xi (A_1 \cos \xi h + A_2 \sin \xi h) + e B_1 = 0,$$

$$c_2 \xi_2 [G_1 \cos \xi_2 (h + 2h_1 + 2h_2) - G_2 \sin \xi_2 (h + 2h_1 + 2h_2)] = 0, \qquad (7.98)$$

$$\frac{e}{\varepsilon} (A_1 \sin \xi h + A_2 \cos \xi h) + B_1 h + B_2 = V,$$

$$\frac{e}{\varepsilon} (-A_1 \sin \xi h + A_2 \cos \xi h) - B_1 h + B_2 = -V. \qquad (7.99)$$

Equations (7.96) through (7.99) are eight linear equations for A_1, A_2, B_1, B_2, F_1, F_2, G_1 and G_2, driven by V. We consider free vibrations with $V = 0$. In this case Eqs. (7.96)–(7.99) become homogeneous. For nontrivial solutions the determinant of the coefficient matrix has to vanish, which gives the frequency equation that determines the resonant frequency ω. The structure of the coefficient matrix of the linear equations shows that the frequency equation is linear in K. We are mainly interested in the case of large K or near perfect bonding. For convenience we introduce the interface effective compliance by

$$J = 1/K. \tag{7.100}$$

Then the frequency equation can be written as:

$$f(\omega) + Jg(\omega) = 0, \tag{7.101}$$

where

$$\begin{aligned}
f(\omega) = -\frac{1}{\varepsilon}\Big[& e^2 c_1 \xi_1 c_2 \xi_2 \sin 2\xi h \cos 2\xi_1 h_1 \sin 2\xi_2 h_2 \\
& + e^2 c_1^2 \xi_1^2 \sin 2\xi h \sin 2\xi_1 h_1 \cos 2\xi_2 h_2 \\
& + 4e^2 c_1 \xi_1 \bar{c} \xi \sin^2(\xi h) \cos 2\xi_1 h_1 \cos 2\xi_2 h_2 \\
& - 4e^2 c_2 \xi_2 \bar{c} \xi \sin^2(\xi h) \sin 2\xi_1 h_1 \sin 2\xi_2 h_2 \\
& + 2h\varepsilon c_2 \xi_2 \bar{c}^2 \xi^2 \sin 2\xi h \sin 2\xi_1 h_1 \sin 2\xi_2 h_2 \\
& - 2h\varepsilon c_1^2 \xi_1^2 \bar{c} \xi \cos 2\xi h \sin 2\xi_1 h_1 \cos 2\xi_2 h_2 \\
& - 2h\varepsilon c_1 \xi_1 c_2 \xi_2 \bar{c} \xi \cos 2\xi h \cos 2\xi_1 h_1 \sin 2\xi_2 h_2 \\
& - 2h\varepsilon c_1 \xi_1 \bar{c}^2 \xi^2 \sin 2\xi h \cos 2\xi_1 h_1 \cos 2\xi_2 h_2 \Big],
\end{aligned} \tag{7.102}$$

$$\begin{aligned}
g(\omega) = \frac{c_1 \xi_1 c_2 \xi_2}{\varepsilon}\Big[& 4e^2 \bar{c} \xi \sin(\xi h) \cos 2\xi_1 h_1 \sin 2\xi_2 h_2 \\
& - 2h\varepsilon \bar{c}^2 \xi^2 \sin 2\xi h \cos 2\xi_1 h_1 \sin 2\xi_2 h_2 \\
& - 2h\varepsilon c_1 \xi_1 \bar{c} \xi \cos 2\xi h \sin 2\xi_1 h_1 \sin 2\xi_2 h_2 \\
& + e^2 c_1 \xi_1 \sin 2\xi h \sin 2\xi_1 h_1 \sin 2\xi_2 h_2 \Big].
\end{aligned} \tag{7.103}$$

The case of near perfect bonding is represented by a small J. In this case, let the frequency for a perfectly bonded interface ($J = 0$) be ω_0 such that $f(\omega_0) = 0$. We also denote $\omega = \omega_0 + \Delta\omega$. Then, to the first order of $\Delta\omega$, Eq. (7.101) can be approximated by

$$\Delta\omega \cong -\frac{g(\omega_0)}{f'(\omega_0)} J. \tag{7.104}$$

Equation (7.104) gives a linear relation between resonator frequency shift and interface compliance and thus provides the foundation for an experimental procedure for measuring J.

As a numerical example, for geometric parameters we choose $h = 0.0005$ m, $h_1 = 0.0005$ m, and $h_2 = 0.0005$ m. For the piezoelectric layer we consider PZT-5H. For the lower elastic layer we consider aluminum with $\rho_1 = 2,800$ kg/m^3, and $c_1 = 30 \times 10^9$ N/m^2. For the upper elastic layer we consider steel with $\rho_2 = 7,850$ kg/m^3, and $c_2 = 80 \times 10^9$ N/m^2. Frequency shift versus J is calculated from Eq. (7.104) and is presented in Fig. 7.7 for the first three resonances. The first mode has a relatively low sensitivity. The second and the third modes show about the same sensitivity.

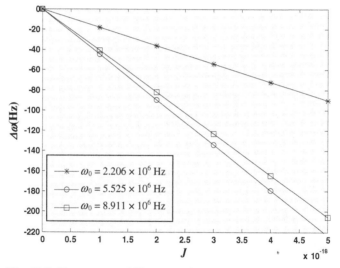

Fig. 7.7. Frequency shift versus interface compliance for the three lowest resonances.

Figure 7.8 shows the frequency shift of the first resonance for different plate thickness. A thicker plate has a lower frequency as expected. The frequency shift is sensitive to the plate thickness. For both Figs. 7.7 and 7.8, the frequency shift depends on the stress distribution along the plate thickness and how much stress the interface experiences. Therefore some design is needed for an optimal sensor.

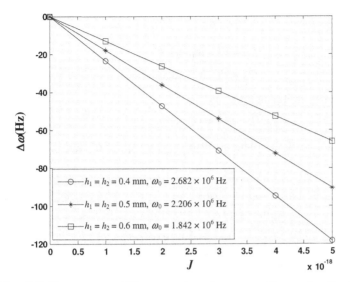

Fig. 7.8. Frequency shift versus interface compliance for the first
resonance with different plate thickness.

7.6. Edge Modes in a Semi-infinite Plate

Consider the semi-infinite plate in Fig. 7.9 [47]. All surfaces are traction-
free. The two major surfaces at $x_2 = \pm h$ are unelectroded. The free-space
electric field is neglected.

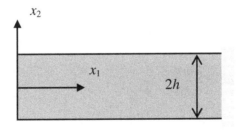

Fig. 7.9. A semi-infinite plate.

The governing equations are

$$\bar{c}\nabla^2 u = \rho\ddot{u}, \quad \nabla^2\psi = 0, \quad \phi = \psi + \frac{e}{\varepsilon}u. \tag{7.105}$$

At the plate major surfaces the traction-free and charge-free boundary conditions are

$$T_{23} = 0, \quad x_2 = \pm h,$$
$$D_2 = 0, \quad x_2 = \pm h,$$

(7.106)

or, equivalently, in terms of u and ψ,

$$u_{,2} = 0, \quad \psi_{,2} = 0, \quad x_2 = \pm h.$$

(7.107)

For the right end of the plate at infinity, we require decaying behavior for edge modes.

It can be verified by direct substitution that the following fields satisfy Eqs. (7.105) and (7.106), and can be classified into waves symmetric or antisymmetric in x_2. For symmetric fields,

$$u = \cos \xi_2 x_2 A \exp(-\xi_1 x_1) \exp(i\omega t),$$
$$\psi = \cos \xi_2 x_2 B \exp(-\xi_2 x_1) \exp(i\omega t),$$

(7.108)

$$\xi_2 = \frac{m\pi}{2h}, \quad m = 0, 2, 4, \cdots,$$

and for antisymmetric fields,

$$u = \sin \xi_2 x_2 A \exp(-\xi_1 x_1) \exp(i\omega t),$$
$$\psi = \sin \xi_2 x_2 B \exp(-\xi_2 x_1) \exp(i\omega t),$$

(7.109)

$$\xi_2 = \frac{m\pi}{2h}, \quad m = 1, 3, 5, \cdots,$$

where A and B are undetermined constants, and

$$\xi_1 = \sqrt{\xi_2^2 - \frac{\rho \omega^2}{\bar{c}}} = \sqrt{\frac{\rho}{\bar{c}}} \sqrt{\left(\frac{m\pi}{2h}\right)^2 - \omega^2 \frac{\bar{c}}{\rho}} = \frac{1}{v_T} \sqrt{\omega_m^2 - \omega^2},$$

(7.110)

$$v_T = \sqrt{\frac{\bar{c}}{\rho}}, \quad \omega_m^2 = \left(\frac{m\pi}{2h}\right)^2 \frac{\bar{c}}{\rho}.$$

ω_m is the cutoff frequency of thickness-twist waves. In particular, $m = 0$ is the face-shear mode. We will not consider this mode because it cannot be an edge mode. The modes corresponding to $m > 0$ may decay exponentially from the free edge at $x_1 = 0$ and therefore are called edge modes.

Equation (7.108) or (7.109) still needs to satisfy the boundary conditions on the traction-free minor surface at $x_1 = 0$. For the electrical boundary conditions at $x_1 = 0$, we consider an electroded surface with the electrode grounded:

$$T_{13} = 0, \quad \phi = 0, \quad x_1 = 0. \tag{7.111}$$

Then, from the fields in Eqs. (7.108) and (7.109), for symmetric or antisymmetric fields, we have the same equations for A and B as below:

$$\bar{c}(-\xi_1)A + e(-\xi_2)B = 0,$$

$$\frac{e}{\varepsilon}A + B = 0. \tag{7.112}$$

For nontrivial solutions we must require that

$$\begin{vmatrix} \bar{c}\xi_1 & e\xi_2 \\ e/\varepsilon & 1 \end{vmatrix} = \bar{c}\xi_1 - \frac{e^2}{\varepsilon}\xi_2 = 0, \tag{7.113}$$

which can be written as

$$\frac{1}{v_T}\sqrt{\omega_m^2 - \omega^2} = \bar{k}^2\xi_2, \tag{7.114}$$

or

$$\omega^2 = \omega_m^2 - v_T^2\bar{k}^4\xi_2^2 = (1 - \bar{k}^4)\frac{\bar{c}}{\rho}\left(\frac{m\pi}{2h}\right)^2, \tag{7.115}$$

$$m = 1, 2, 3, \cdots.$$

Equation (7.115) determines the frequencies of the edge modes. If the minor face at $x_1 = 0$ is unelectroded, or if it is mechanically fixed, no edge modes can be found. We point the resemblance between Eqs. (7.115) and (4.12). In fact, the semi-infinite plate in Fig. 7.9 may be viewed as a slice of the half-space in Fig. 4.1 with a proper thickness depending on the wavelength of the SAW in Sec. 4.1. Then a connection between the edge modes in this section and the surface wave in Sec. 4.1 can be made.

7.7. Mass Sensitivity of Edge Modes

The edges modes in a semi-infinite plate discussed in the previous section can be used to make mass sensors [48]. Consider the semi-infinite piezoelectric plate in Fig. 7.10. The two major surfaces at $x_2 = \pm h$ are traction-free and are unelectroded. The end surface at $x_1 = 0$ is electroded and there is a mass layer on the end surface.

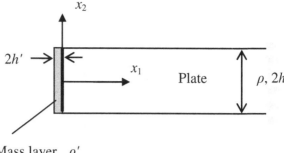

Fig. 7.10. A semi-infinite plate with an end electrode and an end mass layer.

The governing equations and boundary conditions are

$$\overline{c}\nabla^2 u = \rho\ddot{u},$$

$$\nabla^2\psi = 0, \qquad (7.116)$$

$$\phi = \psi + \frac{e}{\varepsilon}u,$$

$$T_{23} = 0, \quad x_2 = \pm h,$$
$$D_2 = 0, \quad x_2 = \pm h, \qquad (7.117)$$

$$T_{13} = \rho'h'\ddot{u}, \quad x_1 = 0,$$
$$\phi = 0, \quad x_1 = 0. \qquad (7.118)$$

For the right end of the plate at infinity, we require decaying behavior for edge modes.

Fields satisfying the above equations and boundary conditions are the same as those in the previous section. They can be classified into waves symmetric or antisymmetric in x_2, and are given by

$$u = \cos\xi_2 x_2 A\exp(-\xi_1 x_1)\exp(i\omega t),$$

$$\psi = \cos\xi_2 x_2 B\exp(-\xi_2 x_1)\exp(i\omega t), \qquad (7.119)$$

$$\xi_2 = \frac{m\pi}{2h}, \quad m = 0, 2, 4, \cdots,$$

$$u_3 = \sin\xi_2 x_2 A\exp(-\xi_1 x_1)\exp(i\omega t),$$

$$\psi = \sin\xi_2 x_2 B\exp(-\xi_2 x_1)\exp(i\omega t), \qquad (7.120)$$

$$\xi_2 = \frac{m\pi}{2h}, \quad m = 1, 3, 5, \cdots,$$

respectively, where

$$\xi_1 = \sqrt{\xi_2^2 - \frac{\rho \omega^2}{\bar{c}}} = \sqrt{\frac{\rho}{\bar{c}}}\sqrt{\left(\frac{m\pi}{2h}\right)^2 - \omega^2 \frac{\bar{c}}{\rho}} = \frac{1}{v_T}\sqrt{\omega_m^2 - \omega^2},$$

(7.121)

$$v_T = \sqrt{\frac{\bar{c}}{\rho}}, \quad \omega_m^2 = \left(\frac{m\pi}{2h}\right)^2 \frac{\bar{c}}{\rho}.$$

A and B are undetermined constants. $m = 0$ is the face-shear mode which is not an edge mode.

Equation (7.119) or (7.120) still needs to satisfy the boundary conditions on the minor surface at $x_1 = 0$. For time-harmonic motions, the mechanical boundary condition in Eq. (7.118) takes the following form:

$$T_{13} = -\omega^2 \rho' h' u.$$

(7.122)

Substituting Eqs. (7.119) or (7.120) into Eq. (7.122) and the electrical boundary condition at $x_1 = 0$ in Eq. (7.118), we obtain

$$\bar{c}(-\xi_1)A + e(-\xi_2)B = -\omega^2 \rho' 2h' A,$$

$$\frac{e}{\varepsilon}A + B = 0.$$

(7.123)

For nontrivial solutions we must require that

$$\begin{vmatrix} \bar{c}\xi_1 - \omega^2 \rho' 2h' & e\xi_2 \\ e/\varepsilon & 1 \end{vmatrix} = \bar{c}\xi_1 - \omega^2 \rho' 2h' - \frac{e^2}{\varepsilon}\xi_2 = 0,$$

(7.124)

which can be written as

$$\frac{1}{v_T}\sqrt{\omega_m^2 - \omega^2} = \bar{k}^2 \xi_2 + \frac{\omega^2 \rho' 2h'}{\bar{c}}.$$

(7.125)

Equation (7.125) determines the frequencies of the edge modes. For small ρ' and h', the root of Eq. (7.125) is approximately given by

$$\omega^2 \cong v_T^2 \xi_2^2 (1 - \bar{k}^4)\left(1 - \bar{k}^2 m\pi \frac{\rho' 2h'}{\rho h}\right),$$

(7.126)

$$m = 1, 2, 3, \cdots.$$

Equation (7.126) is true when the second term in the right pair of parentheses is much smaller than one. Therefore Eq. (7.126) is not valid for higher-order modes with a large m. This is fine because in applications usually lower-order modes with a small m is used. If we denote the unperturbed frequencies when the mass layer is not present by

$$\hat{\omega}^2 = \omega_m^2 - v_T^2 \bar{k}^4 \xi_2^2 = v_T^2 \xi_2^2 (1 - \bar{k}^4) = (1 - \bar{k}^4) \frac{\bar{c}}{\rho} \left(\frac{m\pi}{2h} \right)^2, \qquad (7.127)$$

from Eq. (7.126) we can obtain the frequency shift as

$$\frac{\omega - \hat{\omega}}{\hat{\omega}} \cong -\bar{k}^2 m\pi \frac{\rho' h'}{\rho h}. \qquad (7.128)$$

We make the following observations from Eq. (7.128):

(i) The frequency shift is linear in the layer mass.

(ii) The inertial effect of the mass layer lowers the frequencies as expected.

(iii) Higher-order modes imply higher sensitivity (this is true within a certain range of m).

(iv) The mass-frequency effect is proportional to $\bar{k}^2$ and therefore can only be detected in a piezoelectric plate, but not in an elastic plate. Higher piezoelectric coupling implies higher sensitivity.

7.8. Modes in a Rectangular Plate

Next we study modes in a finite rectangular plate (see Fig. 7.11) [49]. The four faces are traction-free. The boundaries at $x_2 = \pm h$ may be both electroded with shorted electrodes or both unelectroded. $x_1 = \pm a$ are unelectroded.

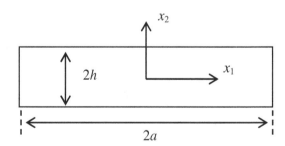

Fig. 7.11. A rectangular plate.

The governing equations are

$$\bar{c}\nabla^2 u = \rho\ddot{u}, \quad \nabla^2 \psi = 0, \quad \phi = \psi + eu / \varepsilon. \qquad (7.129)$$

7.8.1. An unelectroded plate

First consider an unelectroded plate. The boundary conditions are

$$T_{23} = 0, \quad D_2 = 0, \quad x_2 = \pm h,$$
$$T_{13} = 0, \quad D_1 = 0, \quad x_1 = \pm a, \tag{7.130}$$

or, equivalently,

$$u_{,2} = 0, \quad \psi_{,2} = 0, \quad x_2 = \pm h,$$
$$u_{,1} = 0, \quad \psi_{,1} = 0, \quad x_1 = \pm a. \tag{7.131}$$

Consider the possibility of

$$u = A \cos \xi_2 x_2 \cos \xi_1 x_1 \exp(i\omega t),$$
$$\psi = B \cosh \xi_1 x_2 \cos \xi_1 x_1 \exp(i\omega t), \tag{7.132}$$

where A and B are constants. Note that Eq. $(7.132)_2$ satisfies Eq. $(7.129)_2$. For Eq. (7.132) to satisfy Eq. (7.131), we must have

$$\xi_1 a = m\pi / 2, \quad m = 0, 2, 4, \cdots,$$
$$\xi_2 h = n\pi / 2, \quad n = 0, 2, 4, \cdots, \tag{7.133}$$
$$B = 0.$$

For Eq. $(7.132)_1$ to satisfy Eq. $(7.129)_1$ we must have

$$\rho \omega^2 = \bar{c}(\xi_1^2 + \xi_2^2) = \bar{c}\left[\left(\frac{m\pi}{2a}\right)^2 + \left(\frac{n\pi}{2h}\right)^2\right]. \tag{7.134}$$

By similar considerations, it can be concluded that for an unelectroded plate the following are all the modes satisfying Eqs. (7.129) and (7.131):

$$u = \begin{cases} A_1 \cos \dfrac{n\pi}{2h} x_2 \cos \dfrac{m\pi}{2a} x_1 \exp(i\omega t), \\[2mm] A_2 \sin \dfrac{n\pi}{2h} x_2 \cos \dfrac{m\pi}{2a} x_1 \exp(i\omega t), \\[2mm] A_3 \cos \dfrac{n\pi}{2h} x_2 \sin \dfrac{m\pi}{2a} x_1 \exp(i\omega t), \\[2mm] A_4 \sin \dfrac{n\pi}{2h} x_2 \sin \dfrac{m\pi}{2a} x_1 \exp(i\omega t), \end{cases} \tag{7.135}$$

$$\psi = 0,$$

where, corresponding to A_1, A_2, A_3 and A_4, we have $(m, n) =$ (even, even), (even, odd), (odd, even) and (odd, odd), and the frequencies are all given by Eq. (7.134).

7.8.2. An electroded plate

Next consider another case of practical interest in which the boundaries at $x_2 = \pm h$ are electroded and the electrodes are shorted. The boundary conditions are

$$T_{23} = 0, \quad \phi = \psi + eu/\varepsilon = 0, \quad x_2 = \pm h,$$
$$T_{13} = 0, \quad D_1 = 0, \quad x_1 = \pm a. \tag{7.136}$$

In this case it is convenient to classify the modes as symmetric and antisymmetric according to whether their x_2 dependence is even or odd, and discuss symmetric and antisymmetric modes separately.

For symmetric modes consider the possibility of

$$u = A\cos\xi_2 x_2 \cos\xi_1 x_1 \exp(i\omega t),$$
$$\psi = B\cosh\xi_1 x_2 \cos\xi_1 x_1 \exp(i\omega t), \tag{7.137}$$
$$\xi_1 = \frac{m\pi}{2a}, \quad m = 0, 2, 4, \cdots,$$

which satisfy Eq. $(7.129)_2$ and the boundary conditions at $x_1 = \pm a$. For Eq. $(7.137)_1$ to satisfy Eq. $(7.129)_1$, we must have

$$\rho\omega^2 = \bar{c}\left[\left(\frac{m\pi}{2a}\right)^2 + \xi_2^2\right]. \tag{7.138}$$

For Eq. (7.137) to satisfy the boundary conditions at $x_2 = \pm h$, we must have

$$-\bar{c}A\xi_2 \sin\xi_2 h + eB\xi_1 \sinh\xi_1 h = 0,$$
$$\frac{e}{\varepsilon}A\cos\xi_2 h + B\cosh\xi_1 h = 0. \tag{7.139}$$

For nontrivial solutions, the determinant of the coefficient matrix of Eq. (7.139) must vanish, which yields

$$\tan\xi_2 h + \bar{k}^2 \frac{\xi_1}{\xi_2}\tanh\xi_1 h = 0. \tag{7.140}$$

With

$$\xi_1 = \frac{m\pi}{2a}, \quad \xi_2^2 = \frac{\rho\,\omega^2}{c} - \left(\frac{m\pi}{2a}\right)^2, \qquad (7.141)$$

Equation (7.140) can be written as a frequency equation for ω. We note that Eq. (7.140) is the same as Eq. (5.13) except that the wave number in the x_1 direction is known now. Basically the rectangular plate in Fig. 7.11 behaves as a portion of the plate in Fig. 5.1. Similarly,

$$u = A\cos\xi_2 x_2 \sin\xi_1 x_1 \exp(i\omega t),$$

$$\psi = B\cosh\xi_1 x_2 \sin\xi_1 x_1 \exp(i\omega t), \qquad (7.142)$$

$$\xi_1 = \frac{m\pi}{2a}, \quad m = 1, 3, 5, \cdots,$$

are also modes that are even in x_2. The frequency equation is still Eq. (7.140).

For modes antisymmetric in x_2 consider

$$u = A\sin\xi_2 x_2 \cos\xi_1 x_1 \exp(i\omega t),$$

$$\psi = B\sinh\xi_1 x_2 \cos\xi_1 x_1 \exp(i\omega t), \qquad (7.143)$$

$$\xi_1 = \frac{m\pi}{2a}, \quad m = 0, 2, 4, \cdots,$$

which satisfy Eq. $(7.129)_2$ and the boundary conditions at $x_1 = \pm a$. For Eq. $(7.143)_1$ to satisfy Eq. $(7.129)_1$ we still have Eq. (7.138). For Eq. (7.143) to satisfy the boundary conditions at $x_2 = \pm h$, we must have

$$\bar{c}A\xi_2 \cos\xi_2 h + eB\xi_1 \cosh\xi_1 h = 0,$$

$$\frac{e}{\varepsilon}A\sin\xi_2 h + B\sinh\xi_1 h = 0. \qquad (7.144)$$

For nontrivial solutions,

$$\tanh\xi_1 h - \bar{k}^2 \frac{\xi_1}{\xi_2}\tan\xi_2 h = 0, \qquad (7.145)$$

which, with Eq. (7.141), represents the frequency equation. We point out the identity between Eqs. (7.145) and (5.7). Similarly,

$$u = A\sin\xi_2 x_2 \sin\xi_1 x_1 \exp(i\omega t),$$

$$\psi = B\sinh\xi_1 x_2 \sin\xi_1 x_1 \exp(i\omega t), \qquad (7.146)$$

$$\xi_1 = \frac{m\pi}{2a}, \quad m = 1, 3, 5, \cdots,$$

are also allowed by Eqs. (7.129) and (7.136), and are also with Eq. (7.145) as the frequency equation.

7.9. A Rectangular Plate with Thin Films

Consider a rectangular plate with mass layers (see Fig. 7.12) [24]. The four faces of the whole structure are traction-free. The two mass layers are shown by the shaded areas. They are thin with $h' \ll h$ and $h'' \ll h$, and are modeled by the membrane theory of thin films. The membrane theory can describe the lowest order effects of the mass layer inertia and stiffness. The two major surfaces of the plate at $x_2 = \pm h$ can be either unelectroded or with shorted electrodes as shown by the thick lines in the figure, which will be treated separately. The electrodes are assumed to be very thin. They only provide constraints on the electric potential but have no mechanical effects.

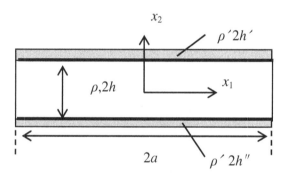

Fig. 7.12. A rectangular plate with thin films.

The governing equations for u, ψ and ϕ are:

$$\bar{c}\nabla^2 u = \rho\ddot{u},$$

$$\nabla^2\psi = 0, \qquad\qquad (7.147)$$

$$\phi = \psi + \frac{e}{\varepsilon}u.$$

The boundary conditions depend on whether the plate is electroded or unelectroded. We discuss these two cases separately below. The results differ from those in Sec. 5.5 only in that the wave number in the x_1 direction is now fixed.

7.9.1. Unelectroded plates

First consider an unelectroded plate. The electric field in the surrounding space is neglected. The boundary conditions are:

$$2h'\mu u_{,11} - T_{23} = \rho' 2h'\ddot{u}, \quad D_2 = 0, \quad x_2 = h,$$

$$2h''\mu u_{,11} + T_{23} = \rho' 2h''\ddot{u}, \quad D_2 = 0, \quad x_2 = -h, \qquad (7.148)$$

$$T_{13} = 0, \quad D_1 = 0, \quad x_1 = \pm a,$$

where μ is the shear elastic constant of the mass layers. Physically, the mechanical boundary conditions at $x_2 = \pm h$ state that the stress gradient in the mass layer and the stress between the mass layer and the plate are responsible for the layer acceleration according to Newton's law. Equation (7.148) is equivalent to

$$2h'\mu u_{,11} - \bar{c} u_{,2} = \rho' 2h'\ddot{u}, \quad \psi_{,2} = 0, \quad x_2 = h,$$

$$2h''\mu u_{,11} + \bar{c} u_{,2} = \rho' 2h''\ddot{u}, \quad \psi_{,2} = 0, \quad x_2 = -h, \qquad (7.149)$$

$$u_{,1} = 0, \quad \psi_{,1} = 0, \quad x_1 = \pm a.$$

We construct the following solution that is symmetric in x_1:

$$u = (A_1 \cos \xi_2 x_2 + A_2 \sin \xi_2 x_2) \cos \xi_1 x_1 \exp(i\omega t),$$
$$\psi = (B_1 \cosh \xi_1 x_2 + B_2 \sinh \xi_1 x_2) \cos \xi_1 x_1 \exp(i\omega t), \qquad (7.150)$$

where A_1, A_2, B_1 and B_2 are undetermined constants,

$$\xi_1 a = m\pi / 2, \quad m = 0, 2, 4, \cdots. \qquad (7.151)$$

Equation (7.150) already satisfies the boundary conditions at $x_1 = \pm a$. Equation $(7.150)_2$ already satisfies Eq. $(7.147)_2$. For Eq. $(7.150)_1$ to satisfy Eq. $(7.147)_1$, we must have

$$\rho \omega^2 = \bar{c}(\xi_1^2 + \xi_2^2) = \bar{c} \left[\left(\frac{m\pi}{2a} \right)^2 + \xi_2^2 \right]. \qquad (7.152)$$

Substitution of Eq. (7.150) into the boundary conditions at $x_2 = \pm h$ gives four linear equations. For nontrivial solutions the determinant of the coefficient matrix has to vanish. This leads to the frequency equation

$$\tan 2\xi_2 h = \frac{\bar{c}\xi_2(\rho'\omega^2 - \mu\xi_1^2)2(h' + h'')}{4(\rho'\omega^2 - \mu\xi_1^2)^2 h'h'' - \bar{c}^2\xi_2^2}. \qquad (7.153)$$

For symmetric mass layers with $h'' = h'$, Eq. (7.153) can be written as

$$\tan 2\xi_2 h = \frac{2\dfrac{\bar{c}\xi_2}{(\rho'\omega^2 - \mu\xi_1^2)2h'}}{1 - \left[\dfrac{\bar{c}\xi_2}{(\rho'\omega^2 - \mu\xi_1^2)2h'}\right]^2} = \frac{2\left[-\dfrac{(\rho'\omega^2 - \mu\xi_1^2)2h'}{\bar{c}\xi_2}\right]}{1 - \left[-\dfrac{(\rho'\omega^2 - \mu\xi_1^2)2h'}{\bar{c}\xi_2}\right]^2}.$$

(7.154)

Comparing Eq. (7.154) with the following trigonometric identity:

$$\tan 2\xi_2 h = \frac{2\tan\xi_2 h}{1 - \left(\tan\xi_2 h\right)^2},$$

(7.155)

we identify the following frequency equations for symmetric mass layers:

$$\tan\xi_2 h = \frac{\bar{c}\xi_2}{(\rho'\omega^2 - \mu\xi_1^2)2h'},$$

(7.156)

$$\text{or} \quad \tan\xi_2 h = -\frac{(\rho'\omega^2 - \mu\xi_1^2)2h'}{\bar{c}\xi_2},$$

For symmetric mass layers, the modes can be classified as antisymmetric and symmetric about x_2, with separate frequency equations as given in Eq. (7.156).

When the mass layers are not present, i.e., $h' = h'' = 0$, or $\rho' = 0$ and $\mu = 0$, we denote the solution to Eq. (7.153) by $\xi_2^{(0)}$ and $\omega^{(0)}$. Then we have $\sin 2\xi_2^{(0)} h = 0$, or $2\xi_2^{(0)} h = n\pi$, $n = 0, 1, 2, \cdots$, and

$$\rho(\omega^{(0)})^2 = \bar{c}[\xi_1^2 + (\xi_2^{(0)})^2] = \bar{c}\left[\left(\frac{m\pi}{2a}\right)^2 + \left(\frac{n\pi}{2h}\right)^2\right],$$

(7.157)

$$m = 0, 2, 4, \cdots, \quad n = 0, 1, 2, \cdots.$$

For thin mass layers, the inertial and stiffness effects of the mass layers in the denominator of Eq. (7.153) (which are second-order) can be neglected. Let

$$\xi_2 = \xi_2^{(0)} + \Delta\xi_2, \quad \omega = \omega^{(0)} + \Delta\omega.$$

(7.158)

Substituting Eqs. (7.157) and (7.158) into Eq. (7.153), to the lowest-order effect, we obtain

$$2(\Delta\xi_2)h \cong \frac{2\left[\rho'(\omega^{(0)})^2 - \mu\xi_1^2\right](h'+h'')}{-\bar{c}\xi_2^{(0)}}, \tag{7.159}$$

which gives the perturbation of ξ_2 due to the mass layers. From Eq. (7.152) we can obtain

$$2\rho\,\omega^{(0)}\Delta\omega = \bar{c}\,2\xi_2^{(0)}\Delta\xi_2, \tag{7.160}$$

which is a relation between the perturbations of ξ_2 and ω. Then, from Eqs. (7.159) and (7.160), the frequency perturbation is determined as

$$\frac{\Delta\omega}{\omega^{(0)}} = -R^{(m)} + R^{(s)}\frac{X^2}{(\Omega^{(0)})^2}$$

$$R^{(m)} = \frac{\rho'(2h'+2h'')}{\rho 2h}, \quad R^{(s)} = \frac{\mu}{\bar{c}}\frac{2h'+2h''}{2h},$$

$$X^2 = \xi_1^2 \Big/ \frac{\pi^2}{4h^2} = m^2\frac{h^2}{a^2}, \quad (\Omega^{(0)})^2 = (\omega^{(0)})^2 \Big/ \frac{\pi^2\bar{c}}{4h^2\rho} = m^2\frac{h^2}{a^2} + n^2,$$

$$\tag{7.161}$$

where $R^{(m)}$ is the mass ratio between the layers and the plate, $R^{(s)}$ is the stiffness ratio, X is the normalized wave number in the x_1 direction, and $\Omega^{(0)}$ is the normalized frequency when the mass layers are not present.

As an example, consider a plate made from PZT-5H. The mass layers are glass (either T-40 or pyrex). Both are isotropic with

	$\rho'(\text{kg/m}^3)$	$\mu(10^{10}\,\text{N/m}^2)$
T - 40	3390	2.26
Pyrex	2320	2.5

Consider the case when the two mass layers are identical, with $h' = h'' = 0.01\,h$. The table below shows the inertial and stiffness effects for the thickness-twist mode with $(m, n) = (2, 1)$.

		$a/h = 6$	$a/h = 8$	$a/h = 10$
T - 40	$R^{(m)} = 0.0090$	$R^{(s)} = 0.0011$	$R^{(s)} = 0.0006$	$R^{(s)} = 0.0004$
Pyrex	$R^{(m)} = 0.0062$	$R^{(s)} = 0.0012$	$R^{(s)} = 0.0007$	$R^{(s)} = 0.0005$

It can be seen that the stiffness effect is smaller than the inertial effect, especially for thin plates.

Since the stiffness effect is small, we examine the inertial effect further below. Still consider the case of symmetric mass layers on an unelectroded plate. For the mode with $(m, n) = (2, 1)$, the exact frequency is determined by Eq. (7.156)$_1$ which can be written as

$$\tan \xi_2 h = \frac{\bar{c} \xi_2}{\left(\rho' \omega^2 - \frac{\mu \pi^2}{a^2} \right) 2h'} , \qquad (7.162)$$

where μ is going to be set to zero in the calculations below. We plot in Fig. 7.13 the frequency shift for the mode versus the mass ratio $R = R^{(m)}$. $h = 1$ mm, $a/h = 10$. The figure shows that for small R the first-order prediction by Eq. (7.161) is close to the result given by Eq. (7.156). When $R = 0.05$, the first-order prediction begins to show deviation from the result of Eq. (7.156). In most applications R is less than 0.05. Therefore the first-order approximation by Eq. (7.161) is usually sufficient.

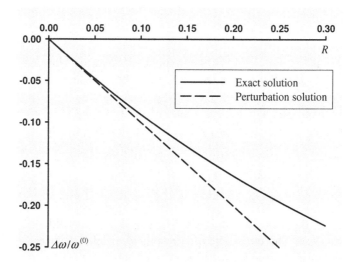

Fig. 7.13. Frequency shift versus $R = R^{(m)}$ when $(m, n) = (2,1)$ for symmetric mass layers.

By similar considerations, it can be shown that modes antisymmetric in x_1 are given by

$$u = (A_1 \cos \xi_2 x_2 + A_2 \sin \xi_2 x_2) \sin \xi_1 x_1 \exp(i\omega t),$$
$$\psi = (B_1 \cosh \xi_1 x_2 + B_2 \sinh \xi_1 x_2) \sin \xi_1 x_1 \exp(i\omega t),$$
(7.163)

where

$$\xi_1 a = m\pi / 2, \quad m = 1, 3, 5, \cdots,$$
(7.164)

and Eqs. (7.153) and (7.161) remain the same.

7.9.2. Electroded plates

Next consider the case when the plate boundaries at $x_2 = \pm h$ are electroded and the electrodes are shorted. The boundary conditions are

$$2h' \mu u_{,11} - T_{23} = \rho' 2h' \ddot{u}, \quad \phi = 0, \quad x_2 = h,$$
$$2h' \mu u_{,11} + T_{23} = \rho'' 2h'' \ddot{u}, \quad \phi = 0, \quad x_2 = -h,$$
(7.165)
$$T_{13} = 0, \quad D_1 = 0, \quad x_1 = \pm a,$$

which are equivalent to

$$2h' \mu u_{,11} - (\bar{c} u_{,2} + e \psi_{,2}) = \rho' 2h' \ddot{u}, \quad \psi + \frac{e}{\varepsilon} u = 0, \quad x_2 = h,$$

$$2h'' \mu u_{,11} + \bar{c} u_{,2} + e \psi_{,2} = \rho'' 2h'' \ddot{u}, \quad \psi + \frac{e}{\varepsilon} u = 0, \quad x_2 = -h,$$
(7.166)

$$u_{,1} = 0, \quad \psi_{,1} = 0, \quad x_1 = \pm a.$$

Since the boundary conditions at $x_1 = \pm a$ are the same as those in Eq. (7.149), the expressions in Eq. (7.150) are still valid. Substitution of Eq. (7.150) into Eq. (7.166) gives four linear equations. For nontrivial solutions the determinant of the coefficient matrix has to vanish. This leads to the frequency equation which is rather lengthy and is not provided. The case of symmetric mass layers is much simpler, which will be examined further below. For symmetric mass layers with $h'' = h'$, the frequency equation splits into two for modes that are symmetric and antisymmetric in x_2:

$$\begin{vmatrix} \bar{c} \xi_2 \sin \xi_2 h + 2h'(-\mu \xi_1^2 + \rho' \omega^2) \cos \xi_2 h & -e \xi_1 \sinh \xi_1 h \\ \dfrac{e}{\varepsilon} \cos \xi_2 h & \cosh \xi_1 h \end{vmatrix} = 0,$$
(7.167)

$$\begin{vmatrix} \overline{c}\xi_2\cos\xi_2 h + 2h'(\mu\xi_1^2 - \rho'\omega^2)\sin\xi_2 h & e\xi_1\cosh\xi_1 h \\ \dfrac{e}{\varepsilon}\sin\xi_2 h & \sinh\xi_1 h \end{vmatrix} = 0. \tag{7.168}$$

Equations (7.167) and (7.168) can be written as

$$\overline{c}\xi_2\tan(\xi_2 h) + \frac{e^2}{\varepsilon}\xi_1\tanh(\xi_1 h) = 2h'(\mu\xi_1^2 - \rho'\omega^2), \tag{7.169}$$

$$\overline{c}\xi_2\cot(\xi_2 h) - \frac{e^2}{\varepsilon}\xi_1\coth(\xi_1 h) = -2h'(\mu\xi_1^2 - \rho'\omega^2). \tag{7.170}$$

When the mass layers are not present, i.e., $\rho' = 0$ and $\mu = 0$, we denote the solution to Eqs. (7.169) and (7.170) by $\xi_2^{(0)}$ and $\omega^{(0)}$ such that

$$\overline{c}\xi_2^{(0)}\tan(\xi_2^{(0)}h) + \frac{e^2}{\varepsilon}\xi_1\tanh(\xi_1 h) = 0, \tag{7.171}$$

$$\overline{c}\xi_2^{(0)}\cot(\xi_2^{(0)}h) - \frac{e^2}{\varepsilon}\xi_1\coth(\xi_1 h) = 0. \tag{7.172}$$

Then, for thin mass layers, we look for a perturbation solution by substituting Eq. (7.158) into Eqs. (7.169) and (7.170). To the lowest-order effect, for a fixed ξ_1, we obtain

$$\overline{c}(\Delta\xi_2)\tan(\xi_2^{(0)}h) + \overline{c}\xi_2^{(0)}\frac{1}{\cos^2(\xi_2^{(0)}h)}(\Delta\xi_2 h) = 2h'[\mu\xi_1^2 - \rho'(\omega^{(0)})^2], \tag{7.173}$$

$$\overline{c}(\Delta\xi_2)\cot(\xi_2^{(0)}h) + \overline{c}\xi_2^{(0)}\frac{-1}{\sin^2(\xi_2^{(0)}h)}(\Delta\xi_2 h) = -2h'[\mu\xi_1^2 - \rho'(\omega^{(0)})^2]. \tag{7.174}$$

From Eqs. (7.173) and (7.160), and Eqs. (7.174) and (7.160), we obtain

$$\frac{\Delta\omega}{\omega^{(0)}} = \frac{-R^{(m)} + R^{(s)}\dfrac{X^2}{(\Omega^{(0)})^2}}{1 + \tan^2(\xi_2^{(0)}h) + \tan(\xi_2^{(0)}h)\big/(\xi_2^{(0)}h)}, \tag{7.175}$$

$$\frac{\Delta\omega}{\omega^{(0)}} = \frac{-R^{(m)} + R^{(s)}\dfrac{X^2}{(\Omega^{(0)})^2}}{1 + \cot^2(\xi_2^{(0)}h) - \cot(\xi_2^{(0)}h)\big/(\xi_2^{(0)}h)}. \tag{7.176}$$

We plot in Fig. 7.14 the frequency shift for the mode with $(m, n) = (2, 1)$ versus the mass ratio $R = R^{(m)}$. $h = 1$ mm, $a/h = 10$. The perturbation solution in Eq. (7.176) is used. The unperturbed frequency is determined from Eq. (7.172). The exact solution is from Eq. (7.170).

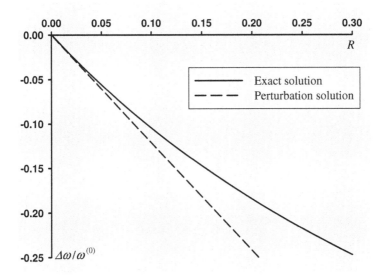

Fig. 7.14. Frequency shift versus $R = R^{(m)}$ when $(m, n) = (2,1)$ for symmetric mass layers.

For modes that are antisymmetric in x_1 as given in Eq. (7.163), Eqs. (7.169) and (7.170) remain the same.

Chapter 8
Free Vibrations in Polar Coordinates

This chapter and Chapter 7 are companion chapters. In addition to plates for which rectangular coordinates are convenient, circular cylindrical bodies are common structures for electromechanical devices. For antiplane motions of these bodies, polar coordinates are usually suitable. In this chapter we mainly discuss free vibrations of finite bodies in polar coordinates. One section is on wedges which are semi-infinite. The last section is in elliptical coordinates.

8.1. Thickness-shear in a Circular Cylinder

Consider an infinite circular cylinder of inner radius a and outer radius b (see Fig. 8.1) [50]. We choose (r, θ, z) to correspond to (1, 2, 3). The inner and outer surfaces are electroded.

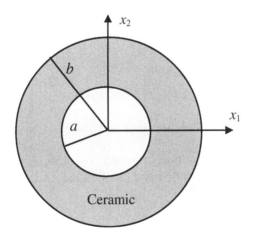

Fig. 8.1. A circular cylindrical ceramic shell.

From Sec. 1.6, the governing equations are:

$$\bar{c}\nabla^2 u = \rho\ddot{u}, \quad a < r < b,$$
$$\nabla^2\psi = 0, \quad a < r < b, \tag{8.1}$$
$$\phi = \psi + \frac{e}{\varepsilon}u, \quad a < r < b.$$

For boundary conditions we consider the following possibilities:

$$u = 0, \quad r = a, b, \quad \text{if the cylindical surfaces are fixed,}$$
$$\text{or} \quad T_{rz} = 0, \quad r = a, b, \quad \text{if the cylindical surfaces are free;}$$
$$\phi(r = a) = \phi(r = b), \quad \text{if the electrodes are shorted,} \tag{8.2}$$
$$\text{or} \quad D_r = 0, \quad r = a, b, \quad \text{if the electrodes are open.}$$

For axisymmetric motions we look for solutions in the following form:

$$u(r,t) = u(r)\exp(i\omega t),$$
$$\psi(r,t) = \psi(r)\exp(i\omega t), \tag{8.3}$$

and drop the exponential factor. The equations for u and ψ are

$$\nabla^2 u = \frac{\partial^2 u}{\partial r^2} + \frac{1}{r}\frac{\partial u}{\partial r} = -\frac{\rho\omega^2}{\bar{c}}u,$$
$$\nabla^2\psi = \frac{\partial^2\psi}{\partial r^2} + \frac{1}{r}\frac{\partial\psi}{\partial r} = 0. \tag{8.4}$$

The general solution to Eq. (8.4) is

$$u = A_1 J_0(\xi r) + A_2 Y_0(\xi r),$$
$$\psi = A_3 \ln\frac{r}{b} + A_4, \tag{8.5}$$

where A_1, A_2, A_3 and A_4 are undetermined constants, J_0 and Y_0 are zero-order Bessel functions of the first and second kind, and

$$\xi^2 = \frac{\rho\omega^2}{\bar{c}}. \tag{8.6}$$

The stress and electric displacement components are

$$\phi = \frac{e}{\varepsilon}[A_1 J_0(\xi r) + A_2 Y_0(\xi r)] + A_3 \ln \frac{r}{b} + A_4,$$

$$T_{rz} = \overline{c}u_{z,r} + e\psi_{,r} = -\overline{c}\xi[A_1 J_1(\xi r) + A_2 Y_1(\xi r)] + e\frac{A_3}{r}, \qquad (8.7)$$

$$D_r = -\varepsilon\psi_{,r} = -\varepsilon\frac{A_3}{r},$$

where $J_0' = -J_1$ and $Y_0' = -Y_1$ have been used.

8.1.1. Clamped and electroded surfaces

First consider the case when the two cylindrical surfaces are fixed and the two electrodes are shorted and grounded. Then we have

$$u = 0, \quad r = a, b,$$
$$\phi = 0, \quad r = a, b, \qquad (8.8)$$

which implies that

$$\psi = 0, \quad r = a, b. \qquad (8.9)$$

Hence

$$A_3 = 0, \quad A_4 = 0, \qquad (8.10)$$

and the frequency equation is

$$\begin{vmatrix} J_0(\xi a) & Y_0(\xi a) \\ J_0(\xi b) & Y_0(\xi b) \end{vmatrix} = 0. \qquad (8.11)$$

8.1.2. Free and unelectroded surfaces

Next consider the case when

$$T_{rz} = 0, \quad r = a, b,$$
$$D_r = 0, \quad r = a, b. \qquad (8.12)$$

Then $A_3 = 0$ and the frequency equation is

$$\begin{vmatrix} J_1(\xi a) & Y_1(\xi a) \\ J_1(\xi b) & Y_1(\xi b) \end{vmatrix} = 0. \qquad (8.13)$$

8.1.3. Free and electroded surfaces

Finally, consider the case when

$$T_{rz} = 0, \quad r = a, b,$$
$$\phi = 0, \quad r = a, b. \tag{8.14}$$

It can be shown that the frequency equation is

$$
\begin{vmatrix} J_1(\xi a) & Y_1(\xi a) \\ J_1(\xi b) & Y_1(\xi b) \end{vmatrix}
$$

$$
= \frac{\overline{k}^2}{\xi b \ln\dfrac{a}{b}} \begin{vmatrix} J_0(\xi a) - J_0(\xi b) & J_1(\xi a) - \dfrac{b}{a} J_1(\xi b) \\ Y_0(\xi a) - Y_0(\xi b) & Y_1(\xi a) - \dfrac{b}{a} Y_1(\xi b) \end{vmatrix}. \tag{8.15}
$$

For large x, Bessel functions can be approximated by

$$J_v(x) \cong \sqrt{\frac{2}{\pi x}} \cos\left(x - \frac{v\pi}{2} - \frac{\pi}{4} \right),$$

$$Y_v(x) \cong \sqrt{\frac{2}{\pi x}} \sin\left(x - \frac{v\pi}{2} - \frac{\pi}{4} \right). \tag{8.16}$$

Then it can be shown that for large a and b, Eq. (8.15) simplifies to

$$\sin\xi(b-a) = \frac{\overline{k}^2}{\xi b \ln\dfrac{a}{b}} \left[\left(1 + \frac{b}{a} \right) \cos\xi(b-a) - 2\sqrt{\frac{b}{a}} \right]. \tag{8.17}$$

Setting $2h = b - a$ and letting $a, b \to \infty$, we have

$$\xi b \ln\frac{a}{b} = \xi b \ln\frac{b-2h}{b} = \xi b \ln\left(1 - \frac{2h}{b} \right) \cong \xi b\left(-\frac{2h}{b} \right) = -\xi 2h,$$

$$\sin\xi(b-a) = \sin(\xi 2h)$$

$$\left(1 + \frac{b}{a} \right)\cos\xi(b-a) - 2\sqrt{\frac{b}{a}} \tag{8.18}$$

$$\cong 2\cos(\xi 2h) - 2 = -2[1 - \cos(\xi 2h)] = -4\sin^2\xi h.$$

Then Eq. (8.17) reduces to

$$\sin \xi h = 0 \quad \text{or} \quad \tan \xi h = \frac{\xi h}{k^2}, \tag{8.19}$$

which is the frequency equation for thickness-shear vibrations of an electroded plate given by Eqs. (7.16) and (7.22).

8.2. A Circular Cylinder with Unattached Electrodes

Next we analyze the case of a cylinder with unattached electrodes and air gaps (see Fig. 8.2) [51]. For free vibrations the voltage will be set to zero.

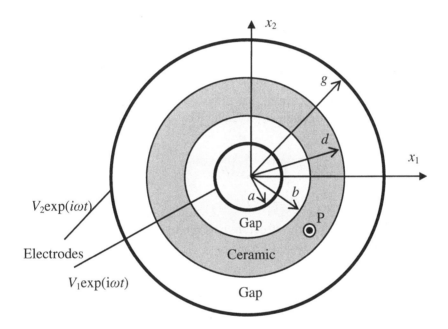

Fig. 8.2. A circular cylindrical shell with unattached electrodes.

The governing equations for the cylinder are

$$v_T^2 \nabla^2 u = \ddot{u}, \quad \nabla^2 \psi = 0, \quad \phi = \psi + \frac{e}{\varepsilon} u. \tag{8.20}$$

In the air gaps the electric filed is governed by

$$\nabla^2 \phi = 0,$$
$$E_r = -\phi_{,r},$$
$$D_r = \varepsilon_0 E_r = -\varepsilon_0 \phi_{,r}.$$

(8.21)

We have the following boundary and continuity conditions:

$$\phi = V_1 \exp(i\omega t), \quad r = a,$$
$$\phi = V_2 \exp(i\omega t), \quad r = g,$$
$$T_{rz}(r = b) = T_{rz}(r = d) = 0,$$
$$\phi(r = b^+) = \phi(r = b^-),$$
$$\phi(r = d^+) = \phi(r = d^-),$$
$$D_r(r = b^+) = D_r(r = b^-),$$
$$D_r(r = d^+) = D_r(r = d^-).$$

(8.22)

The charge on and the current that flows into the outer electrode per unit length in the axial direction are calculated by

$$D_r = \varepsilon_0 E_r,$$
$$Q_e = \int_0^{2\pi} -D_r g\, d\theta,$$
$$I = \dot{Q}_e = i\omega Q_e.$$

(8.23)

Then the impedance of the structure is given by

$$Z = (V_2 - V_1)/I.$$

(8.24)

For axisymmetric motions, in the cylinder, the general solution in the previous section still applies:

$$u = A_1 J_0(\xi r) + A_2 Y_0(\xi r),$$
$$\psi = A_3 \ln r + A_4,$$

(8.25)

$$\phi = \frac{e}{\varepsilon}[A_1 J_0(\xi r) + A_2 Y_0(\xi r)] + A_3 \ln r + A_4,$$

$$T_{rz} = -\bar{c}\xi[A_1 J_1(\xi r) + A_2 Y_1(\xi r)] + e\frac{A_3}{r},$$

(8.26)

$$D_r = -\varepsilon \frac{A_3}{r},$$

where A_1 through A_4 are undetermined constants. J_0 and Y_0 are the zero-order Bessel functions of the first and second kinds, and

$$\xi^2 = \frac{\rho\omega^2}{c}. \tag{8.27}$$

In the inner air gap, we have

$$\phi = F_1 \ln r + G_1,$$

$$D_r = \varepsilon_0 E_r = -\varepsilon_0 \frac{F_1}{r}, \tag{8.28}$$

where F_1 and G_1 are undetermined constants. Similarly, for the outer air gap,

$$\phi = F_2 \ln r + G_2,$$

$$D_r = -\varepsilon_0 \frac{F_2}{r}, \tag{8.29}$$

where F_2 and G_2 are undetermined constants. Imposing the boundary and continuity conditions, we obtain

$$F_1 \ln a + G_1 = V_1,$$

$$F_2 \ln g + G_2 = V_2,$$

$$-\overline{c}\xi[A_1 J_1(\xi b) + A_2 Y_1(\xi b)] + e\frac{A_3}{b} = 0,$$

$$-\overline{c}\xi[A_1 J_1(\xi d) + A_2 Y_1(\xi d)] + e\frac{A_3}{d} = 0,$$

$$F_1 \ln b + G_1 = \frac{e}{\varepsilon}[A_1 J_0(\xi b) + A_2 Y_0(\xi b)] + A_3 \ln b + A_4, \tag{8.30}$$

$$\frac{e}{\varepsilon}[A_1 J_0(\xi d) + A_2 Y_0(\xi d)] + A_3 \ln d + A_4 = F_2 \ln d + G_2,$$

$$-\varepsilon_0 \frac{F_1}{b} = -\varepsilon\frac{A_3}{b},$$

$$-\varepsilon\frac{A_3}{d} = -\varepsilon_0 \frac{F_2}{d}.$$

For free vibration the voltage is zero. For nontrivial solutions of the undetermined constants, the determinant of the coefficient matrix has to vanish. This yields the following equation that determines the resonant frequencies:

$$
\begin{vmatrix} J_1(\xi b) & Y_1(\xi b) \\ J_1(\xi d) & Y_1(\xi d) \end{vmatrix} = \frac{\overline{k}^2}{\xi d[\ln\dfrac{b}{d} - \dfrac{\varepsilon}{\varepsilon_0}(\ln\dfrac{b}{a} - \ln\dfrac{d}{g})]}
$$

$$
\times \begin{vmatrix} J_0(\xi b) - J_0(\xi d) & J_1(\xi b) - \dfrac{d}{b}J_1(\xi d) \\ Y_0(\xi b) - Y_0(\xi d) & Y_1(\xi b) - \dfrac{d}{b}Y_1(\xi d) \end{vmatrix}.
$$

(8.31)

For large x, Bessel functions can be approximated by

$$
J_\nu(x) \cong \sqrt{\frac{2}{\pi x}}\cos(x - \frac{\nu\pi}{2} - \frac{\pi}{4}),
$$

$$
Y_\nu(x) \cong \sqrt{\frac{2}{\pi x}}\sin(x - \frac{\nu\pi}{2} - \frac{\pi}{4}).
$$

(8.32)

Then it can be shown that, for large b and d, Eq. (8.31) simplifies to

$$
\sin\xi(d-b) = \frac{\overline{k}_{15}^2}{\xi d[\ln\dfrac{b}{d} - \dfrac{\varepsilon_{11}}{\varepsilon_0}(\ln\dfrac{b}{a} - \ln\dfrac{d}{g})]}
$$

$$
\times[(1+\frac{d}{b})\cos\xi(d-b) - 2\sqrt{\frac{d}{b}}].
$$

(8.33)

If there are no air gaps in the structure, i.e., $a = b$ and $g = d$, Eqs. (8.31) and (8.33) reduce to

$$
\begin{vmatrix} J_1(\xi b) & Y_1(\xi b) \\ J_1(\xi d) & Y_1(\xi d) \end{vmatrix}
$$

$$
= \frac{\overline{k}^2}{\xi d \ln\dfrac{b}{d}} \begin{vmatrix} J_0(\xi b) - J_0(\xi d) & J_1(\xi b) - \dfrac{d}{b}J_1(\xi d) \\ Y_0(\xi b) - Y_0(\xi d) & Y_1(\xi b) - \dfrac{d}{b}Y_1(\xi d) \end{vmatrix},
$$

(8.34)

$$
\sin\xi(d-b) = \frac{\overline{k}^2}{\xi d \ln\dfrac{b}{d}}[(1+\frac{d}{b})\cos\xi(d-b) - 2\sqrt{\frac{d}{b}}],
$$

which are Eqs. (8.15) and (8.17). For a numerical example, we fix $b = 5$ mm and $d = 8$ mm. We consider the case when the gaps are of the same thickness and denote $g - d = b - a = x$ which represents the common thickness of the gaps. Then we varies x from 0 to b. The first three resonant frequencies are shown in Fig. 8.3. We see that the odd-order modes are more sensitive to air gaps than the even-order modes. For thick air gaps the frequencies become stabilized.

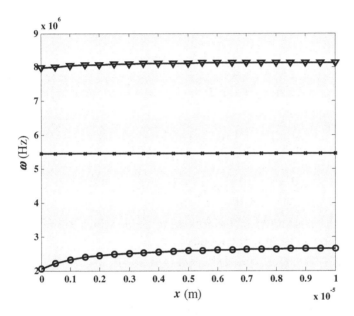

Fig. 8.3. Resonant frequencies versus air gap thickness.

8.3. A Wedge

In this section, we show that exact solutions can be obtained for a ceramic wedge [52]. The results are useful to the understanding and design of contoured piezoelectric resonators. Consider a semi-infinite wedge of ceramics poled in the x_3 direction as shown in Fig. 8.4. We consider the case that the wedge surfaces are traction-free and are unelectroded. Effects of the electric field that may exist in the surrounding free space are neglected.

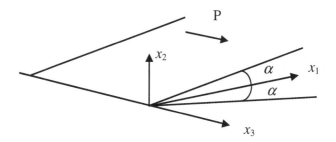

Fig. 8.4. A wedge.

In polar coordinates defined by $x_1 = r\cos\theta$ and $x_2 = r\sin\theta$, the governing equations are

$$v_T^2\left(\frac{\partial^2 u}{\partial r^2} + \frac{1}{r}\frac{\partial u}{\partial r} + \frac{1}{r^2}\frac{\partial^2 u}{\partial \theta^2}\right) = \ddot{u},$$

$$\frac{\partial^2 \psi}{\partial r^2} + \frac{1}{r}\frac{\partial \psi}{\partial r} + \frac{1}{r^2}\frac{\partial^2 \psi}{\partial \theta^2} = 0. \tag{8.35}$$

Consider the possibility of the following fields which are odd in θ and may be called antisymmetric modes:

$$u(r,\theta,t) = u(r)\sin\nu\theta\exp(-i\omega t),$$

$$\psi(r,\theta,t) = \psi(r)\sin\nu\theta\exp(-i\omega t). \tag{8.36}$$

Substitution of Eq. (8.36) into Eq. (8.35) results in

$$\frac{\partial^2 u}{\partial r^2} + \frac{1}{r}\frac{\partial u}{\partial r} + \left(\xi^2 - \frac{\nu^2}{r^2}\right)u = 0,$$

$$\frac{\partial^2 \psi}{\partial r^2} + \frac{1}{r}\frac{\partial \psi}{\partial r} - \frac{\nu^2}{r^2}\psi = 0, \tag{8.37}$$

where we have denoted $\xi = \omega/v_T$. ξ may be viewed as a wave number in the r direction. Equation (8.37)$_1$ can be written as Bessel's equation of order ν. Equation (8.37)$_2$ allows a simpler power function solution. The general solutions can be written as

$$u = [C_1 J_\nu(\xi r) + C_2 Y_\nu(\xi r)]\sin\nu\theta\exp(-i\omega t),$$

$$\psi = [C_3 r^\nu + C_4 r^{-\nu}]\sin\nu\theta\exp(-i\omega t), \tag{8.38}$$

where J_ν and Y_ν are the νth order Bessel functions of the first and second kinds. C_1–C_4 are undetermined constants. Consider the case of $\nu > 0$.

Since Y_ν and $r^{-\nu}$ are singular at the origin, terms associated with C_2 and C_4 have to be dropped. From Eq. (8.38) the stress and electric displacement components needed for boundary conditions can be obtained. We have

$$u = C_1 J_\nu(\xi r) \sin \nu\theta \exp(-i\omega t),$$

$$\psi = C_3 r^\nu \sin \nu\theta \exp(-i\omega t),$$

$$T_{\theta z} = [\bar{c}\frac{\nu}{r}C_1 J_\nu(\xi r) + e\frac{\nu}{r}C_3 r^\nu]\cos \nu\theta \exp(-i\omega t), \qquad (8.39)$$

$$D_\theta = -\varepsilon\frac{\nu}{r}C_3 r^\nu \cos \nu\theta \exp(-i\omega t).$$

At $\theta = \pm\alpha$ we need to impose the following boundary conditions for a free wedge:

$$T_{\theta z} = [\bar{c}\frac{\nu}{r}C_1 J_\nu(\xi r) + e\frac{\nu}{r}C_3 r^\nu]\cos \nu\alpha \exp(-i\omega t) = 0,$$

$$D_\theta = -\varepsilon\frac{\nu}{r}C_3 r^\nu \cos \nu\alpha \exp(-i\omega t) = 0, \qquad (8.40)$$

which implies that

$$\cos \nu\alpha = 0, \quad \nu = \frac{n\pi}{2\alpha}, \quad n = 1, 3, 5\cdots. \qquad (8.41)$$

Equation (8.41) determines the order of the Bessel function.

In a similar way, if the $\sin \nu\theta$ in Eq. (8.38) is replaced by $\cos \nu\theta$, a set of symmetric modes is obtained:

$$u = C_1 J_\nu(\xi r) \cos \nu\theta \exp(-i\omega t),$$

$$\psi = C_3 r^\nu \cos \nu\theta \exp(-i\omega t), \qquad (8.42)$$

$$\nu = \frac{n\pi}{2\alpha}, \quad n = 0, 2, 4, 6\cdots.$$

Note that in the modes given by Eqs. (8.39) and (8.42), ψ is unbounded for large r. If we require boundedness of ψ at large r, C_3 must vanish. Then $\psi = 0$ and the electric potential is given by

$$\phi = \begin{cases} \dfrac{e}{\varepsilon}C_1 J_\nu(\xi r) \sin \nu\theta \exp(-i\omega t), & n = 1, 3, 5\cdots, \\[12pt] \dfrac{e}{\varepsilon}C_1 J_\nu(\xi r) \cos \nu\theta \exp(-i\omega t), & n = 0, 2, 4, \cdots. \end{cases} \qquad (8.43)$$

Since a wedge occupies a semi-infinite region, we have a continuous spectrum. For any given ω, a ξ can be determined from $\xi = \omega / v_T$. Then antisymmetric and symmetric modes are given by Eqs. (8.39) and (8.42), respectively.

For resonator applications, we are interested in long waves in a narrow wedge with a small α. For a narrow wedge, from Eq. (8.41), ν is large. We have Bessel functions with large orders. By long waves we mean that, at a finite r, the wavelength is much larger than the local wedge thickness:

$$\frac{2\pi}{\xi} >> 2\alpha r, \quad \text{or} \quad \xi r << \frac{\pi}{\alpha}. \tag{8.44}$$

In this case ξr has to be small or no more than being finite.

For the lowest symmetric mode with $n = 0$ (face-shear), from Eq. (8.42) $\nu = 0$ irrespective of what α is. Since $J_0(0) = 1$, the tip of the edge is vibrating and this mode is not trapped.

All other modes ($n = 1, 2, 3, 4, \ldots$), symmetric or antisymmetric, may be called thickness-twist modes. For these modes ν is positive. For small arguments, Bessel functions have the following asymptotic expression

$$J_\nu(x) \cong \frac{x^\nu}{2^\nu \Gamma(1 + \nu)}, \tag{8.45}$$

which shows that $J_\nu(0) = 0$ and the modes grow from the tip of the wedge. Therefore all thickness-twist modes show energy trapping because they vanish at the wedge tip and grow away from there. The growing rate is small for a large ν, or a small α. Since Bessel functions decay to zero for large arguments, A trapped mode vanishes both at the wedge tip and at infinity when C_3 is taken to be zero.

The above observations are based on the displacement field. It is also informative to examine the stress, strain, electric field and electric displacement fields. For a general measure of all fields we consider the internal energy density. For example, for the antisymmetric modes in Eq. (8.39), from the real parts of the fields we obtain the internal energy density as

$$U = \frac{1}{2}\bar{c}C_1^2 \left\{ \xi^2 \left[J_\nu'(\xi r) \right]^2 \sin^2 \nu\theta + \frac{\nu^2}{r^2} J_\nu^2(\xi r)\cos^2 \nu\theta \right\} \cos^2 \omega t. \tag{8.46}$$

Equation (8.46) shows that although the displacement vanishes at the wedge tip for any $v > 0$, the internal energy density vanishes at the wedge tip only when $v > 1$. This is not surprising in view of the fact that the energy density depends on the displacement gradient.

8.4. A Circular Cylindrical Panel

Consider a portion of a circular cylindrical shell with inner radius a and outer radius b as shown in Fig. 8.5 [53]. The two radii going through the edges of the shell form an angle 2β. β is less than π and is otherwise arbitrary. Mechanically the boundaries are all traction-free. Electrically the boundaries at $\theta = \pm\beta$ are unelectroded. The boundaries at $r = a, b$ may be either electroded or unelectroded. A cylindrical coordinate system is defined by $x = r\cos\theta$, $y = r\sin\theta$ and $z = z$. In the index notation below, (x, y, z) correspond to $(1, 2, 3)$.

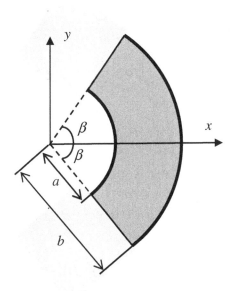

Fig. 8.5. A circular cylindrical panel.

8.4.1. Governing equations and fields

The governing equations are

$$v_T^2 \left(\frac{\partial^2 u}{\partial r^2} + \frac{1}{r}\frac{\partial u}{\partial r} + \frac{1}{r^2}\frac{\partial^2 u}{\partial \theta^2} \right) = \ddot{u},$$

$$\frac{\partial^2 \psi}{\partial r^2} + \frac{1}{r}\frac{\partial \psi}{\partial r} + \frac{1}{r^2}\frac{\partial^2 \psi}{\partial \theta^2} = 0. \tag{8.47}$$

We consider time-harmonic motions. All fields are with an $\exp(i\omega t)$ factor which will be dropped for convenience. According to whether u and ψ are even or odd functions of θ, we classify the modes into symmetric and antisymmetric. First consider the following symmetric modes:

$$u = u(r)\cos v\theta, \quad \psi = \psi(r)\cos v\theta. \tag{8.48}$$

Substitution of Eq. (8.48) into Eq. (8.47) results in

$$\frac{\partial^2 u}{\partial r^2} + \frac{1}{r}\frac{\partial u}{\partial r} + (\xi^2 - \frac{v^2}{r^2})u = 0,$$

$$\frac{\partial^2 \psi}{\partial r^2} + \frac{1}{r}\frac{\partial \psi}{\partial r} - \frac{v^2}{r^2}\psi = 0, \tag{8.49}$$

where we have denoted

$$\xi = \frac{\omega}{v_T}. \tag{8.50}$$

Equation (8.49)$_1$ can be written as Bessel's equation of order v. Equation (8.49)$_2$ allows a simple power function solution. When $v = 0$, we have

$$u = CJ_0(\xi r) + DY_0(\xi r),$$

$$\psi = F\ln r + G. \tag{8.51}$$

$$\phi = \psi + \frac{e}{\varepsilon}u = \frac{e}{\varepsilon}[CJ_0(\xi r) + DY_0(\xi r)] + F\ln r + G,$$

$$T_{rz} = \bar{c}u_{,r} + e\psi_{,r} = -\bar{c}\xi[CJ_1(\xi r) + DY_1(\xi r)] + e\frac{F}{r}, \tag{8.52}$$

$$D_r = -\varepsilon\psi_{,r} = -\varepsilon\frac{F}{r}.$$

When $v \neq 0$,

$$u = [CJ_v(\xi r) + DY_v(\xi r)]\cos v\theta,$$
$$\psi = [Fr^v + Gr^{-v}]\cos v\theta, \tag{8.53}$$

$$\phi = [\frac{e}{\varepsilon}CJ_v(\xi r) + \frac{e}{\varepsilon}DY_v(\xi r) + Fr^v + Gr^{-v}]\cos v\theta,$$

$$T_{rz} = [\bar{c}\xi CJ'_v(\xi r) + \bar{c}\xi DY'_v(\xi r) + evFr^{v-1} - evGr^{-v-1}]\cos v\theta, \tag{8.54}$$

$$T_{\theta z} = -\frac{v}{r}[\bar{c}CJ_v(\xi r) + \bar{c}DY_v(\xi r) + eFr^v + eGr^{-v}]]\sin v\theta,$$

$$D_r = [-\varepsilon vFr^{v-1} + \varepsilon vGr^{-v-1}]\cos v\theta,$$
$$D_\theta = \varepsilon\frac{v}{r}[Fr^v + Gr^{-v}]\sin v\theta, \tag{8.55}$$

where J_v and Y_v are the vth order Bessel functions of the first and second kinds. C, D, F and G are undetermined constants.

For traction-free and unelectroded edges at $\theta = \pm\beta$, we must have $T_{\theta z} = 0$ and $D_\theta = 0$, which implied that

$$u_{,\theta} = 0, \quad \psi_{,\theta} = 0, \quad \theta = \pm\beta. \tag{8.56}$$

From Eq. (8.56) we obtain

$$\sin v\beta = 0, \quad v\beta = m\pi, \quad v = v_m = \frac{m\pi}{\beta}, \quad m = 0, 1, 2, \cdots. \tag{8.57}$$

Similarly, for antisymmetric modes, we have

$$\cos v\beta = 0, \quad v\beta = (m + \frac{1}{2})\pi, \quad v = v_m = (m + \frac{1}{2})\frac{\pi}{\beta}, \quad m = 0, 1, 2, \cdots. \tag{8.58}$$

For antisymmetric modes $v \neq 0$. We will discuss unelectroded and electroded transducers separately below.

8.4.2. An unelectroded transducer

For traction-free and unelectroded boundaries at $r = a, b$, we must have

$$T_{rz} = 0 \quad \text{and} \quad D_r = 0, \tag{8.59}$$

or, equivalently,

$$u_{,r} = 0, \quad \psi_{,r} = 0, \quad r = a, b. \tag{8.60}$$

Then, when $\nu \neq 0$ or $m \neq 0$, we have

$$
\begin{aligned}
&C\xi J'_{\nu_m}(\xi a) + D\xi Y'_{\nu_m}(\xi a) = 0, \\
&F\nu_m a^{\nu_m - 1} - G\nu_m a^{-\nu_m - 1} = 0, \\
&C\xi J'_{\nu_m}(\xi b) + D\xi Y'_{\nu_m}(\xi b) = 0, \\
&F\nu_m b^{\nu_m - 1} - G\nu_m b^{-\nu_m - 1} = 0,
\end{aligned}
\tag{8.61}
$$

or, when $m = 0$:

$$
\begin{aligned}
&-\bar{c}\xi[CJ_1(\xi a) + DY_1(\xi a)] + e\frac{F}{a} = 0, \\
&-\bar{c}\xi[CJ_1(\xi b) + DY_1(\xi b)] + e\frac{F}{b} = 0, \\
&-\varepsilon\frac{F}{a} = 0, \\
&-\varepsilon\frac{F}{b} = 0.
\end{aligned}
\tag{8.62}
$$

For nontrivial solutions of C, D, F or G, the determinant of the coefficient matrix of Eqs. (8.61) or (8.62) must vanish, which gives the frequency equation that determines the resonant frequencies. We note that Eqs. $(8.61)_{1,3}$ are not coupled to Eqs. $(8.61)_{2,4}$. Therefore the frequency equation determined by Eq. (8.61) can be factored into two

$$
\begin{vmatrix}
\xi J'_{\nu_m}(\xi a) & \xi Y'_{\nu_m}(\xi a) \\
\xi J'_{\nu_m}(\xi b) & \xi Y'_{\nu_m}(\xi b)
\end{vmatrix} = 0,
\tag{8.63}
$$

$$
\begin{vmatrix}
\nu_m a^{\nu_m - 1} & -\nu_m a^{-\nu_m - 1} \\
\nu_m b^{\nu_m - 1} & -\nu_m b^{-\nu_m - 1}
\end{vmatrix} = 0.
\tag{8.64}
$$

Equation (8.64) has no solution when $m \neq 0$. In the case of Eq. (8.62), we have $F = 0$ which implies a constant ψ. In practice we are more interested in the case of an electroded plate.

8.4.3. An electroded transducer

When the transducer is electroded at $r = a$, b and the electrodes are shorted and grounded, the electrical boundary conditions at $r = a$, b are that $\phi = 0$. When $m = 0$ we have

$$-\overline{c}\xi[CJ_1(\xi a) + DY_1(\xi a)] + e\frac{F}{a} = 0,$$

$$-\overline{c}\xi[CJ_1(\xi b) + DY_1(\xi b)] + e\frac{F}{b} = 0,$$

$$\frac{e}{\varepsilon}[CJ_0(\xi a) + DY_0(\xi a)] + F\ln a + G = 0, \qquad (8.65)$$

$$\frac{e}{\varepsilon}[CJ_0(\xi b) + DY_0(\xi b)] + F\ln b + G = 0.$$

When $m \neq 0$, we have

$$\frac{e}{\varepsilon}CJ_{v_m}(\xi a) + \frac{e}{\varepsilon}DY_{v_m}(\xi a) + Fa^{v_m} + Ga^{-v_m} = 0,$$

$$\overline{c}\xi CJ'_{v_m}(\xi a) + \overline{c}\xi DY'_{v_m}(\xi a) + ev_m Fa^{v_m-1} - ev_m Ga^{-v_m-1} = 0,$$

$$\frac{e}{\varepsilon}CJ_{v_m}(\xi b) + \frac{e}{\varepsilon}DY_{v_m}(\xi b) + Fb^{v_m} + Gb^{-v_m} = 0, \qquad (8.66)$$

$$\overline{c}\xi CJ'_{v_m}(\xi b) + \overline{c}\xi DY'_{v_m}(\xi b) + ev_m Fb^{v_m-1} - ev_m Gb^{-v_m-1} = 0.$$

The frequency equation is obtained by setting the determinant of the coefficient matrix of the above systems of linear equations to zero.

8.4.4. Numerical results

As a numerical example, for geometric parameters we choose $a = 0.2$ m, $b = 0.21$ m, $\beta = \pi/6$. For the material we consider PZT-7A.

Figure 8.6 shows resonant frequencies versus m. m is effectively the wave number in the θ direction. Therefore, in Fig. 8.6, if the dots are properly connected in the horizontal direction the resulting curves resemble the dispersion curves of SH waves in plates or circular cylindrical shells, but here the wave number and frequencies are discrete. Frequencies with $m = 0$ are called thickness-shear frequencies whose modes have no θ dependence. When m is positive the frequencies are called thickness-twist frequencies whose modes have θ dependence.

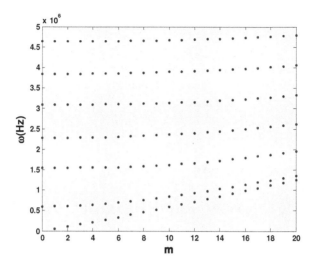

Fig. 8.6. Resonant frequencies.

Figures 8.7 through 8.10 show the displacement and electric potential distributions of the first two modes with $m = 0$. Due to symmetry only the fields in the first quadrant of Fig. 8.5 with $0 \le \theta \le \beta$ are plotted. For a fixed m, modes with more nodal points in the r direction have higher frequencies. Increasing m introduces more nodal points in the θ direction.

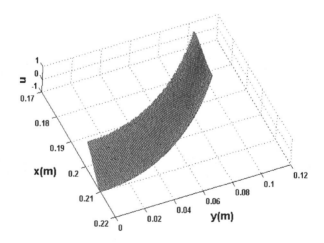

Fig. 8.7. Displacement for $m = 0$, $\omega = 0.60 \times 10^6$ Hz .

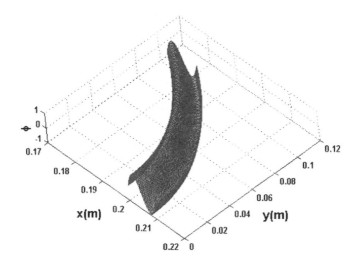

Fig. 8.8. Electric potential for $m = 0$, $\omega = 0.60 \times 10^6$ Hz .

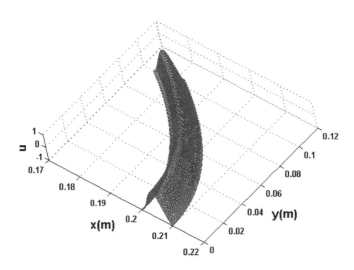

Fig. 8.9. Displacement for $m = 0$, $\omega = 1.55 \times 10^6$ Hz .

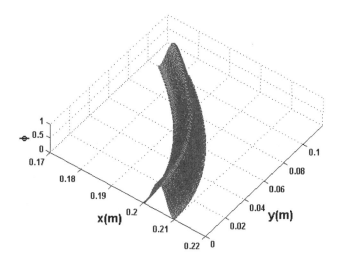

Fig. 8.10. Electric potential for $m = 0$, $\omega = 1.55 \times 10^6$ Hz .

8.5. An Elliptical Cylinder

This section is mainly for posing a problem. The solution to this problem, when obtained, will serve as a good example for the phenomenon of energy trapping. Consider an elliptical cylinder of ceramics poled in the x_3 direction as shown in Fig. 8.11. The surface is traction-free and is unelectroded. The electric field that may exist in the surrounding free space is neglected.

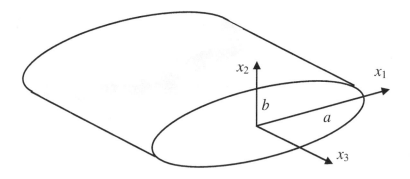

Fig. 8.11. An elliptical cylinder and coordinate system.

The governing equations are

$$v_T^2 \nabla^2 u = \ddot{u}, \quad \nabla^2 \psi = 0, \quad \phi = \psi + \frac{e}{\varepsilon}u \,. \tag{8.67}$$

In elliptical coordinates defined by

$$x_1 = h\cosh\xi\cos\eta, \quad x_2 = h\sinh\xi\sin\eta, \tag{8.68}$$

Eq. (8.67) changes to

$$\frac{\partial^2 u}{\partial \xi^2} + \frac{\partial^2 u}{\partial \eta^2} + 2q(\cosh 2\xi - \cos 2\eta)u = 0, \tag{8.69}$$

and

$$\frac{\partial^2 \psi}{\partial \xi^2} + \frac{\partial^2 \psi}{\partial \eta^2} = 0, \tag{8.70}$$

where $2h$ is the focal distance of the ellipses, and

$$h = \sqrt{a^2 - b^2} \,. \tag{8.71}$$

a and b are the semi-major and -minor axes. The coefficient q is given by

$$q = \frac{\omega^2 h^2}{4v_T^2} \,. \tag{8.72}$$

The solutions to Eqs. (8.69) and (8.70) can be formally obtained. Consider the possibility of the following fields which are odd about the major axis and may be called antisymmetric modes:

$$u(\xi,\eta,t) = ASe_m(\xi,q)se_m(\eta,q)\exp(-i\omega t),$$
$$\psi(\xi,\eta,t) = B\cos N\xi\left(\sinh N\eta + \cosh N\eta\right)\exp(-i\omega t), \tag{8.73}$$

where A and B are undetermined constants. se_m and Se_m are the mth order Mathieu functions and modified Mathieu functions of the sine type. The stress components and the electric displacement components can be obtained accordingly to satisfy boundary conditions. Similarly, symmetric modes can be written as

$$u(\xi,\eta,t) = ACe_m(\xi,q)ce_m(\eta,q)\exp(-i\omega t),$$
$$\psi(\xi,\eta,t) = B\sin N\xi\left(\sinh N\eta + \cosh N\eta\right)\eta\exp(-i\omega t), \tag{8.74}$$

where ce_m and Ce_m are the mth order Mathieu functions and modified Mathieu functions of the cosine type.

Chapter 9
Forced Vibrations in Cartesian Coordinates

In the rest of the book we study forced vibrations. For different applications, a piezoelectric body may be driven electrically or mechanically. For resonators, although free-vibration frequency analysis is fundamental, to calculate the motional capacitance of a resonator an electrically forced vibration analysis is needed. For power handling devices like generators (power harvesters) or transformers, a forced vibration is necessary to obtain the electrical output.

9.1. Thickness-shear in a Plate Driven by a Voltage

Consider an unbounded plate (see Fig. 9.1). The plate surfaces at $x_2 = \pm h$ are traction-free and are electroded. A time-harmonic voltage is applied across the plate thickness.

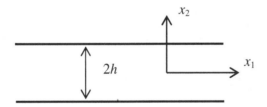

Fig. 9.1. An infinite plate under a driving voltage.

The governing equations and boundary conditions are:

$$c\nabla^2 u + e\nabla^2 \phi = \rho \ddot{u},$$
$$e\nabla^2 u - \varepsilon\nabla^2 \phi = 0,$$
$$T_{23} = 0, \quad x_2 = \pm h,$$
$$\phi(x_2 = h) - \phi(x_2 = -h) = V\exp(i\omega t).$$

(9.1)

From Sec. 7.1 we have (see Eqs. (7.10)–(7.12))

$$u = A_1 \sin \xi x_2 + A_2 \cos \xi x_2,$$
$$\phi = \frac{e}{\varepsilon}(A_1 \sin \xi x_2 + A_2 \cos \xi x_2) + B_1 x_2 + B_2, \tag{9.2}$$

$$T_{23} = \bar{c}(A_1 \xi \cos \xi x_3 - A_2 \xi \sin \xi x_3) + eB_1,$$
$$D_2 = -\varepsilon B_1, \tag{9.3}$$

where B_1, B_2, A_1 and A_2 are integration constants, and

$$\xi^2 = \frac{\rho}{\bar{c}} \omega^2. \tag{9.4}$$

The boundary conditions require that

$$\bar{c}A_1 \xi \cos \xi h - \bar{c}A_2 \xi \sin \xi h + eB_1 = 0,$$
$$\bar{c}A_1 \xi \cos \xi h + \bar{c}A_2 \xi \sin \xi h + eB_1 = 0, \tag{9.5}$$

$$2\frac{e}{\varepsilon} A_1 \sin \xi h + 2B_1 h = V,$$

or, add the first two, and subtract the first two from each other:

$$\bar{c}A_1 \xi \cos \xi h + eB_1 = 0,$$
$$\bar{c}A_2 \xi \sin \xi h = 0, \tag{9.6}$$

$$2\frac{e}{\varepsilon} A_1 \sin \xi h + 2B_1 h = V.$$

We have $A_2 = 0$ which means that the symmetric modes cannot be excited, and

$$A_1 = \frac{\begin{vmatrix} 0 & e \\ V & 2h \end{vmatrix}}{\begin{vmatrix} \bar{c}\xi \cos \xi h & e \\ 2\frac{e}{\varepsilon} \sin \xi h & 2h \end{vmatrix}} = \frac{\div eV}{2\bar{c}\xi h \cos \xi h - 2\frac{e^2}{\varepsilon} \sin \xi h}, \tag{9.7}$$

$$B_1 = \frac{\begin{vmatrix} \bar{c}\xi \cos \xi h & 0 \\ 2\frac{e}{\varepsilon} \sin \xi h & V \end{vmatrix}}{\begin{vmatrix} \bar{c}\xi \cos \xi h & e \\ 2\frac{e}{\varepsilon} \sin \xi h & 2h \end{vmatrix}} = \frac{V\bar{c}\xi \cos \xi h}{2\bar{c}\xi h \cos \xi h - 2\frac{e^2}{\varepsilon} \sin \xi h}. \tag{9.8}$$

Hence

$$D_2 = -\varepsilon B_1 = -\varepsilon \frac{V}{2h} \frac{\xi h}{\xi h - \bar{k}^2 \tan \xi h} = -\sigma, \tag{9.9}$$

where σ is the surface free charge per unit area on the electrode at $x_2 = h$. The denominator of Eq. (9.9), if set to zero, gives the frequency equation for the antisymmetric modes (see Eq. (7.22)). The capacitance per unit area is

$$C = \frac{\sigma}{V} = \frac{\varepsilon}{2h} \frac{\xi h}{\xi h - \bar{k}^2 \tan \xi h}. \tag{9.10}$$

We note the following limits:

$$\lim_{e \to 0} C = \frac{\varepsilon}{2h},$$

$$\lim_{\omega \to 0} C = \frac{\varepsilon}{2h} \frac{1}{1 - \dfrac{k^2}{1 + k^2}} = \frac{\varepsilon}{2h}(1 + k^2). \tag{9.11}$$

9.2. Thickness-shear in a Plate Driven by Traction: A Generator

Consider an unbounded plate as shown in Fig. 9.2. The surfaces of the plate are electroded and the electrodes are connected by a circuit whose impedance is denoted by Z when the motion is time-harmonic. When the plate is driven by a time-harmonic tangential surface traction p, there is an electric response in the circuit. The structure is one of the simplest for converting mechanical work into electric output and may be called a generator or power harvestor.

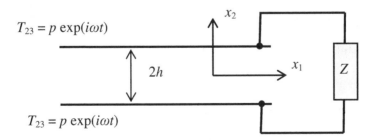

$T_{23} = p \exp(i\omega t)$

$2h$

$T_{23} = p \exp(i\omega t)$

Fig. 9.2. An infinite plate under surface traction.

The governing equations are:

$$c\nabla^2 u + e\nabla^2 \phi = \rho\ddot{u},$$
$$e\nabla^2 u - \varepsilon\nabla^2 \phi = 0. \tag{9.12}$$

The mechanical boundary conditions are

$$T_{23} = p\exp(i\omega t), \quad x_2 = \pm h. \tag{9.13}$$

On each of the two electrodes, the electric potential is spatially constant, varying with time only. We denote the potential difference between the two electrodes by a voltage V

$$\phi(x_2 = h) - \phi(x_2 = -h) = V. \tag{9.14}$$

The free charge per unit area on the electrode at $x_2 = h$ and the current per unit area that flows out of the electrode are

$$Q_e = -D_2, \quad I = -\dot{Q}_e. \tag{9.15}$$

For time-harmonic motions, we use the following complex notation:

$$\{V, Q_e, I\} = \text{Re}\{(\bar{V}, \bar{Q}_e, \bar{I})\exp(i\omega t)\}. \tag{9.16}$$

Then

$$\bar{I} = -i\omega\bar{Q}_e. \tag{9.17}$$

We have the following relation for the output circuit:

$$\bar{I} = \bar{V} / Z. \tag{9.18}$$

From Sec. 7.1 we have (see Eqs. (7.10)–(7.12))

$$u = A_1 \sin\xi x_2 + A_2 \cos\xi x_2,$$
$$\phi = \frac{e}{\varepsilon}(A_1 \sin\xi x_2 + A_2 \cos\xi x_2) + B_1 x_2 + B_2, \tag{9.19}$$

$$T_{23} = \bar{c}(A_1\xi\cos\xi x_3 - A_2\xi\sin\xi x_3) + eB_1,$$
$$D_2 = -\varepsilon B_1, \tag{9.20}$$

where B_1, B_2, A_1 and A_2 are integration constants, and

$$\xi^2 = \frac{\rho}{\bar{c}}\omega^2. \tag{9.21}$$

The boundary conditions require that

$$\bar{c}A_1\xi\cos\xi h - \bar{c}A_2\xi\sin\xi h + eB_1 = p,$$
$$\bar{c}A_1\xi\cos\xi h + \bar{c}A_2\xi\sin\xi h + eB_1 = p. \tag{9.22}$$

Adding and subtracting the two equations in Eq. (9.22), respectively, yield the following:

$$\bar{c}A_1\xi\cos\xi h + eB_1 = p,$$
$$\bar{c}A_2\xi\sin\xi h = 0, \tag{9.23}$$

which implies that $A_2 = 0$. With Eq. (9.20), we can write the charge and current as

$$\bar{Q}_e = \varepsilon B_1, \quad \bar{I} = -i\omega\varepsilon B_1. \tag{9.24}$$

Then the circuit condition in Eq. (9.17) takes the following form:

$$-i\omega\varepsilon B_1 = \bar{V} / Z. \tag{9.25}$$

The voltage across the electrodes is given by

$$\phi(h) - \phi(-h) = 2(\frac{e}{\varepsilon}A_1\sin\xi h + B_1 h) = \bar{V}. \tag{9.26}$$

Equations (9.23)$_1$, (9.25) and (9.26) are three equations for A_1, B_1 and $\bar{V}$. Solving these equations, we obtain

$$A_1 = \frac{p}{\bar{c}}h\frac{1}{\xi h\cot\xi h - \dfrac{\bar{k}^2}{1 + Z/Z_0}}\frac{1}{\sin\xi h},$$

$$B_1 = -\frac{e}{\varepsilon}\frac{p}{\bar{c}}\frac{1}{(1 + Z/Z_0)\xi h\cot\xi h - \bar{k}^2}, \tag{9.27}$$

$$\bar{V} = i\omega e\frac{p}{\bar{c}}\frac{Z}{(1 + Z/Z_0)\xi h\cot\xi h - \bar{k}^2},$$

where

$$Z_0 = \frac{1}{i\omega C_0}, \quad C_0 = \frac{\varepsilon}{2h}. \tag{9.28}$$

In the above, C_0 is the static capacitance per unit area of the plate. From Eq. (9.27)$_1$ the resonant frequencies are determined by:

$$\xi h\cot\xi h - \frac{\bar{k}^2}{1 + Z/Z_0} = 0. \tag{9.29}$$

From Eqs. (9.24) and (9.27) we have

$$\bar{I} = i\omega e \frac{p}{\bar{c}} \frac{1}{(1 + Z/Z_0)\xi h \cot \xi h - \bar{k}^2}. \tag{9.30}$$

With the complex notation and the circuit condition, the average output electrical power per unit plate area over a period is given by

$$P_2 = \frac{1}{4}(\overline{IV}^* + \overline{I}^*\overline{V}) = \frac{1}{4}\overline{I}\,\overline{I}^*(Z + Z^*) = \frac{1}{2}|\overline{I}|^2 \text{Re}\{Z\}$$

$$= \frac{1}{2}p^2\omega^2 \frac{\varepsilon|\bar{k}^2|}{|\bar{c}|} \text{Re}\{Z\}\left|\frac{1}{(1 + Z/Z_0)\xi h \cot \xi h - \bar{k}^2}\right|^2, \tag{9.31}$$

where an asterisk represents complex conjugate. Equation (9.31) shows that the output power depends on the input stress amplitude p quadratically, as expected. The output power also explicitly depends on ω^2, ε and $|\bar{k}^2|$ linearly, and is inversely proportional to $\bar{c}$. Therefore, a large dielectric constant ε, a large electromechanical coupling factor $\bar{k}$, and a small stiffness $\bar{c}$ are helpful for raising the output power. ω and $\bar{k}^2$, and hence p, ε, e and $\bar{c}$ are also present implicitly in the last factor in Eq. (9.31). When the driving frequency is near a resonant frequency, the denominator of this factor is very small, leading to a large output power. We further see from Eq. (9.31) that the output power P_2 depends on $\text{Re}\{Z\}$ linearly for a small load impedance Z, and it diminishes as Z becomes large. The output power P_2 depends only on the real part of the load impedance (resistance), because its imaginary part represents a capacitance or inductance which takes away energy during half of a period and gives the energy back during the other half with no net energy consumption when averaged in a period.

To calculate the mechanical input power we need the velocities at the plate surfaces

$$v_3(x_2 = \pm h) = \pm i\omega A_1 \sin \xi h$$

$$= \pm i\omega \frac{p}{\bar{c}} h \frac{1}{\xi h \cot \xi h - \dfrac{\bar{k}^2}{1 + Z/Z_0}}. \tag{9.32}$$

Therefore the input mechanical power averaged over a period is

$$P_1 = 2\frac{1}{4}(pv_3^* + p^* v_3) = \frac{1}{2} p(v_3^* + v_3) = p\,\mathrm{Re}\{v_3\}$$

$$= \omega p^2 h \frac{\mathrm{Im}\{\overline{c}^*(\dfrac{\overline{k}^2}{1 + Z/Z_0} - \xi h \cot \xi h)^*\}}{\left|\overline{c}\right|^2 \left|\dfrac{\overline{k}^2}{1 + Z/Z_0} - \xi h \cot \xi h\right|^2}. \tag{9.33}$$

The efficiency of the generator, which measures its performance in converting the input mechanical power into the output electric power, is defined as

$$\eta = \frac{P_2}{P_1} = \frac{1}{2h} \frac{\omega \varepsilon \left|\overline{k}^2\right| \mathrm{Re}\{Z\}}{\dfrac{\left|(1 + Z/Z_0)\right|^2}{\left|\overline{c}\right|} \mathrm{Im}\{\overline{c}^*(\dfrac{\overline{k}^2}{1 + Z/Z_0} - \xi h \cot \xi h)^*\}}. \tag{9.34}$$

The efficiency depends on the load impedance Z linearly for a small load. It diminishes with a large Z. It also becomes large at the resonant frequencies. Another quantity of practical interest is the power density defined as the output power per unit volume. In our case, it is given by

$$p_2 = \frac{P_2}{2h}. \tag{9.35}$$

9.3. Thickness-twist in a Plate Driven by Traction: A Generator

Consider a finite rectangular plate under surface mechanical loads (see Fig. 9.3) [54]. The major faces at $x_2 = \pm h$ are electroded and are under a time-harmonic tangential driving traction τ, where τ is real. Note that τ is antisymmetric about the x_2 axis. Because of the antisymmetric τ we use electrodes that break at $x_1 = 0$. Driven by τ there is an output voltage V across the electrodes. The two electrodes with the potential $V/2$ are connected, and the other two with $-V/2$ are connected. The electrodes are joined by an output circuit whose impedance is Z. The two minor faces at $x_1 = \pm a$ are traction-free and are unelectroded. The structure can be used as an example of a generator. More complicated than the generator in the previous section, this generator has fields with x_1 dependence.

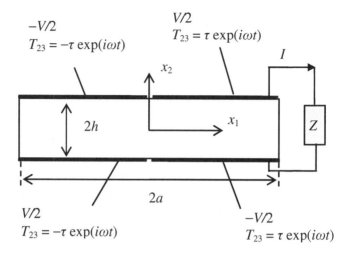

Fig. 9.3. A finite plate under mechanical loads as a generator.

9.3.1. Governing equations

The governing equations for u, ψ and ϕ are

$$\bar{c}\nabla^2 u = \rho\ddot{u},$$
$$\nabla^2\psi = 0,$$
$$\phi = \psi + \frac{e}{\varepsilon}u.$$

(9.36)

The boundary conditions are:

$$T_{23} = -\tau\exp(i\omega t), \quad \phi = -\frac{V}{2}, \quad x_2 = h, \quad -a < x_1 < 0,$$

$$T_{23} = \tau\exp(i\omega t), \quad \phi = \frac{V}{2}, \quad x_2 = h, \quad 0 < x_1 < a,$$

$$T_{23} = -\tau\exp(i\omega t), \quad \phi = \frac{V}{2}, \quad x_2 = -h, \quad -a < x_1 < 0,$$

$$T_{23} = \tau\exp(i\omega t), \quad \phi = -\frac{V}{2}, \quad x_2 = -h, \quad 0 < x_1 < a,$$

$$T_{13} = 0, \quad D_1 = 0, \quad x_1 = \pm a,$$

(9.37)

or, equivalently, in terms of u and ψ,

$$\bar{c}u_{,2} + e\psi_{,2} = -\tau\exp(i\omega t), \quad \psi + \frac{e}{\varepsilon}u = -\frac{V}{2}, \quad x_2 = h, \quad -a < x_1 < 0,$$

$$\bar{c}u_{,2} + e\psi_{,2} = \tau\exp(i\omega t), \quad \psi + \frac{e}{\varepsilon}u = \frac{V}{2}, \quad x_2 = h, \quad 0 < x_1 < a,$$

$$\bar{c}u_{,2} + e\psi_{,2} = -\tau\exp(i\omega t), \quad \psi + \frac{e}{\varepsilon}u = \frac{V}{2}, \quad x_2 = -h, \quad -a < x_1 < 0, \quad (9.38)$$

$$\bar{c}u_{,2} + e\psi_{,2} = \tau\exp(i\omega t), \quad \psi + \frac{e}{\varepsilon}u = -\frac{V}{2}, \quad x_2 = -h, \quad 0 < x_1 < a,$$

$$u_{,1} = 0, \quad \psi_{,1} = 0, \quad x_1 = \pm a.$$

In Eq. (9.38), V is unknown. We also need the equation for the output circuit to determine V. To obtain the circuit equation we first calculate the charge Q_e and the current I on the electrode at $x_2 = h$ with $0 < x_1 < a$, and the electrode at $x_2 = -h$ with $-a < x_1 < 0$. These two electrodes are connected. The total charge on these two electrodes and the total current flowing out of them are:

$$Q_e = 2\int_0^a -D_2\big|_{x_2=h}\,dx_1,$$

$$I = -\dot{Q}_e. \qquad (9.39)$$

For harmonic motions we use the complex notation:

$$(u,\psi,\phi,T_{23},D_2,V,I,Q_e)$$
$$= \mathrm{Re}\{(U,\Psi,\Phi,T,D,\bar{V},\bar{I},\bar{Q}_e)\exp(i\omega t)\}. \qquad (9.40)$$

Then the charge, the current and the circuit equation can be written as

$$\bar{Q}_e = 2\int_0^a -D\big|_{x_2=h}\,dx_1,$$

$$\bar{I} = -i\omega\bar{Q}_e, \qquad (9.41)$$

$$\bar{V} = \bar{I}Z.$$

The output power is given by

$$P_2 = \frac{1}{4}(\bar{I}\bar{V}^* + \bar{I}^*\bar{V}) = \frac{1}{4}(\bar{I}\bar{I}^*Z^* + \bar{I}^*\bar{I}Z)$$

$$= \frac{1}{4}\bar{I}\bar{I}^*(Z^* + Z) = \frac{1}{2}|\bar{I}|^2\,\mathrm{Re}\{Z\}, \qquad (9.42)$$

where an asterisk is for complex conjugate. The input power is given by

$$
\begin{aligned}
P_1 &= 4 \int_0^a \left. (T_{23} \dot{u}) \right|_{x_2=h} dx_1 \\
&= \int_0^a \left. [T(i\omega U)^* + T^*(i\omega U)] \right|_{x_2=h} dx_1 \\
&= \int_0^a \left. -2\tau\omega \mathrm{Im}(U) \right|_{x_2=h} dx_1.
\end{aligned}
\tag{9.43}
$$

The efficiency and the power density are defined by

$$
\eta = \frac{P_2}{P_1}, \quad p = \frac{P_2}{4ah}.
\tag{9.44}
$$

9.3.2. Series solution

For the displacement and potential fields, we only need modes that are antisymmetric in x_1 and x_2. From the modes in Sec. 7.8, we use $\sin \xi_{(m)} x_1$ (where $\xi_{(m)} = (m+1/2)\pi/a$, $m = 0, 1, 2, \cdots$) as base functions in the x_1 direction to construct the following solution:

$$
\begin{aligned}
U &= \sum_{m=0}^{\infty} A^{(m)} \sin \eta_{(m)} x_2 \sin \xi_{(m)} x_1, \\
\Psi &= \sum_{m=0}^{\infty} B^{(m)} \sinh \xi_{(m)} x_2 \sin \xi_{(m)} x_1,
\end{aligned}
\tag{9.45}
$$

where $A_{(m)}$ and $B_{(m)}$ are undetermined constants, and

$$
\eta_{(m)}^2 = \frac{\rho \omega^2}{\overline{c}} - \left(\frac{m+1/2}{a} \pi \right)^2, \quad m = 0, 1, 2, \cdots.
\tag{9.46}
$$

Equation (9.45) satisfies Eq. (9.36) and the boundary conditions at $x_1 = \pm a$. To apply the boundary conditions at $x_2 = \pm h$, we calculate the following:

$$
\begin{aligned}
\Phi &= \sum_{m=0}^{\infty} (B^{(m)} \sinh \xi_{(m)} x_2 + \frac{e}{\varepsilon} A^{(m)} \sin \eta_{(m)} x_2) \sin \xi_{(m)} x_1, \\
D &= -\varepsilon \sum_{m=0}^{\infty} B^{(m)} \xi_{(m)} \cosh \xi_{(m)} x_2 \sin \xi_{(m)} x_1, \\
T &= \sum_{m=0}^{\infty} (\overline{c} A^{(m)} \eta_{(m)} \cos \eta_{(m)} x_2 + e B^{(m)} \xi_{(m)} \cosh \xi_{(m)} x_2) \sin \xi_{(m)} x_1.
\end{aligned}
\tag{9.47}
$$

Substituting Eq. (9.47) into the boundary conditions at $x_2 = h$ in Eq. (9.38) gives the following equations for $A_{(m)}$ and $B_{(m)}$:

$$\sum_{m=0}^{\infty} (B^{(m)} \sinh \xi_{(m)} h + \frac{e}{\varepsilon} A^{(m)} \sin \eta_{(m)} h) \sin \xi_{(m)} x_1$$

$$= \begin{cases} -\bar{V}/2, & -a < x_1 < 0, \\ \bar{V}/2, & 0 < x_1 < a, \end{cases}$$

$$\sum_{m=0}^{\infty} (\bar{c} A^{(m)} \eta_{(m)} \cos \eta_{(m)} h + e B^{(m)} \xi_{(m)} \cosh \xi_{(m)} h) \sin \xi_{(m)} x_1 \tag{9.48}$$

$$= \begin{cases} -\tau, & -a < x_1 < 0, \\ \tau, & 0 < x_1 < a. \end{cases}$$

Due to antisymmetry the boundary conditions at $x_2 = -h$ do not provide new equations other than Eq. (9.48). Multiplying both sides of Eq. (9.48) by $\sin \xi_{(n)} x_1$ and integrating them over $[-a, a]$, with the orthogonality of $\sin \xi_{(m)} x_1$ over $[0, a]$, we obtain

$$(B^{(m)} \sinh \xi_{(m)} h + \frac{e}{\varepsilon} A^{(m)} \sin \eta_{(m)} h) a = \frac{\bar{V}}{\xi_{(m)}},$$

$$(\bar{c} A^{(m)} \eta_{(m)} \cos \eta_{(m)} h + e B^{(m)} \xi_{(m)} \cosh \xi_{(m)} h) a = \frac{2\tau}{\xi_{(m)}}. \tag{9.49}$$

From Eq. (9.49) we solve for

$$A^{(m)} = \alpha_1^{(m)} \bar{V} + \alpha_2^{(m)} \tau,$$

$$B^{(m)} = \beta_1^{(m)} \bar{V} + \beta_2^{(m)} \tau, \tag{9.50}$$

where

$$\alpha_1^{(m)} = \frac{e \cosh \xi_{(m)} h}{a \Delta_{(m)}}, \quad \alpha_2^{(m)} = -\frac{2 \sinh \xi_{(m)} h}{\xi_{(m)} a \Delta_{(m)}},$$

$$\beta_1^{(m)} = -\frac{\bar{c} \eta_{(m)} \cos \eta_{(m)} h}{\xi_{(m)} a \Delta_{(m)}}, \quad \beta_2^{(m)} = \frac{2e \sin \eta_{(m)} h}{\varepsilon \xi_{(m)} a \Delta_{(m)}}, \tag{9.51}$$

$$\Delta_{(m)} = \frac{e^2}{\varepsilon} \xi_{(m)} \sin \eta_{(m)} h \cosh \xi_{(m)} h - \bar{c} \eta_{(m)} \cos \eta_{(m)} h \sinh \xi_{(m)} h.$$

Since $\overline{V}$ is unknown, Eq. (9.50) has not fully determined $A_{(m)}$ and $B_{(m)}$ yet. From Eqs. (9.41)$_{1,2}$, Eq. (9.47)$_2$ and Eq. (9.50)$_2$ we calculate the current and substitute the result into the circuit condition in Eq. (9.41)$_3$ to obtain

$$\frac{\overline{V}}{Z} = \overline{I} = -i\omega 2\varepsilon \sum_{m=0}^{\infty} (\beta_1^{(m)}\overline{V} + \beta_2^{(m)}\tau)\cosh\xi_{(m)}h$$

$$= \beta_1\overline{V} + \beta_2\tau,$$

(9.52)

where

$$\beta_1 = -i\omega 2\varepsilon \sum_{m=0}^{\infty} \beta_1^{(m)}\cosh\xi_{(m)}h,$$

$$\beta_2 = -i\omega 2\varepsilon \sum_{m=0}^{\infty} \beta_2^{(m)}\cosh\xi_{(m)}h.$$

(9.53)

Equation (9.52) determines $\overline{V}$ in terms of τ and completes the solution:

$$\overline{V} = \frac{\beta_2 Z}{1 - \beta_1 Z}\tau.$$

(9.54)

Equation (9.54) shows that the output voltage is linear in τ as a consequence of the linearity of the system. It is also proportional to Z for small Z, and approaches a constant for large Z (voltage saturation at open circuit). Once $\overline{V}$ is known, from the circuit equation we have

$$\overline{I} = \frac{\beta_2}{1 - \beta_1 Z}\tau.$$

(9.55)

Equation (9.55) shows that the output current approaches zero for large Z (open circuit). The input power can be written as

$$P_1 = \int_0^a -2\tau\omega\,\mathrm{Im}[\sum_{m=0}^{\infty} (\alpha_1^{(m)}\overline{V} + \alpha_2^{(m)}\tau)\sin\eta_{(m)}h\sin\xi_{(m)}x_1]dx_1$$

$$= \alpha\tau^2,$$

(9.56)

where

$$\alpha = -2\omega\,\mathrm{Im}[\sum_{m=0}^{\infty} (\alpha_1^{(m)}\frac{\beta_2 Z}{1 - \beta_1 Z} + \alpha_2^{(m)})\frac{\sin\eta_{(m)}h}{\xi_{(m)}}].$$

(9.57)

P_1 is quadratic in τ as expected.

9.3.3. Numerical results

As an example, consider a generator made from PZT-5H. Some damping is introduced by allowing the elastic material constant c to assume complex values, which can represent viscous damping in the material. In our calculations c is replaced by $c(1 + iQ^{-1})$ where Q is a large and real number. For polarized ceramics the value of Q is of the order of 10^2 to 10^3. We fix $Q = 100$. $\tau = 1$ N/m^2 is used in our calculation. The plate thickness is $h = 1$ cm. $a = 10\,h$. The load impedance Z is normalized by $Z_0 = 1 / (i\omega C_0)$, where $C_0 = \varepsilon / (2h)$ is the static capacitance per unit area of the plate when piezoelectric coupling is neglected.

We plot in Fig. 9.4 the output power density versus the driving frequency for different load impedance. The system has infinitely many resonant frequencies. Only the behavior near the first resonance is shown, which is the operating frequency. Near resonances the output has maxima. This shows that a generator is a resonant device operating near a resonant frequency only. Thickness-twist is a so-called high-frequency mode whose frequency depends mainly on the plate thickness, the smallest dimension and hence the highest frequency for a plate.

Output power density versus load is shown in Fig. 9.5 for different driving frequencies near the first resonance. The figure shows that if the driving frequency is not close to a resonant frequency the output power is much less.

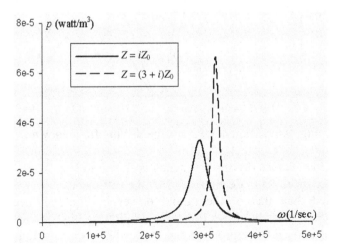

Fig. 9.4. Power density versus the driving frequency ω.

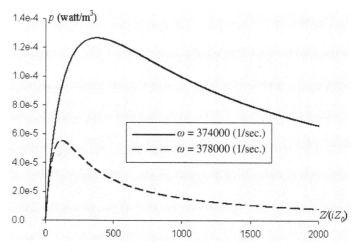

Fig. 9.5. Power density versus the load Z.

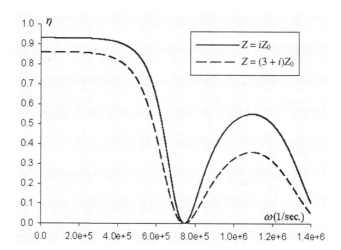

Fig. 9.6. Efficiency versus the driving frequency ω.

Efficiency versus driving frequency is shown in Fig. 9.6 for different load impedance. The dependence of the efficiency on ω is more complicated than that of the output power in Fig. 9.4, because the efficiency is a quotient of two functions. In plotting Figs. 9.4 and 9.6, the load Z was chosen to be proportional to Z_0 which is a function inversely proportional to ω. This has an effect on the behavior of the curves especially for a small ω, and is partially responsible for the different behaviors in Figs. 9.4 and 9.6 for small ω.

Efficiency as a function of Z is plotted in Fig. 9.7. In Figs. 9.6 and 9.7, the highest efficiency seems to be higher than the usual electromechanical coupling factor $\bar{k}$ of the material. This causes no contradiction because $\bar{k}$ is defined for static processes of energy conversion and does not directly apply to a dynamic problem. For time-harmonic motions the concept of a frequency dependent electromechanical coupling coefficient is needed.

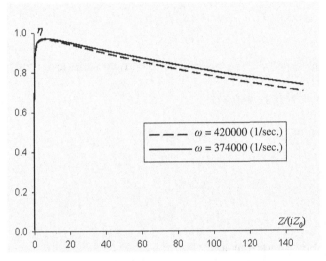

Fig. 9.7. Efficiency versus the load Z.

9.4. A Plate Transformer

Consider a rectangular plate as shown in Fig. 9.8 [55]. We consider unit thickness in the x_3 direction. The major faces at $x_2 = \pm h$ and the minor faces at $x_1 = \pm a$ are all traction-free. The faces at $x_2 = \pm h$ are electroded, with four electrodes shown by the thick lines. The two electrodes with $\pm V_1$ are the input electrodes under a known time-harmonic driving voltage. The two electrodes with $\pm V_2$ are the output electrodes connected by a circuit whose impedance is Z. The minor faces at $x_1 = \pm a$ are unelectroded.

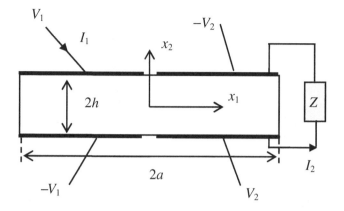

Fig. 9.8. A rectangular plate under electrical loads as a transformer.

9.4.1. Governing equations

The governing equations for u, ψ and ϕ are

$$\bar{c}\nabla^2 u = \rho\ddot{u},$$
$$\nabla^2\psi = 0, \qquad (9.58)$$
$$\phi = \psi + \frac{e}{\varepsilon}u.$$

The boundary conditions are

$$T_{23} = 0, \quad \phi = \pm V_1, \quad x_2 = \pm h, \quad -a < x_1 < 0,$$
$$T_{23} = 0, \quad \phi = \mp V_2, \quad x_2 = \pm h, \quad 0 < x_1 < a, \qquad (9.59)$$
$$T_{13} = 0, \quad D_1 = 0, \quad x_1 = \pm a,$$

or, equivalently, in terms of u and ψ,

$$\bar{c}u_{,2} + e\psi_{,2} = 0, \quad \psi + \frac{e}{\varepsilon}u = \pm V_1, \quad x_2 = \pm h, \quad -a < x_1 < 0,$$
$$\bar{c}u_{,2} + e\psi_{,2} = 0, \quad \psi + \frac{e}{\varepsilon}u = \mp V_2, \quad x_2 = \pm h, \quad 0 < x_1 < a, \qquad (9.60)$$
$$u_{,1} = 0, \quad \psi_{,1} = 0, \quad x_1 = \pm a.$$

To obtain the input and output currents and powers we need the charges and currents on the electrodes. For the input electrode at $x_2 = h$, $-a < x_1 < 0$, we have

$$Q_1 = \int_{-a}^{0} -D_2\big|_{x_2=h} dx_1,$$

(9.61)

$$I_1 = \dot{Q}_1.$$

Similarly, for the output electrode at $x_2 = -h$, $0 < x_1 < a$, we have

$$Q_2 = \int_{0}^{a} D_2\big|_{x_2=-h} dx_1,$$

(9.62)

$$I_2 = -\dot{Q}_2.$$

For harmonic motions we use the complex notation:

$$(u, \psi, \varphi, T_{23}, D_2, V_1, I_1, Q_1, V_2, I_2, Q_2)$$
$$= \text{Re}\{(U, \Psi, \Phi, T, D, \bar{V}_1, \bar{I}_1, \bar{Q}_1, \bar{V}_2, \bar{I}_2, \bar{Q}_2) \exp(i\omega t)\}.$$

(9.63)

We also need the equation for the output circuit:

$$2\bar{V}_2 = \bar{I}_2 Z.$$

(9.64)

The input and output powers are given by

$$P_1 = \frac{1}{4}(\bar{I}_1 2\bar{V}_1^* + \bar{I}_1^* 2\bar{V}_1),$$

$$P_2 = \frac{1}{4}(\bar{I}_2 2\bar{V}_2^* + \bar{I}_2^* 2\bar{V}_2) = \frac{1}{4}(\bar{I}_2 \bar{I}_2^* Z^* + \bar{I}_2^* \bar{I}_2 Z)$$

(9.65)

$$= \frac{1}{4}\bar{I}_2 \bar{I}_2^* (Z^* + Z) = \frac{1}{2}|\bar{I}_2|^2 \text{Re}\{Z\}.$$

The efficiency and power density of the transformer are defined by

$$\eta = \frac{P_2}{P_1}, \qquad p = \frac{P_2}{4ah}.$$

(9.66)

9.4.2. Series solution

We only need modes that are antisymmetric in x_2. In the x_1 direction, both symmetric and antisymmetric modes are needed due to the lack of symmetry about $x_2 = 0$. From the x_1 dependence of the modes in Sec. 7.8, we construct the following solution:

$$U = \hat{A}^{(0)} \sin \hat{\eta}_{(0)} x_2$$

$$+ \sum_{m=1}^{\infty} (A^{(m)} \sin \xi_{(m)} x_1 \sin \eta_{(m)} x_2 + \hat{A}^{(m)} \cos \hat{\xi}_{(m)} x_1 \sin \hat{\eta}_{(m)} x_2),$$

$$\Psi = \hat{B}^{(0)} x_2 \tag{9.67}$$

$$+ \sum_{m=1}^{\infty} (B^{(m)} \sin \xi_{(m)} x_1 \sinh \xi_{(m)} x_2 + \hat{B}^{(m)} \cos \hat{\xi}_{(m)} x_1 \sinh \hat{\xi}_{(m)} x_2),$$

where $A_{(m)}$, $\hat{A}_{(m)}$, $B_{(m)}$ and $\hat{B}_{(m)}$ are undetermined constants, and

$$\xi_{(m)} = \frac{m - 1/2}{a} \pi, \quad \eta_{(m)}^2 = \frac{\rho \omega^2}{\overline{c}} - \left(\frac{m - 1/2}{a} \pi \right)^2, \quad m = 1, 2, 3, \cdots,$$

$$\hat{\xi}_{(m)} = \frac{m}{a} \pi, \quad \hat{\eta}_{(m)}^2 = \frac{\rho \omega^2}{\overline{c}} - \left(\frac{m}{a} \pi \right)^2, \quad m = 1, 2, 3, \cdots, \tag{9.68}$$

$$\hat{\eta}_{(0)}^2 = \frac{\rho \omega^2}{\overline{c}}.$$

$A_{(m)}$ and $B_{(m)}$ are for modes antisymmetric in x_1. $\hat{A}_{(m)}$ and $\hat{B}_{(m)}$ are for modes symmetric in x_1. Equation (9.67) satisfies Eq. (9.58) and the boundary conditions at $x_1 = \pm a$. To apply the boundary conditions at $x_2 = \pm h$, we calculate the following:

$$\Phi = \hat{B}^{(0)} x_2 + \frac{e}{\varepsilon} \hat{A}^{(0)} \sin \hat{\eta}_{(0)} x_2$$

$$+ \sum_{m=1}^{\infty} [(B^{(m)} \sinh \xi_{(m)} x_2 + \frac{e}{\varepsilon} A^{(m)} \sin \eta_{(m)} x_2) \sin \xi_{(m)} x_1 \tag{9.69}$$

$$+ (\hat{B}^{(m)} \sinh \hat{\xi}_{(m)} x_2 + \frac{e}{\varepsilon} \hat{A}^{(m)} \sin \hat{\eta}_{(m)} x_2) \cos \hat{\xi}_{(m)} x_1],$$

$$D = -\varepsilon \sum_{m=1}^{\infty} (B^{(m)} \xi_{(m)} \cosh \xi_{(m)} x_2 \sin \xi_{(m)} x_1$$

$$+ \hat{B}^{(m)} \hat{\xi}_{(m)} \cosh \hat{\xi}_{(m)} x_2 \cos \hat{\xi}_{(m)} x_1), \tag{9.70}$$

$$T = \bar{c}\hat{A}^{(0)}\hat{\eta}_{(0)}\cos\hat{\eta}_{(0)}x_2 + e\hat{B}^{(0)}$$

$$+ \sum_{m=1}^{\infty} [(\bar{c}A^{(m)}\eta_{(m)}\cos\eta_{(m)}x_2 + eB^{(m)}\xi_{(m)}\cosh\xi_{(m)}x_2)\sin\xi_{(m)}x_1 \qquad (9.71)$$

$$+ (\bar{c}\hat{A}^{(m)}\hat{\eta}_{(m)}\cos\hat{\eta}_{(m)}x_2 + e\hat{B}^{(m)}\hat{\xi}_{(m)}\cosh\hat{\xi}_{(m)}x_2)\cos\hat{\xi}_{(m)}x_1].$$

Substituting Eqs. (9.69)–(9.71) into the boundary conditions at $x_2 = h$ in Eq. (9.60) gives the following equations for $A_{(m)}$, $\hat{A}_{(m)}$, $B_{(m)}$ and $\hat{B}_{(m)}$:

$$\hat{B}^{(0)}h + \frac{e}{\varepsilon}\hat{A}^{(0)}\sin\hat{\eta}_{(0)}h$$

$$+ \sum_{m=1}^{\infty} [(B^{(m)}\sinh\xi_{(m)}h + \frac{e}{\varepsilon}A^{(m)}\sin\eta_{(m)}h)\sin\xi_{(m)}x_1$$

$$+ (\hat{B}^{(m)}\sinh\hat{\xi}_{(m)}h + \frac{e}{\varepsilon}\hat{A}^{(m)}\sin\hat{\eta}_{(m)}h)\cos\hat{\xi}_{(m)}x_1] \qquad (9.72)$$

$$= \begin{cases} \overline{V}_1, & -a < x_1 < 0, \\ -\overline{V}_2, & 0 < x_1 < a, \end{cases}$$

$$= \begin{cases} \dfrac{\overline{V}_1 + \overline{V}_2}{2}, & -a < x_1 < 0, \\ -\dfrac{\overline{V}_1 + \overline{V}_2}{2}, & 0 < x_1 < a, \end{cases} + \begin{cases} \dfrac{\overline{V}_1 - \overline{V}_2}{2}, & -a < x_1 < 0, \\ \dfrac{\overline{V}_1 - \overline{V}_2}{2}, & 0 < x_1 < a, \end{cases}$$

$$\bar{c}\hat{A}^{(0)}\hat{\eta}_{(0)}\cos\hat{\eta}_{(0)}h + e\hat{B}^{(0)}$$

$$+ \sum_{m=1}^{\infty} [(\bar{c}A^{(m)}\eta_{(m)}\cos\eta_{(m)}h + eB^{(m)}\xi_{(m)}\cosh\xi_{(m)}h)\sin\xi_{(m)}x_1 \qquad (9.73)$$

$$+ (\bar{c}\hat{A}^{(m)}\hat{\eta}_{(m)}\cos\hat{\eta}_{(m)}h + e\hat{B}^{(m)}\hat{\xi}_{(m)}\cosh\hat{\xi}_{(m)}h)\cos\hat{\xi}_{(m)}x_1] = 0,$$

where we have decomposed the voltages into the sum of an antisymmetric function and a symmetric function of x_1. Due to the antisymmetry in x_2 the boundary conditions at $x_2 = -h$ yield the same conditions as Eqs. (9.72) and (9.73). Integrating both sides of Eqs. (9.72) and (9.73) over $[-a, a]$, also multiplying both sides of Eqs. (9.72) and (9.73) by $\sin\xi_{(n)}x_1$ for $n = 1, 2, 3,\ldots$ and integrating the resulting expressions over $[-a, a]$, and doing the same with $\cos\hat{\xi}_{(n)}x_1$ for $n = 1, 2, 3,\ldots$, we obtain

$$(\hat{B}^{(0)}h + \frac{e}{\varepsilon}\hat{A}^{(0)}\sin\hat{\eta}_{(0)}h)2a = (\bar{V}_1 - \bar{V}_2)a,$$

$$\bar{c}\hat{A}^{(0)}\hat{\eta}_{(0)}\cos\hat{\eta}_{(0)}h + e\hat{B}^{(0)} = 0,$$

$$(B^{(m)}\sinh\xi_{(m)}h + \frac{e}{\varepsilon}A^{(m)}\sin\eta_{(m)}h)a$$

$$= -2\frac{\bar{V}_1 + \bar{V}_2}{2}\frac{1}{\xi_{(m)}}, \quad m = 1, 2, 3, \ldots,$$

$$(\hat{B}^{(m)}\sinh\hat{\xi}_{(m)}h + \frac{e}{\varepsilon}\hat{A}^{(m)}\sin\hat{\eta}_{(m)}h)a$$

$$= 2\frac{\bar{V}_1 - \bar{V}_2}{2}\frac{\sin\hat{\xi}_{(m)}a}{\hat{\xi}_{(m)}} = 0, \quad m = 1, 2, 3, \ldots,$$

$$\bar{c}A^{(m)}\eta_{(m)}\cos\eta_{(m)}h + eB^{(m)}\xi_{(m)}\cosh\xi_{(m)}h = 0, \quad m = 1, 2, 3, \ldots,$$

$$\bar{c}\hat{A}^{(m)}\hat{\eta}_{(m)}\cos\hat{\eta}_{(m)}h + e\hat{B}^{(m)}\hat{\xi}_{(m)}\cosh\hat{\xi}_{(m)}h = 0, \quad m = 1, 2, 3, \ldots.$$

$$(9.74)$$

From Eq. (9.74) we solve for

$$A^{(m)} = \alpha^{(m)}(\bar{V}_1 + \bar{V}_2), \quad m = 1, 2, 3, \ldots,$$

$$B^{(m)} = \beta^{(m)}(\bar{V}_1 + \bar{V}_2), \quad m = 1, 2, 3, \ldots,$$

$$\hat{A}^{(0)} = \hat{\alpha}^{(0)}(\bar{V}_1 - \bar{V}_2),$$

$$\hat{B}^{(0)} = \hat{\beta}^{(0)}(\bar{V}_1 - \bar{V}_2),$$

$$\hat{A}^{(m)} = 0, \quad m = 1, 2, 3, \ldots,$$

$$\hat{B}^{(m)} = 0, \quad m = 1, 2, 3, \ldots,$$

$$(9.75)$$

where

$$\alpha^{(m)} = \frac{e}{\bar{c}}\frac{\xi_{(m)}\cosh\xi_{(m)}h}{\eta_{(m)}\cos\eta_{(m)}h\sinh\xi_{(m)}h - \bar{k}^2\xi_{(m)}\cosh\xi_{(m)}h\sin\eta_{(m)}h}\frac{1}{\xi_{(m)}a},$$

$$m = 1, 2, 3, \ldots,$$

$$\beta^{(m)} = -\frac{\eta_{(m)}\cos\eta_{(m)}h}{\eta_{(m)}\cos\eta_{(m)}h\sinh\xi_{(m)}h - \bar{k}^2\xi_{(m)}\cosh\xi_{(m)}h\sin\eta_{(m)}h}\frac{1}{\xi_{(m)}a},$$

$$m = 1, 2, 3, \ldots,$$

$$(9.76)$$

$$\hat{\alpha}^{(0)} = -\frac{e}{\overline{c}}\frac{1}{2}\frac{1}{\hat{\eta}_{(0)}h\cos\hat{\eta}_{(0)}h - \overline{k}^2\sin\hat{\eta}_{(0)}h},$$

$$\hat{\beta}^{(0)} = \frac{1}{2}\frac{\hat{\eta}_{(0)}\cos\hat{\eta}_{(0)}h}{\hat{\eta}_{(0)}h\cos\hat{\eta}_{(0)}h - \overline{k}^2\sin\hat{\eta}_{(0)}h}. \tag{9.77}$$

Since $\overline{V}_2$ is unknown, Eq. (9.75) has not fully determined $A_{(m)}$, $\hat{A}_{(m)}$, $B_{(m)}$ and $\hat{B}_{(m)}$ yet. From Eqs. (9.62), (9.70) and (9.75) we calculate the output current and substitute the result into the circuit condition in Eq. (9.64) to obtain

$$\frac{2\overline{V}_2}{Z} = \overline{I}_2 = \beta_1\overline{V}_1 + \beta_2\overline{V}_2, \tag{9.78}$$

where

$$\beta_1 = i\omega\varepsilon a\hat{\beta}^{(0)} + i\omega\varepsilon\sum_{m=1}^{\infty}\beta^{(m)}\cosh\xi^{(m)}h,$$

$$\beta_2 = -i\omega\varepsilon a\hat{\beta}^{(0)} + i\omega\varepsilon\sum_{m=1}^{\infty}\beta^{(m)}\cosh\xi^{(m)}h. \tag{9.79}$$

Equation (9.78) determines $\overline{V}_2$ in terms of $\overline{V}_1$ which is known. This completes the solution and yields the transforming ratio n as

$$n = \left|\frac{\overline{V}_2}{\overline{V}_1}\right|, \quad \frac{\overline{V}_2}{\overline{V}_1} = \frac{\beta_1 Z}{2 - \beta_2 Z}. \tag{9.80}$$

Equation (9.80) shows that n is proportional to Z for small Z, and approaches a constant for large Z (voltage saturation at open circuit). Once $\overline{V}_2$ is known, from the circuit equation in Eq. (9.64) we have

$$\overline{I}_2 = \frac{2\beta_1}{2 - \beta_2 Z}\overline{V}_1. \tag{9.81}$$

Equation (9.81) shows that the output current approaches zero for large Z (open circuit). From Eqs. (9.61), (9.70) and (9.75) the input current is

$$\overline{I}_1 = -\beta_2\overline{V}_1 - \beta_1\overline{V}_2. \tag{9.82}$$

With Eq. (9.80), we can write Eq. (9.82) as

$$\frac{\overline{I}_1}{\overline{V}_1} = -\beta_2 - \frac{\beta_1^2 Z}{2 - \beta_2 Z}, \tag{9.83}$$

which is the input admittance of the device.

9.4.3. Numerical results

As an example, consider a transformer made from PZT-5H. Some damping is introduced by using $c(1 + iQ^{-1})$ for c in the calculation. $Q = 100$. For geometric parameters we choose $h = 1$ mm, $a = 10$ mm. We use the following reference frequency and reference impedance to normalize the driving frequency ω and the load impedance Z:

$$\omega_0 = \frac{\pi}{2h}\sqrt{\frac{c}{\rho}}, \quad Z_0 = \frac{1}{i\omega C_0} = \frac{1}{i\omega}\frac{h}{\varepsilon a}. \tag{9.84}$$

In our numerical calculation the convergence of the series seems to be very slow, if it does converge. This may be related to the jump discontinuities of the voltage on the plate surfaces.

Figure 9.9 shows the transforming ratio versus the driving frequency. The transforming ratio assumes its maxima at resonant frequencies as expected. The highest peak is near the first pure thickness-twist frequency with the normalized frequency close to one. The next highest peak is near the third thickness-twist frequency. The second pure thickness-twist mode is symmetric about $x_1 = 0$ and cannot be excited by the applied voltage which is antisymmetric about $x_1 = 0$. Between the first and the third pure thickness modes there are modes that have nodal points along x_1 and therefore the output is low because across a nodal point there is electrical cancellation.

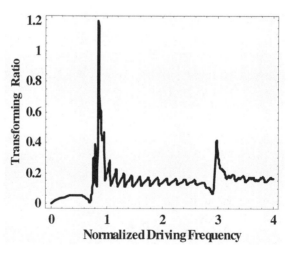

Fig. 9.9. Transforming ratio versus driving frequency ($Z = 2Z_0$).

Figure 9.10 shows the transforming ratio versus the load Z. As the load increases from zero, the transforming ratio increases from zero. Then it exhibits a maximum. For large loads, the transforming ratio approaches a constant, showing saturation. These can also be seen from Eq. (9.80). Physically, for very large values of the load, the output electrodes are essentially open. The output voltage is saturated and the output current essentially vanishes.

Input admittance versus driving frequency is given in Fig. 9.11. The frequency at which the input admittance assumes a minimum is called the antiresonant frequency.

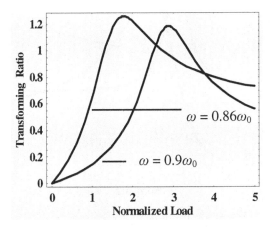

Fig. 9.10. Transforming ratio versus Z/Z_0.

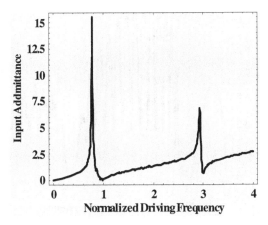

Fig. 9.11. Input admittance versus driving frequency ($Z = 2Z_0$).

Input admittance versus load is plotted in Fig. 9.12 for two values of the driving frequency close to the first resonance. The input admittance exhibits a maximum and saturation, as shown in Eq. (9.83).

Figure 9.13 shows the efficiency versus driving frequency. As a ratio of the input and output powers (two functions with peaks), the behavior of efficiency versus frequency is relatively complicated, with some slowly varying peaks and some relatively sharp peaks. This is related to the behavior of the curves in Fig. 9.9.

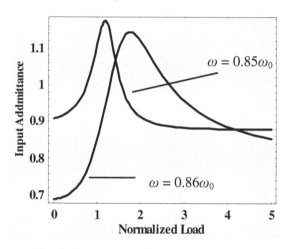

Fig. 9.12. Input admittance versus Z/Z_0.

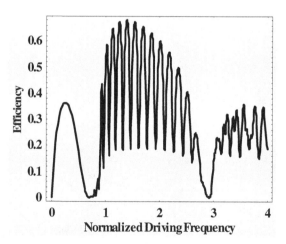

Fig. 9.13. Efficiency versus driving frequency ($Z = 2iZ_0$).

Efficiency versus load is shown in Fig. 9.14 for two values of the driving frequency close to the first resonance. The figure shows that as the load increases from zero, the efficiency first increases from zero and then it reaches a maximum. After the maximum the efficiency goes down monotonically.

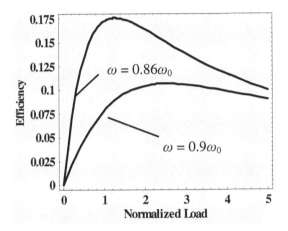

Fig. 9.14. Efficiency versus $Z/(iZ_0)$.

9.5. A Plate with Nonuniform Electrodes

Consider a plate with electrodes of varying thickness (see Fig. 9.15) [56,57]. We consider unit thickness in the x_3 direction. The electrodes are under a time-harmonic driving voltage. The two faces at $x_1 = \pm a$ are traction-free and are unelectroded.

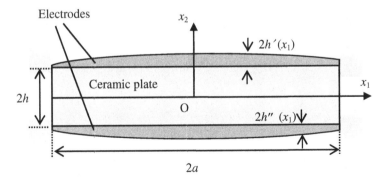

Fig. 9.15. A plate with nonuniform electrodes.

9.5.1. Governing equations and solution

The governing equations for u, ψ and ϕ are

$$\overline{c}\nabla^2 u = \rho\ddot{u},$$
$$\nabla^2\psi = 0, \tag{9.85}$$
$$\phi = \psi + \frac{e}{\varepsilon}u.$$

The electrodes are assumed to be very thin. Only their inertia needs to be considered. Their stiffness is neglected. The boundary conditions are

$$-T_{23} = \rho'2h'(x_1)\ddot{u}, \quad \phi = V(t), \quad x_2 = h,$$
$$T_{23} = \rho'2h''(x_1)\ddot{u}, \quad \phi = -V(t), \quad x_2 = -h, \tag{9.86}$$
$$T_{13} = 0, \quad D_1 = 0, \quad x_1 = \pm a,$$

or, equivalently, in terms of u and ψ,

$$\overline{c}u_{,2} + e\psi_{,2} = -\rho'2h'(x_1)\ddot{u}, \quad \psi + \frac{e}{\varepsilon}u = V(t), \quad x_2 = h,$$
$$\overline{c}u_{,2} + e\psi_{,2} = \rho'2h''(x_1)\ddot{u}, \quad \psi + \frac{e}{\varepsilon}u = -V(t), \quad x_2 = -h, \tag{9.87}$$
$$u_{,1} = 0, \quad \psi_{,1} = 0, \quad x_1 = \pm a.$$

We use the complex notation

$$(u,\psi,\phi,V,T_{23}) = \text{Re}\{(U,\Psi,\Phi,\overline{V},T)\exp(i\omega t)\}. \tag{9.88}$$

From the x_1 dependence of the modes in Sec. 7.8, we construct the following solution:

$$
\begin{aligned}
U &= A_1^{(0)}\cos\eta_{(0)}x_2 + A_2^{(0)}\sin\eta_{(0)}x_2 \\
&+ \sum_{m=2,4,6\cdots}^{\infty}(A_1^{(m)}\cos\eta_{(m)}x_2 + A_2^{(m)}\sin\eta_{(m)}x_2)\cos\xi_{(m)}x_1 \\
&+ \sum_{m=1,3,5\cdots}^{\infty}(A_3^{(m)}\cos\eta_{(m)}x_2 + A_4^{(m)}\sin\eta_{(m)}x_2)\sin\xi_{(m)}x_1, \\
\Psi &= B_1^{(0)} + B_2^{(0)}x_2 \\
&+ \sum_{m=2,4,6\cdots}^{\infty}(B_1^{(m)}\cosh\xi_{(m)}x_2 + B_2^{(m)}\sinh\xi_{(m)}x_2)\cos\xi_{(m)}x_1 \\
&+ \sum_{m=1,3,5\cdots}^{\infty}(B_3^{(m)}\cosh\xi_{(m)}x_2 + B_4^{(m)}\sinh\xi_{(m)}x_2)\sin\xi_{(m)}x_1,
\end{aligned}
\tag{9.89}
$$

where $A_1^{(m)}$ through $A_4^{(m)}$ and $B_1^{(m)}$ through $B_4^{(m)}$ are undetermined constants, and

$$\eta_{(0)}^2 = \frac{\rho \omega^2}{\overline{c}},$$

$$\xi_{(m)} = \frac{m}{2a}\pi, \quad \eta_{(m)}^2 = \frac{\rho \omega^2}{\overline{c}} - \left(\frac{m}{2a}\pi\right)^2, \quad m = 1, 2, 3, \cdots.$$

(9.90)

Equation (9.89) satisfies Eq. (9.85) and the boundary conditions at $x_1 = \pm a$. To apply the boundary conditions at $x_2 = \pm h$, we need:

$$\Phi = B_1^{(0)} + B_2^{(0)}x_2 + \frac{e}{\varepsilon}\left(A_1^{(0)}\cos\eta_{(0)}x_2 + A_2^{(0)}\sin\eta_{(0)}x_2\right)$$

$$+ \sum_{m=2,4,6\cdots}^{\infty}(B_1^{(m)}\cosh\xi_{(m)}x_2 + \frac{e}{\varepsilon}A_1^{(m)}\cos\eta_{(m)}x_2$$

$$+ B_2^{(m)}\sinh\xi_{(m)}x_2 + \frac{e}{\varepsilon}A_2^{(m)}\sin\eta_{(m)}x_2)\cos\xi_{(m)}x_1 \qquad (9.91)$$

$$+ \sum_{m=1,3,5\cdots}^{\infty}(B_3^{(m)}\cosh\xi_{(m)}x_2 + \frac{e}{\varepsilon}A_3^{(m)}\cos\eta_{(m)}x_2$$

$$+ B_4^{(m)}\sinh\xi_{(m)}x_2 + \frac{e}{\varepsilon}A_4^{(m)}\sin\eta_{(m)}x_2)\sin\xi_{(m)}x_1,$$

$$T = \overline{c}(-A_1^{(0)}\eta_{(0)}\sin\eta_{(0)}x_2 + A_2^{(0)}\eta_{(0)}\cos\eta_{(0)}x_2) + eB_2^{(0)}$$

$$+ \sum_{m=2,4,6\cdots}^{\infty}[\overline{c}(-A_1^{(m)}\eta_{(m)}\sin\eta_{(m)}x_2 + A_2^{(m)}\eta_{(m)}\cos\eta_{(m)}x_2)$$

$$+ e(B_1^{(m)}\xi_{(m)}\sinh\xi_{(m)}x_2 + B_2^{(m)}\xi_{(m)}\cosh\xi_{(m)}x_2)]\cos\xi_{(m)}x_1 \qquad (9.92)$$

$$+ \sum_{m=1,3,5\cdots}^{\infty}[\overline{c}(-A_3^{(m)}\eta_{(m)}\sin\eta_{(m)}x_2 + A_4^{(m)}\eta_{(m)}\cos\eta_{(m)}x_2)$$

$$+ e(B_3^{(m)}\xi_{(m)}\sinh\xi_{(m)}x_2 + B_4^{(m)}\xi_{(m)}\cosh\xi_{(m)}x_2)]\sin\xi_{(m)}x_1.$$

Substitution of Eqs. (9.89), (9.91) and (9.92) into the boundary conditions at $x_2 = \pm h$ in Eq. (9.87) gives equations involving $A_1^{(m)}$ through $A_4^{(m)}$ and $B_1^{(m)}$ through $B_4^{(m)}$. These equations need to be multiplied by $\cos\xi_{(n)}x_1$ or $\sin\xi_{(n)}x_1$ and integrated over $[-a, a]$ to obtain linear algebraic equations for the undetermined constants [56,57].

The resulting equations are not included here. Some examples with numerical results are directly provided below.

9.5.2. An example

As an example consider the following electrodes:

$$h''(x_1) = h'(x_1) = \beta\left(1 + \lambda\cos\frac{\pi}{a}x_1\right),$$

$$\beta > 0, \quad |\lambda| \le 1. \tag{9.93}$$

The electrode profile described by Eq. (9.93) for various values of λ is shown in Fig. 9.16. The electrodes can be thicker ($\lambda > 0$) or thinner ($\lambda < 0$) in the middle.

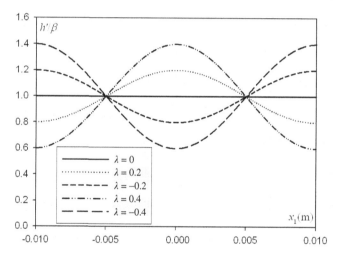

Fig. 9.16. Thickness variations of the electrodes.

For the plate we use polarized ceramics PZT-5H. Some damping is introduced by allowing the elastic material constant to assume complex values. c is replaced by $c(1 + iQ^{-1})$ in the numerical calculation. We fix $Q = 50$ as a representation of all damping that may be present. The driving frequency is normalized by

$$\omega_0 = \frac{\pi}{2h}\sqrt{\frac{\bar{c}}{\rho}}, \tag{9.94}$$

which is the lowest thickness-twist frequency of an unbounded plate. When using this frequency to normalize the driving frequency, we expect thickness-twist resonances to appear close to and smaller than odd integers. We fix h = 1 mm and a = 10 mm. The mass ratio $R = \rho'\beta/(\rho h)$ = 0.02 except in Fig. 9.18.

Figure 9.17 shows the displacement at the center of the upper surface of the plate versus the driving frequency. Major resonances appear near one and three, the first and the third thickness-twist resonant frequencies, as expected. Modes that are symmetric about the middle surface with the normalized frequency approximately being equal to even numbers cannot be excited by the antisymmetric electric field applied. At the major resonances particles of the upper (or lower) surfaces of the plate move essentially in phase. Near the major resonances there also exist minor resonances corresponding to modes with nodal points along the x_1 direction.

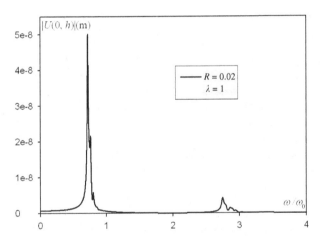

Fig. 9.17. Displacement versus frequency.

Figure 9.18 shows the displacement distribution at the upper surface of the plate when the electrodes are thicker in the middle and the frequency is very close to but is slightly lower than the first resonance. The motion is larger in the central portion of the plate than near the edges. This shows that energy trapping can be achieved in a fully electroded plate by electrodes with nonuniform thickness. The figure also shows that for larger R or electrodes thicker at the center the motion is pushed more toward the center, indicating stronger energy trapping.

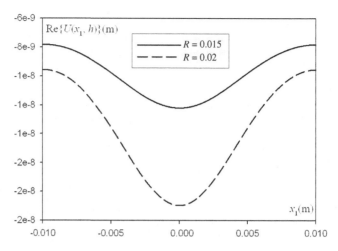

Fig. 9.18. Effect of β on mode shape ($\lambda = 0.4$, $\omega/\omega_0 = 0.72$).

Figure 9.19 shows the effect of λ on the upper surface displacement at a frequency very close to and slightly lower than the first resonance. When $\lambda = 0$, we have flat electrodes without trapping. When λ is positive, the electrodes are thicker in the middle and the motion is trapped in the central portion. A larger λ leads to stronger trapping as expected. When λ is negative, the electrodes are thinner in the middle and the motion is larger near the edges.

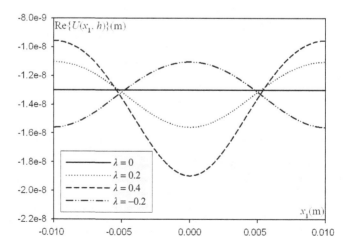

Fig. 9.19. Effect of λ on mode shape ($\omega/\omega_0 = 0.72$).

In Figs. 9.18 and 9.19, only the real part of the complex displacement is plotted. It may be more informative to plot the absolute value of the complex displacement.

9.6. A Multilayered Plate

Consider an N-layered plate of total thickness $2h$ with the x_2 axis normal to the plate (see Fig. 9.20) [58]. The two plate major surfaces and the $N-1$ interfaces are sequentially determined by $x_2 = -h = h_0, h_1, \ldots, h_{N-1}$, and $h_N = h$. The two major surfaces of the plate at $x_2 = \pm h$ are electroded and are under given shear stress and electric potential. The two minor surfaces at $x_1 = \pm a$ are unelectroded and are traction-free.

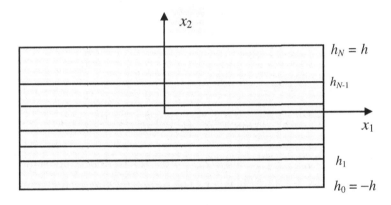

Fig. 9.20. A multilayered plate.

The governing equations for u, ψ and ϕ are

$$\bar{c}\nabla^2 u = \rho\ddot{u},$$
$$\nabla^2 \psi = 0, \qquad (9.95)$$
$$\phi = \psi + \frac{e}{\varepsilon}u,$$

for each layer. At the top and bottom surfaces the boundary conditions are

$$T_{23} = T^{\pm}\exp(i\omega t), \quad x_2 = \pm h,$$
$$\phi = V^{\pm}\exp(i\omega t), \quad x_2 = \pm h. \qquad (9.96)$$

$T^{\pm}$ and $V^{\pm}$ are surface tractions and potentials. At an interface, we require the continuity of u, T_{23}, ϕ and D_2. At the two ends, the boundary conditions are

$$T_{13} = 0, \quad x_1 = \pm a,$$
$$D_1 = 0, \quad x_1 = \pm a. \tag{9.97}$$

Consider the ith layer, with $i = 1, 2, 3, \ldots, N$. For time-harmonic motions we write u, ψ, ϕ, T_{23} and D_2 as

$$(u, \psi, \phi, T_{23}, D_2) = \mathrm{Re}\{(U, \Psi, \Phi, T, D) \exp(i\omega t)\}. \tag{9.98}$$

From the x_1 dependence of the modes in Sec. 7.8, we construct the following solution:

$$U = A_{(0)}^{(i)} \cos\eta_{(0)}^{(i)} x_2 + B_{(0)}^{(i)} \sin\eta_{(0)}^{(i)} x_2$$
$$+ \sum_{m=1}^{\infty} (A_{(m)}^{(i)} \cos\eta_{(m)}^{(i)} x_2 + B_{(m)}^{(i)} \sin\eta_{(m)}^{(i)} x_2) X_{(m)}(x_1),$$
$$\Psi = C_{(0)}^{(i)} + D_{(0)}^{(i)} x_2 \tag{9.99}$$
$$+ \sum_{m=1}^{\infty} (C_{(m)}^{(i)} \cosh\xi_{(m)} x_2 + D_{(m)}^{(i)} \sinh\xi_{(m)} x_2) X_{(m)}(x_1),$$

where $A_{(m)}^{(i)}$, $B_{(m)}^{(i)}$, $C_{(m)}^{(i)}$ and $D_{(m)}^{(i)}$ are undetermined constants. The superscript (i) is for the ith layer. The subscript (m) is for summation within a layer, and

$$\xi_{(m)} = \frac{m}{2a}\pi, \quad X_{(m)}(x_1) = \begin{cases} \sin\xi_{(m)} x_1, & m = 1, 3, 5, \cdots, \\ \cos\xi_{(m)} x_1, & m = 2, 4, 6, \cdots, \end{cases}$$
$$(\eta_{(m)}^{(i)})^2 = \frac{\rho^{(i)} \omega^2}{\bar{c}^{(i)}} - \left(\frac{m}{2a}\pi\right)^2, \quad m = 0, 1, 2, 3, \cdots. \tag{9.100}$$

Equation (9.99) satisfies Eq. (9.95) and the boundary conditions at $x_1 = \pm a$. To apply the interface continuity conditions and the boundary conditions at $x_2 = \pm h$, we calculate the following:

$$\Phi = \frac{e^{(i)}}{\varepsilon^{(i)}} A_{(0)}^{(i)} \cos \eta_{(0)}^{(i)} x_2 + \frac{e^{(i)}}{\varepsilon^{(i)}} B_{(0)}^{(i)} \sin \eta_{(0)}^{(i)} x_2 + C_{(0)}^{(i)} + D_{(0)}^{(i)} x_2$$

$$+ \sum_{m=1}^{\infty} (\frac{e^{(i)}}{\varepsilon^{(i)}} A_{(m)}^{(i)} \cos \eta_{(m)}^{(i)} x_2 + \frac{e^{(i)}}{\varepsilon^{(i)}} B_{(m)}^{(i)} \sin \eta_{(m)}^{(i)} x_2 \tag{9.101}$$

$$+ C_{(m)}^{(i)} \cosh \xi_{(m)} x_2 + D_{(m)}^{(i)} \sinh \xi_{(m)} x_2) X_{(m)}(x_1),$$

$$D = -\varepsilon^{(i)} D_{(0)}^{(i)} + \sum_{m=1}^{\infty} (-\varepsilon^{(i)} C_{(m)}^{(i)} \xi_{(m)} \sinh \xi_{(m)} x_2$$

$$- \varepsilon^{(i)} \xi_{(m)} D_{(m)}^{(i)} \cosh \xi_{(m)} x_2) X_{(m)}(x_1), \tag{9.102}$$

$$T = -\overline{c}^{(i)} A_{(0)}^{(i)} \eta_{(0)}^{(i)} \sin \eta_{(0)}^{(i)} x_2 + \overline{c}^{(i)} B_{(0)}^{(i)} \eta_{(0)}^{(i)} \cos \eta_{(0)}^{(i)} x_2 + e^{(i)} D_{(0)}^{(i)}$$

$$+ \sum_{m=1}^{\infty} (-\overline{c}^{(i)} A_{(m)}^{(i)} \eta_{(m)}^{(i)} \sin \eta_{(m)}^{(i)} x_2 + \overline{c}^{(i)} B_{(m)}^{(i)} \eta_{(m)}^{(i)} \cos \eta_{(m)}^{(i)} x_2 \tag{9.103}$$

$$+ e^{(i)} C_{(m)}^{(i)} \xi_{(m)} \sinh \xi_{(m)} x_2 + e^{(i)} D_{(m)}^{(i)} \xi_{(m)} \cosh \xi_{(m)} x_2) X_{(m)}(x_1).$$

At the interface between the ith and $(i + 1)$th layer at $x_2 = h_i$ (where $i =$ 1, 2, 3,…, $N - 1$), we apply the following continuity conditions

$$U(h_i^-) = U(h_i^+), \tag{9.104}$$

$$\Phi(h_i^-) = \Phi(h_i^+), \tag{9.105}$$

$$D(h_i^-) = D(h_i^+), \tag{9.106}$$

$$T(h_i^-) = T(h_i^+). \tag{9.107}$$

Equations (9.104)–(9.107) imply that

$$\begin{Bmatrix} A_{(0)}^{(i)} \\ B_{(0)}^{(i)} \\ C_{(0)}^{(i)} \\ D_{(0)}^{(i)} \end{Bmatrix} = [T_{(0)}^{(i)}] \begin{Bmatrix} A_{(0)}^{(i+1)} \\ B_{(0)}^{(i+1)} \\ C_{(0)}^{(i+1)} \\ D_{(0)}^{(i+1)} \end{Bmatrix}, \quad i = 1, 2, 3, \cdots, N - 1, \tag{9.108}$$

$$\begin{Bmatrix} A_{(m)}^{(i)} \\ B_{(m)}^{(i)} \\ C_{(m)}^{(i)} \\ D_{(m)}^{(i)} \end{Bmatrix} = [T^{(i)}] \begin{Bmatrix} A_{(m)}^{(i+1)} \\ B_{(m)}^{(i+1)} \\ C_{(m)}^{(i+1)} \\ D_{(m)}^{(i+1)} \end{Bmatrix}, \quad i = 1, 2, 3, \cdots, N - 1, \quad m = 1, 2, 3, \cdots, \tag{9.109}$$

where $[T_{(0)}^{(i)}]$ and $[T^{(i)}]$ are transfer matrices whose expressions are straightforward from Eqs. (9.104)–(9.107) and are too lengthy to be presented here [58]. With Eqs. (9.108) and (9.109) we can express all undetermined constants in terms of those of the Nth layer. In particular, between the first and the last layers, we have

$$\begin{Bmatrix} A_{(0)}^{(1)} \\ B_{(0)}^{(1)} \\ C_{(0)}^{(1)} \\ D_{(0)}^{(1)} \end{Bmatrix} = [T_{(0)}^{(1)}][T_{(0)}^{(2)}]\cdots[T_{(0)}^{(N-1)}] \begin{Bmatrix} A_{(0)}^{(N)} \\ B_{(0)}^{(N)} \\ C_{(0)}^{(N)} \\ D_{(0)}^{(N)} \end{Bmatrix}, \qquad (9.110)$$

$$\begin{Bmatrix} A_{(m)}^{(1)} \\ B_{(m)}^{(1)} \\ C_{(m)}^{(1)} \\ D_{(m)}^{(1)} \end{Bmatrix} = [T^{(1)}][T^{(2)}]\cdots[T^{(N-1)}] \begin{Bmatrix} A_{(m)}^{(N)} \\ B_{(m)}^{(N)} \\ C_{(m)}^{(N)} \\ D_{(m)}^{(N)} \end{Bmatrix}. \qquad (9.111)$$

We also write the boundary data in Eq. (9.96) in series form:

$$T^{\pm}(x_1) = T_{(0)}^{\pm} + \sum_{m=1}^{\infty} T_{(m)}^{\pm} X_{(m)}(x_1),$$

$$V^{\pm}(x_1) = V_{(0)}^{\pm} + \sum_{m=1}^{\infty} V_{(m)}^{\pm} X_{(m)}(x_1), \qquad (9.112)$$

where $T_{(0)}^{\pm}$, $T_{(m)}^{\pm}$, $V_{(0)}^{\pm}$ and $V_{(m)}^{\pm}$ are known expansion coefficients of the applied load. Then the boundary conditions in Eq. (9.96) imply that, for $m = 0$:

$$\bar{c}^{(1)} A_{(0)}^{(1)} \eta_{(0)}^{(1)} \sin\eta_{(0)}^{(1)}h + \bar{c}^{(1)} B_{(0)}^{(1)} \eta_{(0)}^{(1)} \cos\eta_{(0)}^{(1)}h + e^{(1)} D_{(0)}^{(1)} = T_{(0)}^{-}, \qquad (9.113)$$

$$-\bar{c}^{(N)} A_{(0)}^{(N)} \eta_{(0)}^{(N)} \sin\eta_{(0)}^{(N)}h + \bar{c}^{(N)} B_{(0)}^{(N)} \eta_{(0)}^{(N)} \cos\eta_{(0)}^{(N)}h + e^{(N)} D_{(0)}^{(N)} = T_{(0)}^{+}, \qquad (9.114)$$

$$\frac{e^{(1)}}{\varepsilon^{(1)}} A_{(0)}^{(1)} \cos\eta_{(0)}^{(1)}h - \frac{e^{(1)}}{\varepsilon^{(1)}} B_{(0)}^{(1)} \sin\eta_{(0)}^{(1)}h + C_{(0)}^{(1)} - D_{(0)}^{(1)}h = V_{(0)}^{-}, \qquad (9.115)$$

$$\frac{e^{(N)}}{\varepsilon^{(N)}} A_{(0)}^{(N)} \cos\eta_{(0)}^{(N)}h + \frac{e^{(N)}}{\varepsilon^{(N)}} B_{(0)}^{(N)} \sin\eta_{(0)}^{(N)}h + C_{(0)}^{(N)} + D_{(0)}^{(N)}h = V_{(0)}^{+}, \qquad (9.116)$$

and for $m = 1, 2, 3,\ldots$:

$$\bar{c}^{(1)} A_{(m)}^{(1)} \eta_{(m)}^{(1)} \sin \eta_{(m)}^{(1)} h + \bar{c}^{(1)} B_{(m)}^{(1)} \eta_{(m)}^{(1)} \cos \eta_{(m)}^{(1)} h$$
$$- e^{(1)} C_{(m)}^{(1)} \xi_{(m)} \sinh \xi_{(m)} h + e^{(1)} D_{(m)}^{(1)} \xi_{(m)} \cosh \xi_{(m)} h = T_{(m)}^{-}, \tag{9.117}$$

$$- \bar{c}^{(N)} A_{(m)}^{(N)} \eta_{(m)}^{(N)} \sin \eta_{(m)}^{(N)} h + \bar{c}^{(N)} B_{(m)}^{(N)} \eta_{(m)}^{(N)} \cos \eta_{(m)}^{(N)} h$$
$$+ e^{(N)} C_{(m)}^{(N)} \xi_{(m)} \sinh \xi_{(m)} h + e^{(N)} D_{(m)}^{(N)} \xi_{(m)} \cosh \xi_{(m)} h = T_{(m)}^{+}, \tag{9.118}$$

$$\frac{e^{(1)}}{\varepsilon^{(1)}} A_{(m)}^{(1)} \cos \eta_{(m)}^{(1)} h - \frac{e^{(1)}}{\varepsilon^{(1)}} B_{(m)}^{(1)} \sin \eta_{(m)}^{(1)} h$$
$$+ C_{(m)}^{(1)} \cosh \xi_{(m)} h - D_{(m)}^{(1)} \sinh \xi_{(m)} h = V_{(m)}^{-}, \tag{9.119}$$

$$\frac{e^{(N)}}{\varepsilon^{(N)}} A_{(m)}^{(N)} \cos \eta_{(m)}^{(N)} h + \frac{e^{(N)}}{\varepsilon^{(N)}} B_{(m)}^{(N)} \sin \eta_{(m)}^{(N)} h$$
$$+ C_{(m)}^{(N)} \cosh \xi_{(m)} h + D_{(m)}^{(N)} \sinh \xi_{(m)} h = V_{(m)}^{+}. \tag{9.120}$$

Equations (9.110) and (9.113)–(9.116) are eight equations for $A_{(0)}^{(1)}$, $B_{(0)}^{(1)}$, $C_{(0)}^{(1)}$, $D_{(0)}^{(1)}$, $A_{(0)}^{(N)}$, $B_{(0)}^{(N)}$, $C_{(0)}^{(N)}$ and $D_{(0)}^{(N)}$. Equations (9.111) and (9.117)–(9.120) are eight equations for $A_{(m)}^{(1)}$, $B_{(m)}^{(1)}$, $C_{(m)}^{(1)}$, $D_{(m)}^{(1)}$, $A_{(m)}^{(N)}$, $B_{(m)}^{(N)}$, $C_{(m)}^{(N)}$ and $D_{(m)}^{(N)}$ for $m = 1, 2, 3,\ldots$.

As an example, consider a two-layered plate resonator driven electrically as shown in Fig. 9.21. There is one pair of electrodes which are shown by the thick lines in the figure. The two ceramics have significantly different material properties.

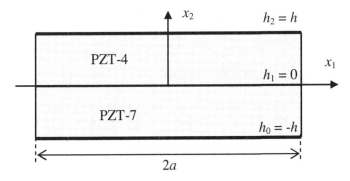

Fig. 9.21. A two-layered plate as a resonator driven electrically.

The boundary conditions are:

$$T_{23}(x_1) = 0, \quad x_2 = \pm h,$$

$$\phi(x_1) = V_0^{\pm} \exp(i\omega t), \quad x_2 = \pm h,$$

(9.121)

where $V_0^{\pm}$ are constants. Some damping is introduced by allowing the elastic material constant c to assume complex values. In our calculations c is replaced by $c(1 + iQ^{-1})$. We fix $Q = 100$. We choose $h = 1$ mm, and $a = 10$ mm. The linear algebraic equations are solved numerically on a computer. In Fig. 9.22 we plot the displacement at the center of the upper surface of the plate versus the driving frequency which is normalized by

$$\omega_0 = \frac{\pi}{2h} \sqrt{\frac{c}{\rho}} .$$

(9.122)

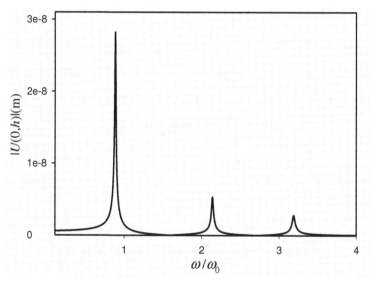

Fig. 9.22. Resonances in a two-layered resonator.

The figure shows the locations of the resonant frequencies. In this example, due to the uniform driving voltage the plate is in fact vibrating like an infinite plate without variation in the x_1 direction. Modes corresponding to different resonant frequencies have different wave numbers in the x_2 direction. These modes are thickness-shear modes in the x_3 direction. Note that for a one-layered plate modes symmetric about the middle plane, e.g., the second resonance, cannot be excited by a thickness electric field. A two-layered plate does not have symmetry about the middle plane and its modes cannot be separated into symmetric and antisymmetric modes and all modes can be excited. Displacement distribution along the plate thickness (mode shape) near the first resonance is shown in Fig. 9.23, which is the fundamental thickness mode.

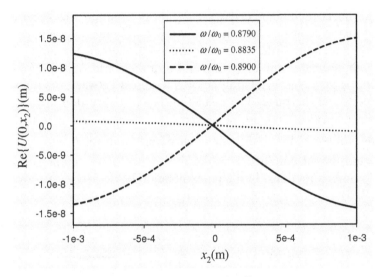

Fig. 9.23. Mode shapes near the first resonance.

9.7. Power Transmission through an Elastic Plate

There is current and growing interest in periodically recharging batteries that power electronic devices operating in a sealed armor or other hazardous environments, such as nuclear storage facilities, into which the physical access is prohibited. It has been proposed that piezoelectric transducers are used to generate acoustic waves propagating through a sealed armor for transmitting a small amount of power to the electronic devices inside. The procedure involves the generation and propagation of acoustic waves and energy harvesting from the waves using piezoelectric transducers. There are also other possible applications of the technology like data transmission through a wall by acoustic waves.

For power transmission through a wall let us consider an elastic plate of thickness $2h$ and length $2L$ as shown in Fig. 9.24 [59]. We consider a unit thickness of the plate in the z direction. The two ends of the plate at $x = \pm L$ are traction-free. Two piezoelectric transducers represented by the shaded areas are attached to the plate. There are no limitations on the length of the transducers except that they are shorter than the plate in the x direction ($a < L$). We assume that the transducers are placed at the center of the plate. This does not cause much loss of generality because, as to be shown, due to energy trapping or vibration localization, the acoustic waves will be confined within and very close to the transducer region and decay rapidly away from the transducer edges. Therefore the acoustic waves in fact will not feel much of the plate edges at $x = \pm L$. The transducers are electroded at their major surfaces. The plate can be made from either a metal or a dielectric. When the plate is metallic, a very thin insulating layer is assumed between the transducer electrodes and the plate. The insulating layer is very thin, its thickness and mechanical effects are neglected. $2V_1$ is a known, time-harmonic driving voltage. $2V_2$ is the unknown output voltage. I_1 and I_2 are the input and output currents. Z is the impedance of the output circuit in time-harmonic motions. In the index notation below, (x, y, z) correspond to $(1, 2, 3)$.

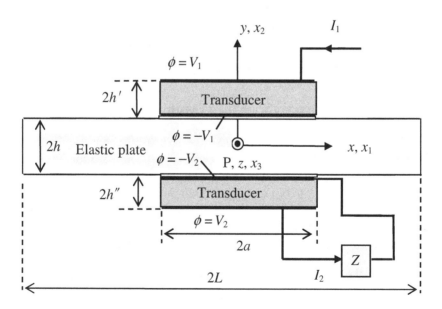

Fig. 9.24. An elastic wall with two piezoelectric transducers.

9.7.1. Governing equations

The transducers are governed by

$$v_T^2 \nabla^2 u = \ddot{u}, \quad \nabla^2 \psi = 0, \quad \phi = \psi + \frac{e}{\varepsilon} u. \tag{9.123}$$

The elastic plate is of an isotropic elastic material. The equation of motion takes the following form:

$$c_2^2 \nabla^2 u = \ddot{u}, \tag{9.124}$$

where $c_2^2 = \mu / \rho_1$. ρ_1 is the mass density. μ is the shear elastic constant. To calculate the charge and current on the upper electrode of the upper transducer, we need

$$Q_1 = \int_{-a}^{a} -D_2\big|_{y=h+2h'} \, dx, \tag{9.125}$$

$$I_1 = \dot{Q}_1 = i\omega Q_1.$$

For the charge and current on the lower electrode of the lower transducer, we have

$$Q_2 = \int_{-a}^{a} D_2\big|_{-h-2h''} \, dx, \tag{9.126}$$

$$I_2 = -\dot{Q}_2 = -i\omega Q_2.$$

We also have the following output circuit condition for time-harmonic motions:

$$I_2 = 2V_2 / Z. \tag{9.127}$$

The boundary and continuity conditions for the upper transducer, the elastic plate, and the bottom transducer are, respectively,

$$\phi(h+2h') = V_1, \quad |x| < a,$$
$$T_{23}(h+2h') = 0, \quad |x| < a,$$
$$\phi(h^-) = -V_1, \quad |x| < a, \tag{9.128}$$
$$u(h^-) = u(h^+), \quad |x| < a,$$

$$T_{23}(h^-) = \begin{cases} T_{23}(h^+), & |x| < a, \\ 0, & a < |x| < L, \end{cases}$$

$$T_{23}(-h^+) = \begin{cases} T_{23}(-h^-), & |x| < a, \\ 0, & a < |x| < L, \end{cases} \tag{9.129}$$

$$\phi(-h^+) = -V_2, \quad |x| < a,$$
$$u(-h^+) = u(-h^-), \quad |x| < a,$$
$$\phi(-h - 2h'') = V_2, \quad |x| < a, \tag{9.130}$$
$$T_{23}(-h - 2h'') = 0, \quad |x| < a.$$

In addition, there are traction-free boundary conditions at the edges of the plate ($x = \pm L$) and the transducers ($x = \pm a$). For the transducers, the edges are unelectroded and are charge free.

9.7.2. Series solution

For the elastic plate, the general solution for u symmetric in x can be written as

$$u = A_0 \sin(\eta_0 y) + B_0 \cos(\eta_0 y)$$
$$+ \sum_{n=1}^{\infty} [A_n \sin(\eta_n y) + B_n \cos(\eta_n y)] \cos\frac{n\pi x}{L}, \tag{9.131}$$

$$T_{23} = \mu\eta_0 A_0 \cos(\eta_0 y) - \mu\eta_0 B_0 \sin(\eta_0 y)$$
$$+ \sum_{n=1}^{\infty} [\mu\eta_n A_n \cos(\eta_n y) - \mu\eta_n B_n \sin(\eta_n y)] \cos\frac{n\pi x}{L}, \tag{9.132}$$

where A_n and B_n are undetermined constants, and

$$\eta_n = \left[\left(\frac{\omega}{c_2}\right)^2 - \left(\frac{n\pi}{L}\right)^2 \right]^{1/2}. \tag{9.133}$$

For the upper transducer, the general solution symmetric in x and satisfying the transducer edge conditions can be written as

$$u = F_0 \sin(\xi_0 y) + G_0 \cos(\xi_0 y)$$
$$+ \sum_{m=1}^{\infty} [F_m \sin(\xi_m y) + G_m \cos(\xi_m y)] \cos\frac{m\pi x}{a}, \tag{9.134}$$

$$\psi = H_0 y + K_0$$
$$+ \sum_{m=1}^{\infty} [H_m \sinh\frac{m\pi y}{a} + K_m \cosh\frac{m\pi y}{a}] \cos\frac{m\pi x}{a}, \tag{9.135}$$

$$\phi = H_0 y + K_0 + \frac{e}{\varepsilon} F_0 \sin(\xi_0 y) + \frac{e}{\varepsilon} G_0 \cos(\xi_0 y)$$

$$+ \sum_{m=1}^{\infty} \left[\frac{e}{\varepsilon} F_m \sin(\xi_m y) + \frac{e}{\varepsilon} G_m \cos(\xi_m y) \right. \tag{9.136}$$

$$\left. + H_m \sinh \frac{m\pi y}{a} + K_m \cosh \frac{m\pi y}{a} \right] \cos \frac{m\pi x}{a},$$

$$T_{23} = \overline{c} \xi_0 F_0 \cos(\xi_0 y) - \overline{c} \xi_0 G_0 \sin(\xi_0 y) + e H_0$$

$$+ \sum_{m=1}^{\infty} \left[\overline{c} \xi_m F_m \cos(\xi_m y) - \overline{c} \xi_m G_m \sin(\xi_m y) \right. \tag{9.137}$$

$$\left. + e \frac{m\pi}{a} H_m \cosh \frac{m\pi y}{a} + e \frac{m\pi}{a} K_m \sinh \frac{m\pi y}{a} \right] \cos \frac{m\pi x}{a},$$

$$D_2 = -\varepsilon H_0 + \sum_{m=1}^{\infty} \left[-\varepsilon \frac{m\pi}{a} H_m \cosh \frac{m\pi y}{a} \right.$$

$$\left. - \varepsilon \frac{m\pi}{a} K_m \sinh \frac{m\pi y}{a} \right] \cos \frac{m\pi x}{a}, \tag{9.138}$$

where F_m, G_m, H_m and K_m are undetermined constants, and

$$\xi_m = \left[\left(\frac{\omega}{v_T^2} \right)^2 - \left(\frac{m\pi}{a} \right)^2 \right]^{1/2}. \tag{9.139}$$

Similarly, for the lower transducer, we have

$$u = P_0 \sin(\xi_0 y) + Q_0 \cos(\xi_0 y)$$

$$+ \sum_{m=1}^{\infty} \left[P_m \sin(\xi_m y) + Q_m \cos(\xi_m y) \right] \cos \frac{m\pi x}{a}, \tag{9.140}$$

$$\psi = N_0 y + R_0 + \sum_{m=1}^{\infty} \left[N_m \sinh \frac{m\pi y}{a} + R_m \cosh \frac{m\pi y}{a} \right] \cos \frac{m\pi x}{a},$$

$$\phi = N_0 y + R_0 + \frac{e}{\varepsilon} P_0 \sin(\xi_0 y) + \frac{e}{\varepsilon} Q_0 \cos(\xi_0 y)$$

$$+ \sum_{m=1}^{\infty} \left[\frac{e}{\varepsilon} P_m \sin(\xi_m y) + \frac{e}{\varepsilon} Q_m \cos(\xi_m y) \right. \tag{9.141}$$

$$\left. + N_m \sinh \frac{m\pi y}{a} + R_m \cosh \frac{m\pi y}{a} \right] \cos \frac{m\pi x}{a},$$

$$T_{23} = \bar{c}\xi_0 P_0 \cos(\xi_0 y) - \bar{c}\xi_0 Q_0 \sin(\xi_0 y) + eN_0$$

$$+ \sum_{m=1}^{\infty} [\bar{c}\xi_m P_m \cos(\xi_m y) - \bar{c}\xi_m Q_m \sin(\xi_m y) \qquad (9.142)$$

$$+ e\frac{m\pi}{a} N_m \cosh\frac{m\pi y}{a} + e\frac{m\pi}{a} R_m \sinh\frac{m\pi y}{a}]\cos\frac{m\pi x}{a},$$

$$D_2 = -\varepsilon N_0 + \sum_{m=1}^{\infty} [-\varepsilon\frac{m\pi}{a} N_m \cosh\frac{m\pi y}{a}$$

$$- \varepsilon\frac{m\pi}{a} R_m \sinh\frac{m\pi y}{a}]\cos\frac{m\pi x}{a}, \qquad (9.143)$$

where P_m, Q_m, N_m and R_m are undetermined constants.

Substitution of the above fields into the boundary and continuity conditions in Eqs. (9.128)–(9.130) gives equations involving the undetermined constants. The equations from Eqs. (9.128) and (9.130) need to be multiplied by $\cos p\pi x/a$ and integrated from $-a$ to a for $p = 0, 1, 2, 3,\ldots$ so that linear algebraic equations for the undetermined constants can be obtained. The equations from Eq. (9.129) need to be multiplied by $\cos p\pi x/L$ and integrated from $-L$ to L for $p = 0, 1, 2, 3,\ldots$ in order to obtain linear algebraic equations for the undetermined constants. The linear algebraic equations for the undetermined constants are lengthy and therefore are not presented here [59]. In addition, the circuit condition in Eq. (9.127) is needed and it takes the following form:

$$-i\omega \int_{-a}^{a} \{-\varepsilon N_0 + \sum_{m=0}^{\infty} [-\varepsilon\frac{m\pi}{a} N_m \cosh\frac{m\pi(-h-2h'')}{a}$$

$$- \varepsilon\frac{m\pi}{a} R_m \sinh\frac{m\pi(-h-2h'')}{a}]\cos\frac{m\pi x}{a}\}dx = \frac{2V_2}{Z}. \qquad (9.144)$$

We note that when a assumes a set of special values of $a = L/k$, where $k = 1, 2, 3,\ldots$ is a positive integer, we have

$$\frac{m\pi}{a} = \frac{km\pi}{L}. \qquad (9.145)$$

This introduces some simplification in the algebra because in this case one set of the expansion base functions is a subset of the other set of base functions. Therefore in fact the same set of base functions is used with orthogonality among the base functions.

9.7.3. Numerical results

As a numerical example, for geometric parameters we choose a = 0.05 m, L = $3a$, h = 0.01 m. For the elastic layer we consider steel with ρ_1 = 7850 kg/m^3 and μ = 80 × 10^9 N/m^2. For the upper and lower piezoelectric layers we consider PZT-5H. Damping in the system is introduced by allowing the elastic material constant c (and μ) to assume complex values. In our calculations c is replaced by $c(1 + iQ^{-1})$ and μ is treated similarly. We fix Q = 20. The linear algebraic equations for the undetermined coefficients are solved numerically using MATLAB on a computer. To describe the relatively simple displacement distribution over a few wavelengths, relatively few terms in the series expansions of u and ψ are needed. When twenty terms in the expansion of u are kept, the error of the maximum displacement is only 0.04%. Therefore twenty terms are kept in our calculations. It takes about 1–2 minutes to form and solve the equations as well as to calculate the data for various fields. We also run numerical tests further which did not show any signs of ill-conditioning of the equations when up to one hundred terms in each series for u and ψ are kept, while at the same time no observable difference in u and ψ results.

For basic electrical behaviors of power transmission through a wall please see Sec. 10.3. Below we focus on vibration localization or energy trapping. Figure 9.25 shows the displacement distribution (absolute value) along the upper surface of the elastic plate at the first resonance for different transducer thickness. In the figure $n = |V_2 / V_1|$ is the relative output voltage, and

$$\omega_0 = \frac{\pi}{2h}\sqrt{\frac{\mu}{\rho_1}}, \quad C_0 = \frac{\varepsilon a}{h'}, \quad Z_0 = \frac{1}{i\omega C_0}, \tag{9.146}$$

where μ is real. At the first resonance the plate is essentially vibrating at the fundamental thickness-twist mode. Due to the inertial effect of the transducer mass the resonant frequencies are a little less that ω_0 as expected. The most important behavior is that the vibration is large under the transducers and decays rapidly away from the transducers. This is the so-called energy trapping phenomenon. In the structure we are analyzing energy trapping which is mainly due to the transducer mass or inertia effect and therefore is sensitive to the transducer thickness. When energy trapping exists, the vibration is localized near the transducer region and does not cause global vibration of the whole structure, which is desirable in many applications. We note that the vibration does not completely decay to zero outside the transducer region. This is because that although the two transducers are of equal thickness, they are under different voltages. Therefore the whole structure is not exactly vibrating at the pure fundamental thickness-twist mode which is antisymmetric in y and can be trapped completely. A small component of the face-shear mode which is symmetric in y is involved which cannot be trapped. We also note that for thin transducers the output voltage is very low. For practical needs we need to increase the transducer thickness for a higher voltage output.

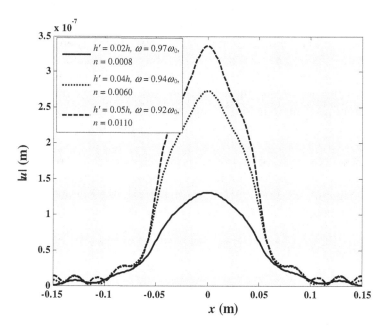

Fig. 9.25. Displacement distribution ($h = 0.01$ m, $h' = h''$, $Z = iZ_0$, $V_1 = 220$ volt).

9.8. A Transducer on an Elastic Plate

In this section we analyze the electrically forced vibration of an elastic plate with a finite piezoelectric actuator. The same actuator also works as a sensor to detect the motion of the elastic plate. This problem has applications in nondestructive evaluation (NDE) of structures. Consider the structure in Fig. 9.26 [60]. The elastic plate is isotropic and is of length $a + b$ and thickness $2h$. The piezoelectric transducer is of length $2L$ and thickness $2h'$. The transducer is electroded at its top and bottom surfaces, with the electrodes shown by the thick lines. The electrodes are very thin. Their mechanical effects are negligible. The transducer is under a driving voltage $2V$ which later will be assumed time-harmonic. The elastic plate is a dielectric (insulator). If a metal plate is considered, a very thin insulating layer is assumed between the bottom electrode of the transducer and the elastic plate.

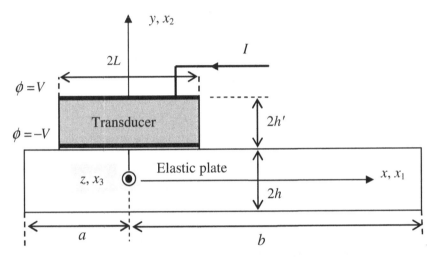

Fig. 9.26. An elastic plate with a piezoelectric transducer.

9.8.1. Governing equations

The transducer is governed by

$$v_T^2 \nabla^2 u = \ddot{u}, \quad \nabla^2 \psi = 0, \quad \phi = \psi + \frac{e}{\varepsilon} u. \tag{9.147}$$

The elastic plate is of an isotropic elastic material. The equation of motion takes the following form:

$$c_2^2 \nabla^2 u = \ddot{u}, \tag{9.148}$$

where $c_2^2 = \mu / \rho_1$. ρ_1 is the mass density. μ is the shear elastic constant. To calculate the charge Q_e and current I on the upper electrode as well as the admittance Y of the transducer, we need

$$Q_e = \int_{-L}^{L} -D_2\big|_{y=h+2h'} dx,$$

$$I = \dot{Q}_e = i\omega Q_e = Y 2V,$$

(9.149)

where a unit dimension in the x_3 direction is considered. At the top and bottom surfaces as well as the interface we have

$$\phi(h + 2h') = V, \quad |x| < L,$$

$$T_{23}(h + 2h') = 0, \quad |x| < L,$$

$$\phi(h^+) = -V, \quad |x| < L,$$

$$u(h^-) = u(h^+), \quad |x| < L,$$

(9.150)

$$T_{23}(h^-) = \begin{cases} T_{23}(h^+), & |x| < L, \\ 0, & -a < x < -L, \quad L < x < b, \end{cases}$$

$$T_{23}(-h) = 0 \quad -a < x < b.$$

(9.151)

In addition, there are traction-free boundary conditions at the left and right edges of the plate:

$$T_{13} = 0, \quad x_1 = -a, b.$$

(9.152)

For the transducer, the left and right edges are traction-free, and are unelectroded and charge-free:

$$T_{13} = 0, \quad D_1 = 0, \quad x_1 = \pm L.$$

(9.153)

9.8.2. Series solution

We look for solutions in the elastic plate and the transducer separately and then connect them by interface conditions. We use the complex notation. All fields are with an $\exp(i\omega t)$ factor which will be dropped for convenience. For the elastic plate we consider the following fields from separation of variables:

$$u = A_0 \sin(\eta_0 y) + B_0 \cos(\eta_0 y) + \sum_{n=1}^{\infty} [A_n \sin(\eta_n y) + B_n \cos(\eta_n y)]$$

$$\times (\cos\frac{bn\pi}{a+b}\cos\frac{n\pi x}{a+b} + \sin\frac{bn\pi}{a+b}\sin\frac{n\pi x}{a+b}),$$

(9.154)

$$T_{23} = \mu\eta_0 A_0 \cos(\eta_0 y) - \mu\eta_0 B_0 \sin(\eta_0 y) + \sum_{n=1}^{\infty} [\mu\eta_n A_n \cos(\eta_n y)$$

$$- \mu\eta_n B_n \sin(\eta_n y)](\cos\frac{bn\pi}{a+b}\cos\frac{n\pi x}{a+b} + \sin\frac{bn\pi}{a+b}\sin\frac{n\pi x}{a+b}),$$

(9.155)

where A_n and B_n are undetermined constants, and

$$\eta_n = \left[\left(\frac{\omega}{c_2}\right)^2 - \left(\frac{n\pi}{a+b}\right)^2\right]^{1/2}.$$

(9.156)

Equation (9.154) satisfies Eq. (9.148) and the edge conditions at $x = -a$ and b.

For the transducer, by separation of variables, the general solution can be written as

$$u = F_0 \sin(\xi_0 y) + G_0 \cos(\xi_0 y)$$

$$+ \sum_{m=2,4,6,\cdots}^{\infty} [F_m \sin(\xi_m y) + G_m \cos(\xi_m y)]\cos\frac{m\pi x}{2L}$$

(9.157)

$$+ \sum_{m=1,3,5,\cdots}^{\infty} [F_m \sin(\xi_m y) + G_m \cos(\xi_m y)]\sin\frac{m\pi x}{2L},$$

$$\psi = H_0 y + K_0$$

$$+ \sum_{m=2,4,6,\cdots}^{\infty} [H_m \sinh\frac{m\pi y}{2L} + K_m \cosh\frac{m\pi y}{2L}]\cos\frac{m\pi x}{2L}$$

(9.158)

$$+ \sum_{m=1,3,5,\cdots}^{\infty} [H_m \sinh\frac{m\pi y}{2L} + K_m \cosh\frac{m\pi y}{2L}]\sin\frac{m\pi x}{2L},$$

$$\phi = H_0 y + K_0 + \frac{e}{\varepsilon} F_0 \sin(\xi_0 y) + \frac{e}{\varepsilon} G_0 \cos(\xi_0 y)$$

$$+ \sum_{m=2,4,6,\cdots}^{\infty} [\frac{e}{\varepsilon} F_m \sin(\xi_m y) + \frac{e}{\varepsilon} G_m \cos(\xi_m y)$$

$$+ H_m \sinh\frac{m\pi y}{2L} + K_m \cosh\frac{m\pi y}{2L}]\cos\frac{m\pi x}{2L} \qquad (9.159)$$

$$+ \sum_{m=1,3,5,\cdots}^{\infty} [\frac{e}{\varepsilon} F_m \sin(\xi_m y) + \frac{e}{\varepsilon} G_m \cos(\xi_m y)$$

$$+ H_m \sinh\frac{m\pi y}{2L} + K_m \cosh\frac{m\pi y}{2L}]\sin\frac{m\pi x}{2L},$$

$$T_{23} = \bar{c}\xi_0 F_0 \cos(\xi_0 y) - \bar{c}\xi_0 G_0 \sin(\xi_0 y) + eH_0$$

$$+ \sum_{m=2,4,6,\cdots}^{\infty} [\bar{c}\xi_m F_m \cos(\xi_m y) - \bar{c}\xi_m G_m \sin(\xi_m y)$$

$$+ e\frac{m\pi}{2L} H_m \cosh\frac{m\pi y}{2L} + e\frac{m\pi}{2L} K_m \sinh\frac{m\pi y}{2L}]\cos\frac{m\pi x}{2L} \qquad (9.160)$$

$$+ \sum_{m=1,3,5,\cdots}^{\infty} [\bar{c}\xi_m F_m \cos(\xi_m y) - \bar{c}\xi_m G_m \sin(\xi_m y)$$

$$+ e\frac{m\pi}{2L} H_m \cosh\frac{m\pi y}{2L} + e\frac{m\pi}{2L} K_m \sinh\frac{m\pi y}{2L}]\sin\frac{m\pi x}{2L},$$

$$D_2 = \sum_{m=2,4,6,\cdots}^{\infty} [-\varepsilon\frac{m\pi}{2L} H_m \cosh\frac{m\pi y}{2L} - \varepsilon\frac{m\pi}{2L} K_m \sinh\frac{m\pi y}{2L}]$$

$$\times \cos\frac{m\pi x}{2L}$$

$$\qquad (9.161)$$

$$+ \sum_{m=1,3,5,\cdots}^{\infty} [-\varepsilon\frac{m\pi}{2L} H_m \cosh\frac{m\pi y}{2L} - \varepsilon\frac{m\pi}{2L} K_m \sinh\frac{m\pi y}{2L}]$$

$$\times \cos\frac{m\pi x}{2L},$$

where F_m, G_m, H_m and K_m are undetermined constants, and

$$\xi_m = \left[\left(\frac{\omega}{v_T^2}\right)^2 - \left(\frac{m\pi}{2L}\right)^2\right]^{1/2}. \qquad (9.162)$$

Equations (9.157)–(9.161) already satisfy Eq. (9.147) and the edge conditions at $x_1 = \pm L$.

Substitution of the above fields into the boundary and continuity conditions in Eqs. (9.150) and (9.151) gives equations involving the undetermined constants. The equations resulted from Eq. (9.150) need to be multiplied by

$$\cos\frac{p\pi x}{2L} \quad \text{or} \quad \sin\frac{p\pi x}{2L} \qquad (9.163)$$

and integrated from $-L$ to L for $p = 0, 2, 4,\ldots$ and $p = 1, 3, 5,\ldots$, respectively, to obtain linear algebraic equations of the undetermined constants. Equation (9.151) needs to be multiplied by

$$\cos\frac{bp\pi}{a+b}\cos\frac{p\pi x}{a+b} + \sin\frac{bp\pi}{a+b}\sin\frac{p\pi x}{a+b} \qquad (9.164)$$

and integrated from $-a$ to b for $p = 1, 2, 3,\ldots$ to obtain linear algebraic equations of the undetermined coefficients.

9.8.3. Numerical results

As a numerical example, consider a transducer made from PZT-5H. Damping is introduced by allowing the relevant elastic constants to assume complex values. In our calculations, c is replaced by $c(1 + iQ^{-1})$ and μ is replaced by $\mu(1 + iQ^{-1})$. We fix $Q = 20$. For the elastic layer, we consider steel with $\rho_1 = 7850\,\text{kg/m}^3$ and $\mu = 80\times10^9\,\text{N/m}^2$. We fix $a = b = 10\,L = 0.5$ m, $h = 0.01$ m, and $V = 220$ volt. We use the following frequency ω_0 as a normalizing frequency

$$\omega_0^2 = \frac{\mu\pi^2}{\rho_1(2h)^2}. \qquad (9.165)$$

Figure 9.27 shows the displacement (real part) of the middle point of the upper surface of the elastic plate versus the driving frequency. At certain frequencies the displacement becomes large (resonance).

In Figs. 9.28 and 9.29 we plot the displacement (real part) distribution along the upper surface of the elastic plate for the first two resonances. Since the excitation is on one side of the plate only, modes symmetric and antisymmetric about the middle plane of the plate are all excited. Depending on the driving frequency, certain modes may

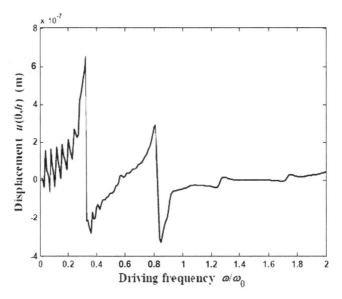

Fig. 9.27. $u(0, h)$ versus driving frequency ($h' = h$).

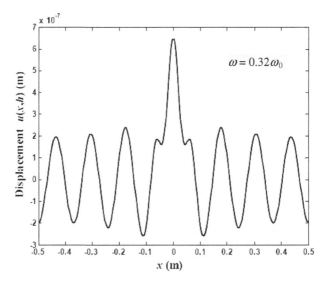

Fig. 9.28. $u(x, h)$ at the first resonance ($h' = h$, $a = b$, $\omega = 0.32\ \omega_0$).

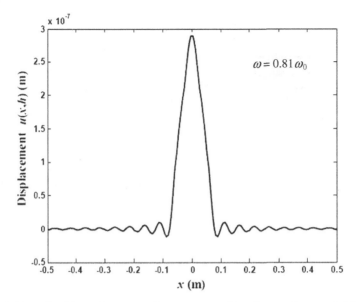

Fig. 9.29. $u(x, h)$ at the second resonance ($h' = h$, $a = b$, $\omega = 0.81\,\omega_0$).

dominate. As combinations of face-shear and thickness-twist modes, Figs. 9.28 and 9.29 sometimes show vibrations essentially confined under or close to the piezoelectric actuator, or essentially all over the plate. These are useful for different applications. In NDE we need modes that can feel the entire plate. Then no matter where a defect is the resonant frequency or the electric admittance is affected.

9.9. Two Transducers on an Elastic Plate

Consider the structure in Fig. 9.30 [61]. The elastic plate is isotropic, and is of length L and thickness $2h$. There are two piezoelectric transducers represented by the dotted areas in the figure. Their lengths are L' and L'', respectively. Their thicknesses are $2h'$ and $2h''$, respectively. The thick lines at the tops and bottoms of the transducers are metal electrodes. The electrodes are very thin. Their mechanical effects like inertia and stiffness are negligible. Electrically the electrodes provide electrical constraints on the electric potential which is a constant on an electrode. The elastic plate is a dielectric (insulator). If a metal plate is considered, a very thin insulating layer is assumed between the bottom electrodes of the transducers and the elastic plate. The transducer on the right is under a given driving voltage $2V_1$ which later will be assumed time-harmonic. The current I_1 flowing into the top driving electrode is unknown. The output transducer on the left is for receiving or detection of the acoustic waves excited by the transducer on the right. For the transducer on the left, both the voltage V_2 and the current I_2 are unknown. The output electrodes are connected by a circuit whose impedance is Z when the motion is time-harmonic.

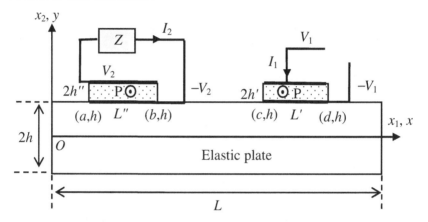

Fig. 9.30. An elastic plate with two piezoelectric transducers.

9.9.1. Governing equations

The transducers are governed by

$$v_T^2 \nabla^2 u = \ddot{u}, \quad \nabla^2 \psi = 0, \quad \phi = \psi + \frac{e}{\varepsilon} u. \qquad (9.166)$$

The elastic plate is of an isotropic material. The equation of motion takes the following form:

$$c_2^2 \nabla^2 u = \ddot{u}, \qquad (9.167)$$

where $c_2^2 = \mu / \rho_1$. ρ_1 is the mass density. μ is the shear elastic constant. The charges and currents on the electrodes of the transducers are given by

$$Q_1 = \int_c^d -D_2\big|_{y=h+2h'} dx,$$

$$I_1 = \dot{Q}_1 = i\omega Q_1, \qquad (9.168)$$

$$Q_2 = \int_a^b -D_2\big|_{y=h+2h''}\, dx,$$

(9.169)

$$I_2 = -\dot{Q}_2 = -i\omega Q_2.$$

The output circuit condition for the left transducer is

$$I_2 = 2V_2/Z.$$

(9.170)

9.9.2. Fields in different regions

We look for the solutions in the elastic plate and the transducers separately and then apply boundary, continuity and circuit conditions. For the elastic plate, from separation of variables, we obtain

$$u = A_0 \sin(\eta_0 y) + B_0 \cos(\eta_0 y)$$
$$+ \sum_{n=1}^{\infty} [A_n \sin(\eta_n y) + B_n \cos(\eta_n y)] \cos\frac{n\pi x}{L},$$

(9.171)

$$T_{23} = \mu\eta_0 A_0 \cos(\eta_0 y) - \mu\eta_0 B_0 \sin(\eta_0 y)$$
$$+ \sum_{n=1}^{\infty} [\mu\eta_n A_n \cos(\eta_n y) - \mu\eta_n B_n \sin(\eta_n y)] \cos\frac{n\pi x}{L},$$

(9.172)

where A_n and B_n are undetermined constants, and

$$\eta_n = \left[\left(\frac{\omega}{c_2}\right)^2 - \left(\frac{n\pi}{L}\right)^2 \right]^{1/2}.$$

(9.173)

Equation (9.171) satisfies Eq. (9.167) and the traction-free edge conditions of $u_{,1} = 0$ at $x = 0$ and $x = L$.

For the right transducer, the general solution can be written as

$$u = F_0 \sin(\xi_0 y) + G_0 \cos(\xi_0 y)$$
$$+ \sum_{m=1}^{\infty} [F_m \sin(\xi_m y) + G_m \cos(\xi_m y) \cos\frac{m\pi(x-c)}{L'},$$

$$\psi = H_0 y + K_0$$
$$+ \sum_{m=1}^{\infty} [H_m \sinh\frac{m\pi y}{L'} + K_m \cosh\frac{m\pi y}{L'}] \cos\frac{m\pi(x-c)}{L'},$$

(9.174)

$$\phi = H_0 y + K_0 + \frac{e}{\varepsilon} F_0 \sin(\xi_0 y) + \frac{e}{\varepsilon} G_0 \cos(\xi_0 y)$$

$$+ \sum_{m=1}^{\infty} [\frac{e}{\varepsilon} F_m \sin(\xi_m y) + \frac{e}{\varepsilon} G_m \cos(\xi_m y) \tag{9.175}$$

$$+ H_m \sinh \frac{m\pi y}{L'} + K_m \cosh \frac{m\pi y}{L'}] \cos \frac{m\pi(x-c)}{L'},$$

$$T_{23} = \overline{c} \xi_0 F_0 \cos(\xi_0 y) - \overline{c} \xi_0 G_0 \sin(\xi_0 y) + eH_0$$

$$+ \sum_{m=1}^{\infty} [\overline{c} \xi_m F_m \cos(\xi_m y) - \overline{c} \xi_m G_m \sin(\xi_m y) \tag{9.176}$$

$$+ e \frac{m\pi}{L'} H_m \cosh \frac{m\pi y}{L'} + e \frac{m\pi}{L'} K_m \sinh \frac{m\pi y}{L'}] \cos \frac{m\pi(x-c)}{L'},$$

$$D_2 = -\varepsilon H_0 + \sum_{m=1}^{\infty} [-\varepsilon \frac{m\pi}{L'} H_m \cosh \frac{m\pi y}{L'}$$

$$- \varepsilon \frac{m\pi}{L'} K_m \sinh \frac{m\pi y}{L'}] \cos \frac{m\pi(x-c)}{L'}, \tag{9.177}$$

where F_m, G_m, H_m and K_m are undetermined constants, and

$$\xi_m = \left[\left(\frac{\omega}{v_T} \right)^2 - \left(\frac{m\pi}{L'} \right)^2 \right]^{1/2}. \tag{9.178}$$

Equation (9.174) satisfies Eq. (9.166) and the boundary conditions at the transducer edges which are traction-free and are unelectroded.

Similarly, for the left transducer we have

$$u = P_0 \sin(\xi_0 y) + Q_0 \cos(\xi_0 y)$$

$$+ \sum_{m=1}^{\infty} [P_m \sin(\xi_m y) + Q_m \cos(\xi_m y)] \cos \frac{m\pi(x-a)}{L''},$$

$$\psi = N_0 y + R_0 \tag{9.179}$$

$$+ \sum_{m=1}^{\infty} [N_m \sinh \frac{m\pi y}{L''} + R_m \cosh \frac{m\pi y}{L''}] \cos \frac{m\pi(x-a)}{L''},$$

$$\phi = N_0 y + R_0 + \frac{e}{\varepsilon} P_0 \sin(\xi_0 y) + \frac{e}{\varepsilon} Q_0 \cos(\xi_0 y)$$

$$+ \sum_{m=1}^{\infty} \left[\frac{e}{\varepsilon} P_m \sin(\xi_m y) + \frac{e}{\varepsilon} Q_m \cos(\xi_m y) \right. \tag{9.180}$$

$$\left. + N_m \sinh\frac{m\pi y}{L''} + R_m \cosh\frac{m\pi y}{L''} \right] \cos\frac{m\pi(x-a)}{L''},$$

$$T_{23} = \bar{c}\,\xi_0 P_0 \cos(\xi_0 y) - \bar{c}\,\xi_0 Q_0 \sin(\xi_0 y) + e N_0$$

$$+ \sum_{m=1}^{\infty} \left[\bar{c}\,\xi_m P_m \cos(\xi_m y) - \bar{c}\,\xi_m Q_m \sin(\xi_m y) \right. \tag{9.181}$$

$$\left. + e\frac{m\pi}{L''} N_m \cosh\frac{m\pi y}{L''} + e\frac{m\pi}{L''} R_m \sinh\frac{m\pi y}{L''} \right] \cos\frac{m\pi(x-a)}{L''},$$

$$D_2 = -\varepsilon N_0 + \sum_{m=1}^{\infty} \left[-\varepsilon \frac{m\pi}{L''} N_m \cosh\frac{m\pi y}{L''} \right.$$

$$\left. - \varepsilon \frac{m\pi}{L''} R_m \sinh\frac{m\pi y}{L''} \right] \cos\frac{m\pi(x-a)}{L''}, \tag{9.182}$$

where P_m, Q_m, N_m and R_m are undetermined constants. Equation (9.179) satisfies Eq. (9.166) and the boundary conditions at the transducer edges.

9.9.3. Boundary, continuity and circuit conditions

For the two transducers we have

$$\begin{aligned}
\phi(h + 2h') &= V_1, & c < x < d, \\
T_{23}(h + 2h') &= 0, & c < x < d, \\
\phi(h^+) &= -V_1, & c < x < d, \\
u(h^+) &= u(h^-), & c < x < d,
\end{aligned} \tag{9.183}$$

$$\begin{aligned}
\phi(h + 2h'') &= V_2, & a < x < b, \\
T_{23}(h + 2h'') &= 0, & a < x < b, \\
\phi(h^+) &= -V_2, & a < x < b, \\
u(h^+) &= u(h^-), & a < x < b.
\end{aligned} \tag{9.184}$$

For the plate surfaces we impose

$$T_{23}(h^-) = \begin{cases} T_{23}(h^+), & c < x < d, \\ T_{23}(h^+), & a < x < b, \\ 0, & 0 < x < a, \ b < x < c, \ d < x < L, \end{cases} \tag{9.185}$$

$$T_{23}(-h) = 0, \quad 0 < x < L. \tag{9.186}$$

Corresponding to Eq. (9.183), we have

$$H_0(h + 2h') + K_0 + \frac{e}{\varepsilon} F_0 \sin(\xi_0(h + 2h')) + \frac{e}{\varepsilon} G_0 \cos(\xi_0(h + 2h'))$$

$$+ \sum_{m=1}^{\infty} [\frac{e}{\varepsilon} F_m \sin(\xi_m(h + 2h')) + \frac{e}{\varepsilon} G_m \cos(\xi_m(h + 2h'))$$

$$+ H_m \sinh \frac{m\pi(h + 2h')}{L'} + K_m \cosh \frac{m\pi(h + 2h')}{L'}] \tag{9.187}$$

$$\times \cos \frac{m\pi(x - c)}{L'} = V_1,$$

$$\overline{c} \xi_0 F_0 \cos(\xi_0(h + 2h')) - \overline{c} \xi_0 G_0 \sin(\xi_0(h + 2h')) + e H_0$$

$$+ \sum_{m=1}^{\infty} [\overline{c} \xi_m F_m \cos(\xi_m(h + 2h')) - \overline{c} \xi_m G_m \sin(\xi_m(h + 2h'))$$

$$+ e \frac{m\pi}{L'} H_m \cosh \frac{m\pi(h + 2h')}{L'} + e \frac{m\pi}{L'} K_m \sinh \frac{m\pi(h + 2h')}{L'}] \tag{9.188}$$

$$\times \cos \frac{m\pi(x - c)}{L'} = 0,$$

$$H_0 h + K_0 + \frac{e}{\varepsilon} F_0 \sin(\xi_0 h) + \frac{e}{\varepsilon} G_0 \cos(\xi_0 h)$$

$$+ \sum_{m=1}^{\infty} [\frac{e}{\varepsilon} F_m \sin(\xi_m h) + \frac{e}{\varepsilon} G_m \cos(\xi_m h)$$

$$+ H_m \sinh \frac{m\pi h}{L'} + K_m \cosh \frac{m\pi h}{L'}] \cos \frac{m\pi(x - c)}{L'} = -V_1, \tag{9.189}$$

$$A_0 \sin(\eta_0 h) + B_0 \cos(\eta_0 h)$$

$$+ \sum_{n=1}^{\infty} [A_n \sin(\eta_n h) + B_n \cos(\eta_n h)] \cos\frac{n\pi x}{L}$$

$$= F_0 \sin(\xi_0 h) + G_0 \cos(\xi_0 h) \tag{9.190}$$

$$+ \sum_{m=1}^{\infty} [F_m \sin(\xi_m h) + G_m \cos(\xi_m h)] \cos\frac{m\pi(x-c)}{L'},$$

which are to be multiplied by $\cos p\pi(x-c)/L'$ and integrated from c to d for $p = 0, 1, 2, 3,\ldots$ so that linear algebraic equations for the undetermined constants will result.

Similarly, corresponding to Eq. (9.184), we have

$$N_0(h + 2h'') + R_0 + \frac{e}{\varepsilon}P_0 \sin(\xi_0(h + 2h'')) + \frac{e}{\varepsilon}Q_0 \cos(\xi_0(h + 2h''))$$

$$+ \sum_{m=1}^{\infty} [\frac{e}{\varepsilon}P_m \sin(\xi_m(h + 2h'')) + \frac{e}{\varepsilon}Q_m \cos(\xi_m(h + 2h''))$$

$$+ N_m \sinh\frac{m\pi(h + 2h'')}{L''} + R_m \cosh\frac{m\pi(h + 2h'')}{L''}] \tag{9.191}$$

$$\times \cos\frac{m\pi(x-a)}{L''} = V_2,$$

$$\overline{c}\xi_0 P_0 \cos(\xi_0(h + 2h'')) - \overline{c}\xi_0 Q_0 \sin(\xi_0(h + 2h'')) + eN_0$$

$$+ \sum_{m=1}^{\infty} [\overline{c}\xi_m P_m \cos(\xi_m(h + 2h'')) - \overline{c}\xi_m Q_m \sin(\xi_m(h + 2h''))$$

$$+ e\frac{m\pi}{L''}N_m \cosh\frac{m\pi(h + 2h'')}{L''} + e\frac{m\pi}{L''}R_m \sinh\frac{m\pi(h + 2h'')}{L''}] \tag{9.192}$$

$$\times \cos\frac{m\pi(x-a)}{L''} = 0,$$

$$N_0 h + R_0 + \frac{e}{\varepsilon}P_0 \sin(\xi_0 h) + \frac{e}{\varepsilon}Q_0 \cos(\xi_0 h)$$

$$+ \sum_{m=1}^{\infty} [\frac{e}{\varepsilon}P_m \sin(\xi_m h) + \frac{e}{\varepsilon}Q_m \cos(\xi_m h) \tag{9.193}$$

$$+ N_m \sinh\frac{m\pi h}{L''} + R_m \cosh\frac{m\pi h}{L''}] \cos\frac{m\pi(x-a)}{L''} = -V_2,$$

$$A_0 \sin(\eta_0 h) + B_0 \cos(\eta_0 h)$$

$$+ \sum_{n=1}^{\infty} [A_n \sin(\eta_n h) + B_n \cos(\eta_n h)] \cos \frac{n \pi x}{L}$$

$$= P_0 \sin(\xi_0 h) + Q_0 \cos(\xi_0 h) \tag{9.194}$$

$$+ \sum_{m=1}^{\infty} [P_m \sin(\xi_m h) + Q_m \cos(\xi_m h)] \cos \frac{m \pi (x-a)}{L''},$$

which are to be multiplied by $\cos p\pi(x-a)/L''$ and integrated from a to b for $p = 0, 1, 2, 3, \ldots.$

Finally, corresponding to Eqs. (9.185) and (9.186), we have

$$\mu \eta_0 A_0 \cos(\eta_0 h) - \mu \eta_0 B_0 \sin(\eta_0 h)$$

$$+ \sum_{n=1}^{\infty} [\mu \eta_n A_n \cos(\eta_n h) - \mu \eta_n B_n \sin(\eta_n h)] \cos \frac{n \pi x}{L}$$

$$= \begin{cases} \bar{c}\xi_0 F_0 \cos(\xi_0 h) - \bar{c}\xi_0 G_0 \sin(\xi_0 h) + eH_0 \\ \qquad + \sum_{m=1}^{\infty} [\bar{c}\xi_m F_m \cos(\xi_m h) - \bar{c}\xi_m G_m \sin(\xi_m h) \\ \qquad + e\dfrac{m\pi}{L'} H_m \cosh \dfrac{m\pi h}{L'} + e\dfrac{m\pi}{L'} K_m \sinh \dfrac{m\pi h}{L'}] \\ \qquad \times \cos \dfrac{m\pi(x-c)}{L'}, \quad c < x < d, \\ \bar{c}\xi_0 P_0 \cos(\xi_0 h) - \bar{c}\xi_0 Q_0 \sin(\xi_0 h) + eN_0 \\ \qquad + \sum_{m=1}^{\infty} [\bar{c}\xi_m P_m \cos(\xi_m h) - \bar{c}\xi_m Q_m \sin(\xi_m h) \\ \qquad + e\dfrac{m\pi}{L''} N_m \cosh \dfrac{m\pi h}{L''} + e\dfrac{m\pi}{L''} R_m \sinh \dfrac{m\pi h}{L''}] \\ \qquad \times \cos \dfrac{m\pi(x-a)}{L''}, \quad a < x < b, \\ 0, \quad 0 < x < a, \ b < x < c, \text{ and } d < x < L, \end{cases} \tag{9.195}$$

$$\mu\eta_0 A_0 \cos(\eta_0 h) + \mu\eta_0 B_0 \sin(\eta_0 h)$$

$$+ \sum_{n=1}^{\infty} [\mu\eta_n A_n \cos(\eta_n h) + \mu\eta_n B_n \sin(\eta_n h)] \qquad (9.196)$$

$$\times \cos\frac{n\pi x}{L} = 0, \quad 0 < x < L,$$

which are to be multiplied by $\cos q\pi x / L$ and integrated from 0 to L for $q = 0, 1, 2, 3,\ldots.$ In addition, the circuit condition in Eq. (9.170) provides the last linear equation for the undetermined constants and V_2:

$$-i\omega \int_a^b -\{-\varepsilon N_0 + \sum_{m=0}^{\infty} [-\varepsilon \frac{m\pi}{L''} N_m \cosh\frac{m\pi(h+2h'')}{L''}$$

$$-\varepsilon \frac{m\pi}{L''} R_m \sinh\frac{m\pi(h+2h'')}{L''}]\cos\frac{m\pi(x-a)}{L''}\}dx = \frac{2V_2}{Z}. \qquad (9.197)$$

9.10. A Transducer on an Elastic Half-space

In this section we consider a transducer on a semi-infinite half-space (see Fig. 9.31) [62,63]. The transducer is under a driving voltage.

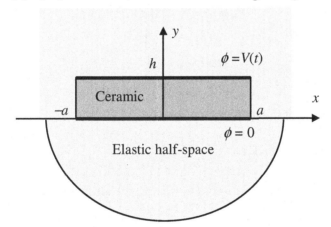

Fig. 9.31. A transducer on an elastic half-space.

9.10.1. Governing equations

The transducer is governed by

$$v_T^2 \nabla^2 u = \ddot{u}, \quad \nabla^2 \psi = 0, \quad \phi = \psi + \frac{e}{\varepsilon} u. \tag{9.198}$$

The elastic substrate is of an isotropic elastic material. The equation of motion takes the following form:

$$c_2^2 \nabla^2 u = \ddot{u}, \tag{9.199}$$

where $c_2^2 = \mu / \rho_1$. ρ_1 is the mass density. μ is the shear elastic constant. To calculate the charge Q_e on and the current I flowing into the upper electrode as well as the admittance Y of the structure, we need

$$Q_e = \int_{-a}^{a} -D_2 \big|_{y=h} \, dx,$$
$$I = \dot{Q}_e = i\omega Q_e = YV, \tag{9.200}$$

where a unit dimension in the x_3 direction is considered. The transducer edges at $x = \pm a$ are traction-free and unelectroded. At the top and bottom surfaces of the transducer we have

$$\phi(h) = V, \quad |x| < a,$$
$$T_{23}(h) = 0, \quad |x| < a,$$
$$\phi(0^+) = 0, \quad |x| < a, \tag{9.201}$$
$$u(0^+) = u(0^-), \quad |x| < a.$$

On the surface of the half-space we have

$$T_{23}(0^-) = \begin{cases} T_{23}(0^+), & |x| < a, \\ 0, & |x| > a. \end{cases} \tag{9.202}$$

We also require that at infinity ($x_2 = y \to -\infty$) all fields are bounded and outgoing.

9.10.2. Fields in different regions

The fields in the transducer are symmetric about $x = 0$ and can be taken from Sec. 9.7. From Eqs. (9.134) through (9.138) we have

$$u = F_0 \sin(\xi_0 y) + G_0 \cos(\xi_0 y)$$

$$+ \sum_{m=1}^{\infty} [F_m \sin(\xi_m y) + G_m \cos(\xi_m y)] \cos\frac{m\pi x}{a},$$

$$\psi = H_0 y + K_0 \tag{9.203}$$

$$+ \sum_{m=1}^{\infty} [H_m \sinh\frac{m\pi y}{a} + K_m \cosh\frac{m\pi y}{a}] \cos\frac{m\pi x}{a},$$

$$\phi = H_0 y + K_0 + \frac{e}{\varepsilon} F_0 \sin(\xi_0 y) + \frac{e}{\varepsilon} G_0 \cos(\xi_0 y)$$

$$+ \sum_{m=1}^{\infty} [\frac{e}{\varepsilon} F_m \sin(\xi_m y) + \frac{e}{\varepsilon} G_m \cos(\xi_m y) \tag{9.204}$$

$$+ H_m \sinh\frac{m\pi y}{a} + K_m \cosh\frac{m\pi y}{a}] \cos\frac{m\pi x}{a},$$

$$T_{23} = \bar{c}\xi_0 F_0 \cos(\xi_0 y) - \bar{c}\xi_0 G_0 \sin(\xi_0 y) + eH_0$$

$$+ \sum_{m=1}^{\infty} [\bar{c}\xi_m F_m \cos(\xi_m y) - \bar{c}\xi_m G_m \sin(\xi_m y) \tag{9.205}$$

$$+ e\frac{m\pi}{a} H_m \cosh\frac{m\pi y}{a} + e\frac{m\pi}{a} K_m \sinh\frac{m\pi y}{a}] \cos\frac{m\pi x}{a},$$

$$D_2 = -\varepsilon H_0 + \sum_{m=1}^{\infty} [-\varepsilon\frac{m\pi}{a} H_m \cosh\frac{m\pi y}{a}$$

$$- \varepsilon\frac{m\pi}{a} K_m \sinh\frac{m\pi y}{a}] \cos\frac{m\pi x}{a}, \tag{9.206}$$

where F_m, G_m, H_m and K_m are undetermined constants, and

$$\xi_m = \left[\left(\frac{\omega}{v_T^2}\right)^2 - \left(\frac{m\pi}{a}\right)^2 \right]^{1/2}. \tag{9.207}$$

Equation (9.203) satisfies Eq. (9.198) and the transducer edge conditions. For the elastic half-space governed by Eq. (9.199), we use Fourier transform in the x direction. Denoting

$$U(\xi, x_2) = \int_{-\infty}^{+\infty} u(x_1, x_2) \exp(-i\xi x_1) dx_1, \tag{9.208}$$

from Eq. (9.199), for time-harmonic motions, we obtain

$$c_2^2(U_{,22} - \xi^2 U) = -\omega^2 U .$$ (9.209)

The solution to Eq. (9.209) is

$$U(\xi, x_2) = A(\xi)\exp(-i\eta x_2) ,$$ (9.210)

where $A(\xi)$ is an undetermined function, and

$$\eta = \sqrt{\frac{\omega^2}{c_2^2} - \xi^2} .$$ (9.211)

In this problem, following [62], we consider the time-harmonic factor to be $\exp(-i\omega t)$, where $\omega > 0$ and $t > 0$. Since

$$\begin{aligned}
\exp(-i\eta x_2)\exp(-i\omega t) &= \exp[-i(\omega t + \eta x_2)] \\
&= \exp[-i(\omega t + \eta^R x_2 + i\eta^I x_2)] \\
&= \exp[-i(\omega t + \eta^R x_2)]\exp(\eta^I x_2),
\end{aligned}$$ (9.212)

for bounded and outgoing behavior at infinity, we require that

$$\eta^R \geq 0, \quad \eta^I \geq 0 .$$ (9.213)

Then u is formally given by the inverse Fourier transform as

$$\begin{aligned}
u(x_1, x_2) &= \frac{1}{2\pi} \int_{-\infty}^{+\infty} U(\xi, x_2)\exp(i\xi x_1)d\xi \\
&= \frac{1}{2\pi} \int_{-\infty}^{+\infty} A(\xi)\exp i(\xi x_1 - \eta x_2)d\xi.
\end{aligned}$$ (9.214)

The corresponding shear stress needed for boundary conditions is

$$T_{23} = \mu u_{,2} = \mu \frac{1}{2\pi} \int_{-\infty}^{+\infty} -i\eta A(\xi)\exp i(\xi x_1 + \eta x_2)d\xi.$$ (9.215)

9.10.3. Boundary and continuity conditions

Substituting the relevant fields into Eqs. (9.201) and (9.202), we obtain

$$H_0 h + K_0 + \frac{e}{\varepsilon} F_0 \sin(\xi_0 h) + \frac{e}{\varepsilon} G_0 \cos(\xi_0 h)$$

$$+ \sum_{m=1}^{\infty} \left[\frac{e}{\varepsilon} F_m \sin(\xi_m h) + \frac{e}{\varepsilon} G_m \cos(\xi_m h) \right. \tag{9.216}$$

$$+ H_m \sinh\frac{m\pi h}{a} + K_m \cosh\frac{m\pi h}{a} \left] \cos\frac{m\pi x}{a} = V, \quad |x| < a, \right.$$

$$\overline{c}\xi_0 F_0 \cos(\xi_0 h) - \overline{c}\xi_0 G_0 \sin(\xi_0 h) + e H_0$$

$$+ \sum_{m=1}^{\infty} [\overline{c}\xi_m F_m \cos(\xi_m h) - \overline{c}\xi_m G_m \sin(\xi_m h)$$

$$+ e\frac{m\pi}{a} H_m \cosh\frac{m\pi h}{a} + e\frac{m\pi}{a} K_m \sinh\frac{m\pi h}{a}] \tag{9.217}$$

$$\times \cos\frac{m\pi x}{a} = 0, \quad |x| < a,$$

$$K_0 + \frac{e}{\varepsilon} G_0 + \sum_{m=1}^{\infty} \left[\frac{e}{\varepsilon} G_m + K_m \right] \cos\frac{m\pi x}{a} = 0, \quad |x| < a, \tag{9.218}$$

$$G_0 + \sum_{m=1}^{\infty} G_m \cos\frac{m\pi x}{a} = \frac{1}{2\pi} \int_{-\infty}^{+\infty} A(\xi) \exp i(\xi x_1) d\xi, \quad |x| < a, \tag{9.219}$$

$$\mu \frac{1}{2\pi} \int_{-\infty}^{+\infty} -i\eta A(\xi) \exp i(\xi x_1) d\xi$$

$$= \begin{cases} \overline{c}\xi_0 F_0 + e H_0 \\ \quad + \sum_{m=1}^{\infty} [\overline{c}\xi_m F_m + e\frac{m\pi}{a} H_m] \cos\frac{m\pi x}{a}, \quad |x| < a, \\ \quad 0, \quad |x| > a. \end{cases} \tag{9.220}$$

Equation (9.220) can be inverted to obtain $A(\xi)$ in terms of F_m and H_m which is substituted back into Eq. (9.219). We then multiply Eqs. (9.216) through (9.219) by $\cos n\pi x/a$ and integrate them from $-a$ to a for $n = 0$, 1, 2, 3,... to obtain linear algebraic equations for the undetermined constants.

Chapter 10
Forced Vibrations in Polar Coordinates

Following the forced vibration analyses in Chapter 9, in this chapter we study forced vibrations in polar coordinates. The structures treated include single and multilayered circular cylindrical shells as well panels. They operate as transducers for power transmission, NDE, generators, and transformers.

10.1. A Shell Generator

The thickness-shear generator in Sec. 9.2 is probably the simplest example of a piezoelectric generator. Its theoretical value is that it allows a simple analytical solution showing the basic behaviors of the device and the effects of various parameters. For real applications, other structural shapes and operating modes need to be studied. In this section we consider a circular cylindrical ceramic shell generator as shown in Fig. 10.1 [64].

The ceramic is poled in the axial or 3 direction. The shell is electroded at its inner and outer surfaces, with electrodes shown by the thick lines in the figure. The electrodes are connected to a circuit whose impedance is Z_L when driven harmonically. The outer surface of the shell is rigidly attached to a fixed wall. The inner surface is driven harmonically by a pressure p at a given frequency ω. Since the ceramic layer is piezoelectric, across the thickness of the shell there exists a harmonic output voltage and a current due to the mechanically driven vibration of the shell.

Assume that $H \gg b - a$ so that edge effects can be neglected. Consider axisymmetric motions independent of z. In polar coordinates, let (r, θ, z) correspond to $(1, 2, 3)$, we have

$$u_z = u(r,t), \quad u_r = u_\theta = 0,$$
$$\phi = \phi(r,t).$$
(10.1)

The strain and electric field components are

$$2S_{rz} = u_{,r}, \quad E_r = -\phi_{,r}.$$
(10.2)

303

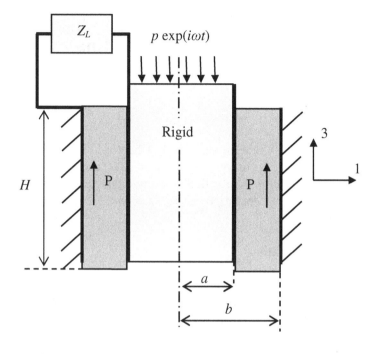

Fig. 10.1. A piezoelectric ceramic shell for power harvesting.

The stress and electric displacement components are

$$T_{rz} = cu_{,r} + e\phi_{,r}, \quad D_r = eu_{,r} - \varepsilon\phi_{,r}. \tag{10.3}$$

The equation of motion and the charge equation of electrostatics take the following form:

$$T_{rz,r} + \frac{1}{r}T_{rz} = \frac{1}{r}(rT_{rz})_{,r} = \rho\ddot{u},$$

$$D_{r,r} + \frac{1}{r}D_r = \frac{1}{r}(rD_r)_{,r} = 0. \tag{10.4}$$

Equation $(10.4)_2$ can be integrated to yield

$$D_r = eu_{,r} - \varepsilon\phi_{,r} = \frac{c_1(t)}{r}, \tag{10.5}$$

where c_1 is an integration constant. Substitution of Eq. (10.5) into Eqs. $(10.3)_1$ and $(10.4)_1$ gives

$$T_{rz} = \overline{c}u_{,r} - \frac{e}{\varepsilon}\frac{c_1}{r},$$

$$\frac{1}{r}(r\overline{c}u_{,r} - \frac{e}{\varepsilon}c_1)_{,r} = \rho\ddot{u},$$ (10.6)

where

$$\overline{c} = c(1+k^2), \quad k^2 = \frac{e^2}{\varepsilon c}.$$ (10.7)

For time-harmonic motions let

$$\{u,\phi,c_1,c_2\} = \text{Re}\{\{U,\Phi,C_1,C_2\}\exp(i\omega t)\}.$$ (10.8)

Then Eq. $(10.6)_2$ becomes

$$U_{,rr} + \frac{1}{r}U_{,r} + \frac{\rho\omega^2}{\overline{c}}U = 0.$$ (10.9)

The general solution to Eq. (10.9) is

$$U = C_3 J_0(\xi r) + C_4 Y_0(\xi r),$$ (10.10)

where J_0 and Y_0 are the first and second kind Bessel functions of order zero, and

$$\xi^2 = \frac{\rho\omega^2}{\overline{c}}.$$ (10.11)

Then,

$$\Phi = \frac{e}{\varepsilon}[C_3 J_0(\xi r) + C_4 Y_0(\xi r)] - \frac{C_1}{\varepsilon}\ln r - \frac{C_2}{\varepsilon},$$

$$T_{rz} = \overline{c}[C_3\xi J_0'(\xi r) + C_4\xi Y_0'(\xi r)] - \frac{e}{\varepsilon}\frac{C_1}{r},$$ (10.12)

$$D_r = \frac{C_1}{r},$$

which will be needed for prescribing boundary conditions. In Eq. (10.12), a prime represents differentiation with respect to the whole argument of the Bessel functions, i.e., ξr. For boundary conditions we also need the charge or current on the electrodes. For the outer electrode at $r = b$, we have the charge as the following integral over the electrode and the current as the time derivative of the charge:

$$Q_e = \int -D_r dA = -c_1 2\pi H,$$

$$I = -\dot{Q}_e.$$ (10.13)

In the complex notation, the current flowing out of this electrode is given by

$$\bar{I} = i\omega C_1 2\pi H .$$ (10.14)

We denote the voltage across the shell by V and its complex amplitude by $\bar{V}$. The electromechanical boundary conditions and the equation of the output circuit are

$$T_{rz}(a) = \bar{c}[C_3 \xi J_0'(\xi a) + C_4 \xi Y_0'(\xi a)] - \frac{e}{\varepsilon}\frac{C_1}{a} = \frac{p\pi a^2}{2\pi aH} = \frac{pa}{2H},$$

$$U(b) = C_3 J_0(\xi b) + C_4 Y_0(\xi b) = 0,$$

$$\Phi(b) - \Phi(a) = \frac{e}{\varepsilon}[C_3 J_0(\xi b) + C_4 Y_0(\xi b)$$ (10.15)

$$- C_3 J_0(\xi a) - C_4 Y_0(\xi a)] - \frac{C_1}{\varepsilon}\ln\frac{b}{a} = \bar{V},$$

$$\bar{I} = i\omega C_1 2\pi H = \frac{\bar{V}}{Z_L},$$

which are four equations for C_1, C_3, C_4 and $\bar{V}$, with p as the driving force. The solution to Eq. (10.15) can be found as

$$C_1 = \frac{pa^2 \varepsilon e[J_0(\xi b)Y_0(\xi a) - J_0(\xi a)Y_0(\xi b)]}{\Delta},$$

$$C_3 = \frac{pa^2 \varepsilon Y_0(\xi b)\ln(b/a)(1 + Z_L/Z_0)}{\Delta},$$

$$C_4 = -\frac{pa^2 \varepsilon J_0(\xi b)\ln(b/a)(1 + Z_L/Z_0)}{\Delta},$$ (10.16)

$$\bar{V} = \frac{pa^2 e[J_0(\xi b)Y_0(\xi a) - J_0(\xi a)Y_0(\xi b)]\ln(b/a)Z_L/Z_0}{\Delta},$$

where

$$\Delta = 2H\{e^2[J_0(\xi a)Y_0(\xi b) - J_0(\xi b)Y_0(\xi a)]$$
$$+ a\bar{c}\varepsilon\xi\ln(b/a)$$
$$\times[Y_0(\xi b)J_0'(\xi a) - J_0(\xi b)Y_0'(\xi a)(\xi a)](1 + Z_L/Z_0)\},$$ (10.17)

$$Z_0 = \frac{1}{i\omega C_0}, \quad C_0 = \frac{\varepsilon 2\pi H}{\ln(b/a)}.$$

$\Delta = 0$ yields an equation that determines the resonant frequencies. The output electrical power is

$$P_2 = \frac{1}{4}(\overline{I}\,\overline{V}^* + \overline{I}^*\overline{V}),$$ (10.18)

where an asterisk represents complex conjugate. The total shear force acting on the inner surface is:

$$F = p\pi a^2 \exp(i\omega t).$$ (10.19)

The velocity of the inner surface is

$$\dot{u} = i\omega U(a)\exp(i\omega t)$$
$$= i\omega[C_3 J_0(\xi a) + C_4 Y_0(\xi a)]\exp(i\omega t).$$ (10.20)

Therefore, the input mechanical power is

$$P_1 = -\frac{1}{4}p\pi a^2\{[i\omega U(a)]^* + i\omega U(a)\}.$$ (10.21)

The efficiency of the device is given by

$$\eta = \frac{P_2}{P_1}.$$ (10.22)

The power density is

$$p_2 = \frac{P_2}{\pi b^2 H}.$$ (10.23)

The basic behavior of this shell generator is qualitatively similar to what is shown in Figs. 9.4 through 9.7 of Sec. 9.3.

10.2. A Shell Transformer

Consider a piezoelectric circular cylinder of polarized ceramics (see Fig. 10.2) [65]. A polar coordinate system is defined by $x_1 = r\cos\theta$ and $x_2 = r\sin\theta$. The inner and outer faces at $r = a, b$ are traction-free. There are three electrodes at $r = a$, b, and c. The electrode at $r = c$ is grounded as a reference. A known, time-harmonic driving voltage V_1 is applied across the electrodes at $r = a$ and c. Due to the particular material orientation, the cylinder is driven into axial thickness-shear vibration and an output voltage V_2 can be picked up across the electrodes at $r = c$ and b which are joined by a load circuit whose impedance is Z. We consider unit thickness in the x_3 direction.

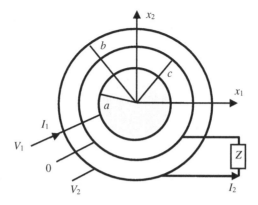

Fig. 10.2. A two-layered cylinder with three electrodes as a transformer.

10.2.1. Governing equations

Consider axisymmetric motions independent of θ. Although the material of the cylinder is homogeneous, due to the presence of an interior electrode, the electric field may have a jump. Therefore we treat the cylinder as two layers. The governing equations, boundary and continuity conditions are

$$\bar{c}\nabla^2 u = \rho\ddot{u}, \quad a < r < c,$$

$$\nabla^2\psi = 0, \quad a < r < c, \tag{10.24}$$

$$\phi = \psi + \frac{e}{\varepsilon}u, \quad a < r < c,$$

$$\bar{c}\nabla^2 u = \rho\ddot{u}, \quad c < r < b,$$

$$\nabla^2\psi = 0, \quad c < r < b, \tag{10.25}$$

$$\phi = \psi + \frac{e}{\varepsilon}u, \quad c < r < b,$$

$$T_{r3} = 0, \quad \phi = V_1, \quad r = a,$$

$$T_{r3} = 0, \quad \phi = V_2, \quad r = b, \tag{10.26}$$

$$u(c^-) = u(c^+), \quad T(c^-) = T(c^+), \quad \phi(c^-) = \phi(c^+) = 0,$$

where, for axisymmetric motions, in polar coordinates,

$$\nabla^2 = \frac{\partial^2}{\partial r^2} + \frac{1}{r}\frac{\partial}{\partial r}. \tag{10.27}$$

To obtain the input and output currents and powers we need the charges and currents on the electrodes. For the input electrode at $r = a$, we have

$$Q_1 = \int_0^{2\pi} D_r\big|_{r=a} \, a d\theta,$$

$$I_1 = \dot{Q}_1.$$
(10.28)

Similarly, for the output electrode at $r = b$, we have

$$Q_2 = \int_0^{2\pi} -D_r\big|_{r=b} \, b d\theta,$$

$$I_2 = -\dot{Q}_2.$$
(10.29)

For harmonic motions we use the complex notation:

$$(u, \psi, \phi, T_{r3}, D_r, V_1, I_1, Q_1, V_2, I_2, Q_2)$$
$$= \mathrm{Re}\{(U, \Psi, \Phi, T, D, \bar{V}_1, \bar{I}_1, \bar{Q}_1, \bar{V}_2, \bar{I}_2, \bar{Q}_2) \exp(i\omega t)\}.$$
(10.30)

We also need the following equation for the output circuit:

$$\bar{V}_2 = \bar{I}_2 Z .$$
(10.31)

The input and output powers are given by

$$P_1 = \frac{1}{4}(\bar{I}_1 \bar{V}_1^* + \bar{I}_1^* \bar{V}_1),$$

$$P_2 = \frac{1}{4}(\bar{I}_2 \bar{V}_2^* + \bar{I}_2^* \bar{V}_2) = \frac{1}{4}(\bar{I}_2 \bar{I}_2^* Z^* + \bar{I}_2^* \bar{I}_2 Z)$$
(10.32)

$$= \frac{1}{4}\bar{I}_2 \bar{I}_2^*(Z^* + Z) = \frac{1}{2}|\bar{I}_2|^2 \, \mathrm{Re}\{Z\}.$$

The efficiency and power density of the transformer are defined as

$$\eta = \frac{P_2}{P_1}, \qquad p = \frac{P_2}{\pi b^2}$$
(10.33)

10.2.2. Solution

For time-harmonic motions, the equations for U and Ψ are

$$\frac{\partial^2 U}{\partial r^2} + \frac{1}{r}\frac{\partial U}{\partial r} = -\frac{\rho\omega^2}{\bar{c}}U,$$

$$\frac{\partial^2 \Psi}{\partial r^2} + \frac{1}{r}\frac{\partial \Psi}{\partial r} = 0.$$
(10.34)

The general solution to Eq. (10.34) is

$$U = A_1 J_0(\xi r) + A_2 Y_0(\xi r), \quad a < r < c,$$
$$\Psi = A_3 \ln r + A_4, \quad a < r < c, \tag{10.35}$$

$$U = B_1 J_0(\xi r) + B_2 Y_0(\xi r), \quad c < r < b,$$
$$\Psi = B_3 \ln r + B_4, \quad c < r < b, \tag{10.36}$$

where A_1 through A_4 and B_1 through B_4 are undetermined constants, J_0 and Y_0 are zero-order Bessel functions of the first and second kind, and

$$\xi^2 = \frac{\rho \omega^2}{\bar{c}}. \tag{10.37}$$

Hence, for $a < r < c$,

$$\Phi = \frac{e}{\varepsilon}[A_1 J_0(\xi r) + A_2 Y_0(\xi r)] + A_3 \ln r + A_4,$$

$$T = -\bar{c}\xi[A_1 J_1(\xi r) + A_2 Y_1(\xi r)] + e\frac{A_3}{r}, \tag{10.38}$$

$$D = -\varepsilon\frac{A_3}{r},$$

where $J_0' = -J_1$ and $Y_0' = -Y_1$ have been used. For $c < r < b$, expressions similar to Eq. (10.38) can be written with A_1 through A_4 replaced by B_1 through B_4. Substitution of Eqs. (10.35) and (10.38) as well as the similar expressions for $c < r < b$ into the boundary and continuity conditions in Eq. (10.26) yields

$$-\bar{c}\xi[A_1 J_1(\xi a) + A_2 Y_1(\xi a)] + e\frac{A_3}{a} = 0,$$

$$\frac{e}{\varepsilon}[A_1 J_0(\xi a) + A_2 Y_0(\xi a)] + A_3 \ln a + A_4 = \bar{V}_1, \tag{10.39}$$

$$-\bar{c}\xi[B_1 J_1(\xi b) + B_2 Y_1(\xi b)] + e\frac{B_3}{b} = 0,$$

$$\frac{e}{\varepsilon}[B_1 J_0(\xi b) + B_2 Y_0(\xi b)] + B_3 \ln b + B_4 = \bar{V}_2, \tag{10.40}$$

$$A_1 J_0(\xi c) + A_2 Y_0(\xi c) = B_1 J_0(\xi c) + B_2 Y_0(\xi c)$$

$$- \bar{c}\,\xi[A_1 J_1(\xi c) + A_2 Y_1(\xi c)] + e\frac{A_3}{c}$$

$$= -\bar{c}\,\xi[B_1 J_1(\xi c) + B_2 Y_1(\xi c)] + e\frac{B_3}{c}, \qquad (10.41)$$

$$\frac{e}{\varepsilon}[A_1 J_0(\xi c) + A_2 Y_0(\xi c)] + A_3 \ln c + A_4 = 0,$$

$$\frac{e}{\varepsilon}[B_1 J_0(\xi c) + B_2 Y_0(\xi c)] + B_3 \ln c + B_4 = 0.$$

Formally Eqs. (10.39)–(10.41) are eight equations for A_1 through A_4 and B_1 through B_4, but the output voltage $\bar{V}_2$ is unknown. The additional equation needed is the output circuit equation given in Eq. (10.31). From Eqs. (10.31), (10.29) and the expression for D we calculate

$$\frac{\bar{V}_2}{Z} = \bar{I}_2 = -i\omega \bar{Q}_2 = i\omega \int_0^{2\pi} D\big|_{r=b}\, b d\theta,$$

$$= i\omega \int_0^{2\pi} -\varepsilon\frac{B_3}{b} b d\theta = -i\omega\varepsilon 2\pi B_3, \qquad (10.42)$$

or

$$\bar{V}_2 + i\omega\varepsilon 2\pi Z B_3 = 0, \qquad (10.43)$$

which is the additional equation that completes the mathematical problem. The nine equations in Eqs. (10.39)–(10.41) and (10.43) determine A_1–A_4, B_1–B_4 and $\bar{V}_2$. To calculate the input power we also need

$$\bar{I}_1 = i\omega \int_0^{2\pi} D\big|_{r=a}\, a d\theta$$

$$= i\omega \int_0^{2\pi} -\varepsilon\frac{A_3}{a} a d\theta = -i\omega\varepsilon 2\pi A_3. \qquad (10.44)$$

10.2.3. Numerical results

As an example, consider a transformer made from Pz24. Some damping is introduced by allowing the elastic material constants c_{44} to assume complex values. In our calculations c_{44} is replaced by $c_{44}(1 + iQ^{-1})$. We fix $Q = 100$, $a = 10$ mm, $b = 30$ mm, $c = 20$ mm. We normalize the frequency and load by

$$\omega_0^2 = \frac{\pi^2 \bar{c}}{4\rho(b-a)^2}, \quad Z_0 = \frac{1}{i\omega C_0}, \quad C_0 = \frac{\varepsilon 2\pi}{\ln(b/c)}. \tag{10.45}$$

The basic electrical behavior of the transformer is qualitatively similar to the transformer in Sec. 9.4. Therefore, in this section, we examine the mechanical fields in the transformer only which we did not do in Sec. 9.4. The distribution of mechanical fields, displacement and stress, are shown in Figs. 10.3 and 10.4, respectively. These fields are not directly related to transformer performance but they need to be considered in transformer design. The displacement has a nodal point near the interface. The stress is large near the interface.

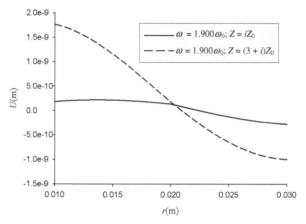

Fig. 10.3. Displacement distribution (u_3).

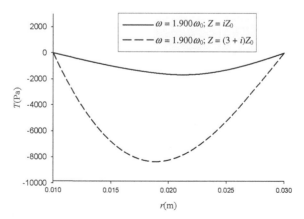

Fig. 10.4. Stress distribution (T_{r3}).

10.3. Power Transmission through an Elastic Shell

Consider a three-layered circular cylindrical shell as shown in Fig. 10.5 [66]. A polar coordinate system is defined by $x_1 = r\cos\theta$ and $x_2 = r\sin\theta$. The middle layer $c < r < d$ is an elastic shell which we assume to be metallic. The inner layer $a < r < c$ and the outer layer $d < r < b$ are piezoelectric transducers. We consider unit dimension in the x_3 direction. The inner and outer surfaces at $r = a$ and b are traction-free. The surfaces of the piezoelectric layers at $r = a$, b, c and d are all electroded. The elastic shell is grounded as a reference for the electric potential. A known, time-harmonic driving voltage V_1 is applied across the electrodes at $r = b$ and d. Due to the particular material orientation, the cylinder is driven into axial thickness-shear vibration, and an output voltage V_2 can be picked up across the electrodes at $r = a$ and c which are joined by a load circuit whose impedance is Z.

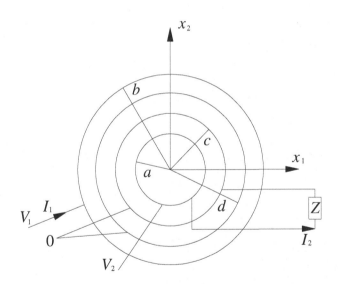

Fig. 10.5. A three-layered circular cylindrical shell.

10.3.1. Governing equations

For axisymmetric motions independent of θ, the governing equations, boundary and continuity conditions are:

$$\overline{c}\nabla^2 u = \rho\ddot{u}, \quad a < r < c \quad \text{or} \quad d < r < b,$$
$$\nabla^2\psi = 0, \quad a < r < c \quad \text{or} \quad d < r < b, \qquad (10.46)$$
$$c_{44}^0\nabla^2 u = \rho_0\ddot{u}, \quad c < r < d,$$

$$T_{r3} = 0, \quad \phi = V_2, \quad r = a,$$
$$T_{r3} = 0, \quad \phi = V_1, \quad r = b,$$
$$u(c^-) = u(c^+), \quad u(d^-) = u(d^+), \qquad (10.47)$$
$$T_{r3}(c^-) = T_{r3}(c^+), \quad T_{r3}(d^-) = T_{r3}(d^+),$$
$$\phi(c^-) = \phi(d^+) = 0,$$

where c_{44}^0 is the shear constant of the elastic shell. To obtain the input and output powers, we need the charges and currents on the electrodes. For the input electrode at $r = b$, we have:

$$Q_1 = -\int_0^{2\pi} D_r\big|_{r=b}\, b(d\theta),$$
$$I_1 = \dot{Q}_1. \qquad (10.48)$$

Similarly, for the output electrode at $r = a$, we have:

$$Q_2 = -\int_0^{2\pi} D_r\big|_{r=a}\, a(d\theta),$$
$$I_2 = \dot{Q}_2. \qquad (10.49)$$

For harmonic motions we use the complex notation:

$$(u,\psi,\phi,T_{r3},D_r,V_1,I_1,Q_1,V_2,I_2,Q_2)$$
$$= \text{Re}\{(U,\Psi,\Phi,T,D,\overline{V}_1,\overline{I}_1,\overline{Q}_1,\overline{V}_2,\overline{I}_2,\overline{Q}_2)\exp(i\omega t)\}. \qquad (10.50)$$

We also need the following equation for the output circuit:

$$\overline{V}_2 = \overline{I}_2 Z. \qquad (10.51)$$

The input and output powers are given by:

$$P_1 = \frac{1}{4}(\overline{I}_1\overline{V}_1^* + \overline{I}_1^*\overline{V}_1),$$

$$P_2 = \frac{1}{4}(\overline{I}_2\overline{V}_2^* + \overline{I}_2^*\overline{V}_2) = \frac{1}{4}(\overline{I}_2\overline{I}_2^* Z^* + \overline{I}_2^*\overline{I}_2 Z) \qquad (10.52)$$

$$= \frac{1}{4}\overline{I}_2\overline{I}_2^*(Z^* + Z) = \frac{1}{2}|\overline{I}_2|^2\, \text{Re}\{Z\}.$$

where an asterisk indicates complex conjugate. The efficiency of the device is defined as:

$$\eta = \frac{P_2}{P_1}. \tag{10.53}$$

10.3.2. Solution

For time-harmonic motions, the equations for U and Ψ are:

$$\frac{\partial^2 U}{\partial r^2} + \frac{1}{r}\frac{\partial U}{\partial r} = -\frac{\rho\omega^2}{\overline{c}}U, \quad a < r < c \quad \text{or} \quad d < r < b,$$

$$\frac{\partial^2 \Psi}{\partial r^2} + \frac{1}{r}\frac{\partial \Psi}{\partial r} = 0, \quad a < r < c \quad \text{or} \quad d < r < b, \tag{10.54}$$

$$\frac{\partial^2 U}{\partial r^2} + \frac{1}{r}\frac{\partial U}{\partial r} = -\frac{\rho_0\omega^2}{c_{44}^0}U, \quad c < r < d.$$

The general solution to Eq. (10.54) is:

$$U = A_1 J_0(\xi r) + A_2 Y_0(\xi r), \quad a < r < c,$$

$$\Psi = A_3 \ln r + A_4, \quad a < r < c,$$

$$U = B_1 J_0(\xi r) + B_2 Y_0(\xi r), \quad d < r < b, \tag{10.55}$$

$$\Psi = B_3 \ln r + B_4, \quad d < r < b,$$

$$U = C_1 J_0(\xi' r) + C_2 Y_0(\xi' r), \quad c < r < d,$$

where A_1 through A_4, B_1 through B_4 and C_1 through C_2 are undetermined constants. J_0 and Y_0 are the zero-order Bessel functions of the first and second kind, and:

$$\xi^2 = \frac{\rho\omega^2}{\overline{c}}, \quad \xi'^2 = \frac{\rho_0\omega^2}{c_{44}^0}. \tag{10.56}$$

Hence,

$$\Phi = \frac{e}{\varepsilon}[A_1 J_0(\xi r) + A_2 Y_0(\xi r)] + A_3 \ln r + A_4, \quad a < r < c,$$

$$T = -\overline{c}\xi[A_1 J_1(\xi r) + A_2 Y_1(\xi r)] + e\frac{A_3}{r}, \quad a < r < c, \tag{10.57}$$

$$D = -\varepsilon\frac{A_3}{r}, \quad a < r < c,$$

$$\Phi = \frac{e}{\varepsilon}[B_1 J_0(\xi r) + B_2 Y_0(\xi r)] + B_3 \ln r + B_4, \quad d < r < b,$$

$$T = -\bar{c}\xi[B_1 J_1(\xi r) + B_2 Y_1(\xi r)] + e\frac{B_3}{r}, \quad d < r < b, \quad (10.58)$$

$$D = -\varepsilon\frac{B_3}{r}, \quad d < r < b,$$

$$T = -c_{44}^0 \xi'[C_1 J_1(\xi' r) + C_2 Y_1(\xi' r)], \quad c < r < d, \quad (10.59)$$

where $J_0' = -J_1$ and $Y_0' = -Y_1$ have been used. Substitution of Eqs. (10.55) and (10.57)–(10.59) into the boundary and continuity conditions in Eq. (10.47) yields:

$$-\bar{c}\xi[A_1 J_1(\xi a) + A_2 Y_1(\xi a)] + e\frac{A_3}{a} = 0,$$

$$\frac{e}{\varepsilon}[A_1 J_0(\xi a) + A_2 Y_0(\xi a)] + A_3 \ln a + A_4 = \bar{V}_2, \quad (10.60)$$

$$-\bar{c}\xi[B_1 J_1(\xi b) + B_2 Y_1(\xi b)] + e\frac{B_3}{b} = 0,$$

$$\frac{e}{\varepsilon}[B_1 J_0(\xi b) + B_2 Y_0(\xi b)] + B_3 \ln b + B_4 = \bar{V}_1, \quad (10.61)$$

$$A_1 J_0(\xi c) + A_2 Y_0(\xi c) = C_1 J_0(\xi' c) + C_2 Y_0(\xi' c),$$

$$-\bar{c}\xi[A_1 J_1(\xi c) + A_2 Y_1(\xi c)] + e\frac{A_3}{c}$$

$$= -c_{44}^0 \xi'[C_1 J_1(\xi' c) + C_2 Y_1(\xi' c)], \quad (10.62)$$

$$\frac{e}{\varepsilon}[A_1 J_0(\xi c) + A_2 Y_0(\xi c)] + A_3 \ln c + A_4 = 0,$$

$$C_1 J_0(\xi' d) + C_2 Y_0(\xi' d) = B_1 J_0(\xi d) + B_2 Y_0(\xi d),$$

$$-c_{44}^0 \xi'[C_1 J_1(\xi' d) + C_2 Y_1(\xi' d)]$$

$$= -\bar{c}\xi[B_1 J_1(\xi d) + B_2 Y_1(\xi d)] + e\frac{B_3}{d}, \quad (10.63)$$

$$\frac{e}{\varepsilon}[B_1 J_0(\xi d) + B_2 Y_0(\xi d)] + B_3 \ln d + B_4 = 0.$$

Formally, Eqs. (10.60)–(10.63) represent ten equations for A_1 through A_4, B_1 through B_4 and C_1 through C_2, but the output voltage $\bar{V}_2$ is unknown.

The additional equation needed is the output circuit equation in Eq. (10.51), which needs to be written in terms of the above-mentioned eleven unknowns. From Eqs. (10.51), (10.49) and the expression for D we calculate:

$$\frac{\overline{V}_2}{Z} = \overline{I}_2 = -i\omega\overline{Q}_2 = -i\omega \int_0^{2\pi} D\big|_{r=a} a(d\theta)$$

$$= i\omega \int_0^{2\pi} \varepsilon \frac{A_3}{b} b(d\theta) = i\omega\varepsilon 2\pi A_3, \tag{10.64}$$

or

$$\overline{V}_2 - i\omega\varepsilon 2\pi Z A_3 = 0, \tag{10.65}$$

which is the additional equation that completes the mathematical problem. The eleven equations in Eqs. (10.60)–(10.63) and (10.65) determine A_1 through A_4, B_1 through B_4, C_1 through C_2 and $\overline{V}_2$. To calculate the input power we also need:

$$\overline{I}_1 = -i\omega \int_0^{2\pi} D\big|_{r=a} b(d\theta)$$

$$= i\omega \int_0^{2\pi} \varepsilon \frac{B_3}{b} b(d\theta) = i\omega\varepsilon 2\pi B_3. \tag{10.66}$$

10.3.3. Numerical results

As an example, consider transducers made from polarized ceramics PZT-5H. Damping is introduced by allowing the elastic constant c_{44} to assume complex values. In our calculations c_{44} is replaced by $c_{44}(1 + iQ^{-1})$. c_{44}^0 is replaced by $c_{44}^0(1 + iQ^{-1})$. We fix $Q = 100$. $A = 10$ mm, $b = 40$ mm, $c = 20$ mm, and $d = 30$ mm are used. We normalize the frequency and load by:

$$\omega_0^2 = \frac{\pi^2 \overline{c}}{4\rho(b-a)^2}, \quad Z_0 = \frac{1}{i\omega C_0}, \quad C_0 = \frac{\varepsilon 2\pi}{\ln(c/a)}. \tag{10.67}$$

Note that in defining ω_0 real material constants are used. For the elastic shell, we consider iron (Fe) with $\rho_0 = 7800$ kg/m^3 and $c_{44}^0 = 11.7 \times 10^{10}$ N/m^2. The input voltage V_1 is fixed to be 3 volt.

In Fig. 10.6, we plot the normalized output voltage $|\overline{V}_2 / \overline{V}_1|$ versus the driving frequency ω. Two different loads are used in the calculation.

One is a pure resistor load. The other is a complex load impedance. The output voltage assume maxima at the resonant frequencies as expected. It seems that if $|Z|$ is larger the output is higher. The two values of Z used in the figure are so chosen that they are reasonably large (away from the short-circuit condition) for a significant voltage output and at the same time they are not close to open-circuit condition so that the output still varies with Z. In principle the load can also affect the resonant frequencies and in turn affect the behavior near resonance. Therefore the results in Fig. 10.6 are load-sensitive in general.

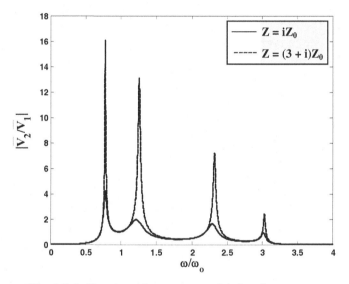

Fig. 10.6. Output voltage versus driving frequency.

Figure 10.7 shows the output voltage $|\bar{V}_2 / \bar{V}_1|$ as a function of a resistor load Z. As the load increases from zero, the output voltage increases from zero essentially linearly. For large loads, the output voltage is essentially a constant, exhibiting saturation. Physically, for very large loads, the output electrodes are essentially open. The output voltage is saturated and the output current essentially vanishes.

Input admittance as a function of the driving frequency is given in Fig. 10.8 for two values of the load Z. The frequency at which the input admittance assumes minimum is called the antiresonant frequency.

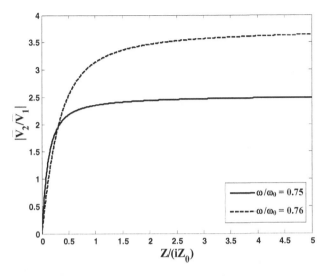

Fig. 10.7. Output voltage versus load near resonance.

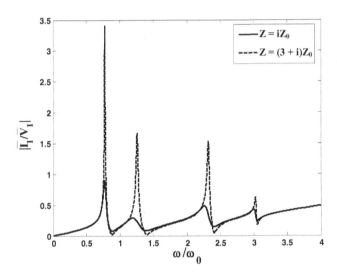

Fig. 10.8. Input admittance (in $1/\Omega$) versus driving frequency.

Figure 10.9 shows the input admittance versus the load for two driving frequencies near the first resonant frequency. For large loads the input admittance approaches a constant that is frequency dependent.

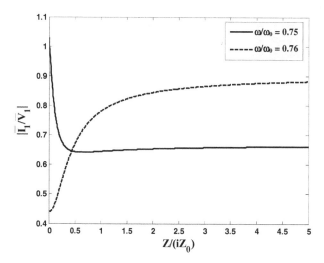

Fig. 10.9. Input admittance (in $1/\Omega$) versus load near resonance.

Figure 10.10 shows the dependence of efficiency on driving frequency. As a ratio between the input and output powers (two functions with peaks), the behavior of efficiency versus frequency is without sharp peaks. It is evident from comparison with Fig. 10.6 that the efficiency does not simply assume maxima at the resonant frequencies.

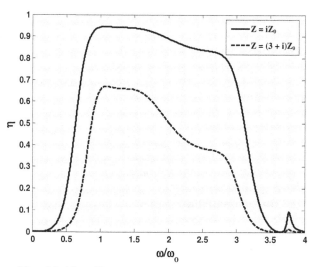

Fig. 10.10. Efficiency versus driving frequency.

Efficiency as a function of the load is shown in Fig. 10.11 for two values of the driving frequency close to the first resonant frequency. As the load increases from zero, the efficiency first increases from zero linearly, then it reaches a maximum. After the maximum, the efficiency decreases monotonically.

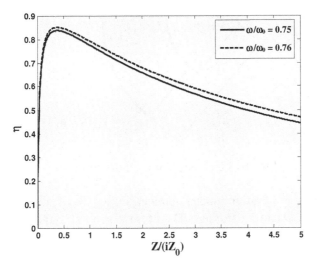

Fig. 10.11. Efficiency versus load near resonance.

10.4. A Circular Cylindrical Panel

In this section we perform a forced vibration analysis on the circular cylindrical panel (see Fig. 10.12) [53] whose free vibration was treated in Sec. 8.4. Mechanically the boundaries are all traction-free. Electrically the boundaries at $\theta = \pm\beta$ are unelectroded. The boundaries at $r = a, b$ are electroded and a driving voltage is applied across the electrodes. A cylindrical coordinate system is defined by $x = r\cos\theta$, $y = r\sin\theta$ and $z = z$. In the index notation below, (x, y, z) correspond to $(1, 2, 3)$.

The driving voltage applied across the electrodes at $r = a, b$ is time-harmonic and is given by

$$\phi = \begin{cases} -V\exp(i\omega t), & r = a, \\ V\exp(i\omega t), & r = b. \end{cases} \qquad (10.68)$$

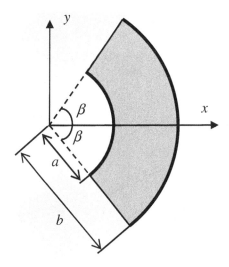

Fig. 10.12. A circular cylindrical panel.

Among the free vibration modes obtained in Sec. 8.4, only those that are symmetric in θ can be excited. Therefore a general symmetric solution satisfying the transducer edge conditions at $\theta = \pm\beta$ is constructed as [53]

$$u = C_0 J_0(\xi r) + D_0 Y_0(\xi r)$$

$$+ \sum_{m=1}^{\infty} [C_m J_{\nu_m}(\xi r) + D_m Y_{\nu_m}(\xi r)] \cos \nu_m \theta, \qquad (10.69)$$

$$\psi = F_0 \ln r + G_0 + \sum_{m=1}^{\infty} [F_m r^{\nu_m} + G_m r^{-\nu_m}] \cos \nu_m \theta.$$

The stress, electric displacement and electric potential needed for boundary conditions are obtained as follows:

$$\phi = \psi + \frac{e}{\varepsilon} u$$

$$= \frac{e}{\varepsilon}[C_0 J_0(\xi r) + D_0 Y_0(\xi r)] + F_0 \ln r + G_0$$

$$+ \sum_{m=1}^{\infty} [\frac{e}{\varepsilon} C_m J_{\nu_m}(\xi r) + \frac{e}{\varepsilon} D_m Y_{\nu_m}(\xi r)$$

$$+ F_m r^{\nu_m} + G_m r^{-\nu_m}] \cos \nu_m \theta, \qquad (10.70)$$

$$T_{rz} = \bar{c}u_{,2} + e\psi_{,2}$$

$$= -\bar{c}\xi[C_0 J_1(\xi r) + D_0 Y_1(\xi r)] + e\frac{F_0}{r}$$

$$+ \sum_{m=1}^{\infty} [\bar{c}\xi C_m J'_{v_m}(\xi r) + \bar{c}\xi D_m Y'_{v_m}(\xi r)$$

$$+ ev_m F_m r^{v_m - 1} - ev_m G_m r^{-v_m - 1}]\cos v_m \theta,$$

(10.71)

$$D_r = -\varepsilon\psi_{,r}$$

$$= -\varepsilon\frac{F_0}{r} + \sum_{m=1}^{\infty} [-\varepsilon v_m F_m r^{v_m - 1} + \varepsilon v_m G_m r^{-v_m - 1}]\cos v_m \theta.$$

(10.72)

The boundary conditions at $r = a, b$ require that

$$\frac{e}{\varepsilon}[C_0 J_0(\xi a) + D_0 Y_0(\xi a)] + F_0 \ln a + G_0$$

$$+ \sum_{m=1}^{\infty} [\frac{e}{\varepsilon}C_m J_{v_m}(\xi a) + \frac{e}{\varepsilon}D_m Y_{v_m}(\xi a)$$

$$+ F_m a^{v_m} + G_m a^{-v_m}]\cos v_m \theta = -V,$$

(10.73)

$$-\bar{c}\xi[C_0 J_1(\xi a) + D_0 Y_1(\xi a)] + e\frac{F_0}{a}$$

$$+ \sum_{m=1}^{\infty} [\bar{c}\xi C_m J'_{v_m}(\xi a) + \bar{c}\xi D_m Y'_{v_m}(\xi a)$$

$$+ ev_m F_m a^{v_m - 1} - ev_m G_m a^{-v_m - 1}]\cos v_m \theta = 0,$$

(10.74)

$$\frac{e}{\varepsilon}[C_0 J_0(\xi b) + D_0 Y_0(\xi b)] + F_0 \ln b + G_0$$

$$+ \sum_{m=1}^{\infty} [\frac{e}{\varepsilon}C_m J_{v_m}(\xi b) + \frac{e}{\varepsilon}D_m Y_{v_m}(\xi b)$$

$$+ F_m b^{v_m} + G_m b^{-v_m}]\cos v_m \theta = V,$$

(10.75)

$$-\bar{c}\xi[C_0 J_1(\xi b) + D_0 Y_1(\xi b)] + e\frac{F_0}{b}$$

$$+ \sum_{m=1}^{\infty} [\bar{c}\xi C_m J'_{v_m}(\xi b) + \bar{c}\xi D_m Y'_{v_m}(\xi b)$$

$$+ ev_m F_m b^{v_m - 1} - ev_m G_m b^{-v_m - 1}]\cos v_m \theta = 0.$$

(10.76)

We multiply Eqs. (10.73)–(10.76) by $\cos v_p \theta$ and integrate the resulting equation from $-\beta$ to β for $p = 0, 1, 2, 3,\ldots$ Then, from the orthogonality of $\cos v_p \theta$, the summations in these equations disappear. Only θ-independent terms are left because the applied voltage does not vary with θ. To calculate the charge on and the current flowing into the electrode at $r = b$, we need

$$D_r = e_{15} 2 S_{rz} + \varepsilon_{11} E_r,$$

$$Q = \int_{-\beta}^{\beta} -D_r\big|_{r=b}\, bd\theta, \tag{10.77}$$

$$I = \dot{Q} = i\omega Q.$$

Then the impedance of the transducer is given by

$$Z = 2V / I. \tag{10.78}$$

With Eq. (10.72), the impedance can be written in the following form:

$$Z = \frac{2V}{i\omega \int_{-\beta}^{\beta} \varepsilon \dfrac{F_0}{b} bd\theta} = \frac{V}{i\omega \varepsilon F_0 \beta}. \tag{10.79}$$

In the numerical calculation for the forced vibration, the elastic constant c is replaced by $c(1 + iQ^{-1})$ to take into viscous damping into consideration. $Q = 20$ is used in the calculation.

Figure 10.13 shows the real part of the impedance as a function of the driving frequency. Only the modes with $m = 0$ can be excited due to the uniformity of the applied voltage. Among the modes with $m = 0$, in the r direction, odd modes are roughly antisymmetric about the middle of the transducer where $r = (a + b)/2$ and even modes are roughly symmetric. Therefore the peaks in Fig. 10.13 correspond to the first and the third free-vibration resonant frequencies with $m = 0$ in Fig. 8.6.

Displacement and electric potential distributions at the first two resonances in Fig. 10.13 are shown in Figs. 10.14 to 10.17. What is shown is the real parts of the complex fields. These are modes with $m = 0$. Modes with higher frequencies have more variations along r.

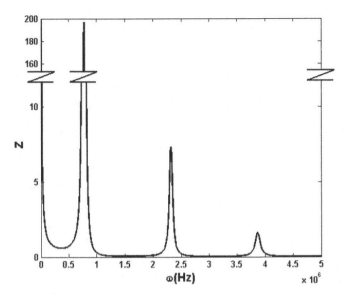

Fig. 10.13. Impedance versus driving frequency.

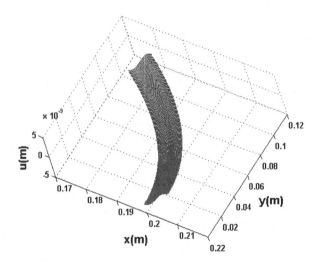

Fig. 10.14. Displacement when $\omega = 0.774 \times 10^6$ Hz .

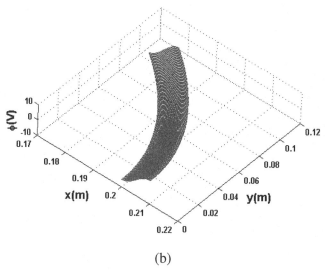

(b)

Fig. 10.15. Electric potential when $\omega = 0.774 \times 10^6$ Hz .

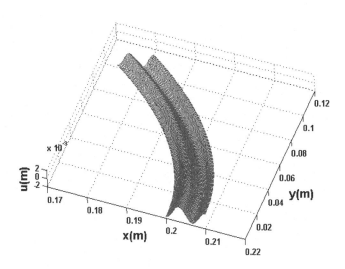

Fig. 10.16. Displacement when $\omega = 2.32 \times 10^6$ Hz .

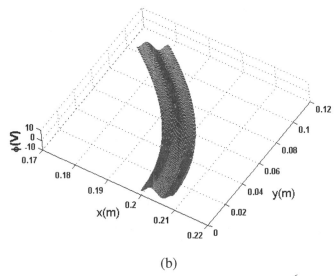

(b)

Fig. 10.17. Electric potential when $\omega = 2.32 \times 10^6$ Hz .

10.5. Power Transmission with Finite Transducers

In the analysis of power transmission through a circular cylindrical shell in Sec. 10.3, the transducers inside and out of the shell are also closed circular cylindrical shells. The whole structure vibrates in pure thickness modes with radial variations only. In this section we consider the more practical situation of a closed circular cylindrical elastic wall with finite piezoelectric patches on both sides [67]. Consider a circular cylindrical elastic shell of inner radius b and outer radius c as shown in Fig. 10.18. The shell is unbounded in the z direction. We consider unit dimension of the shell in the z direction. Two piezoelectric transducers represented by the shaded areas are attached to the shell. The two radii going through the edges of the transducers form an angle 2β. The inner transducer is electroded at $r = a, b$. The outer transducer is electroded at $r = c, d$. The four electrodes are shown by the thick lines in the figure. Under an applied voltage $2V_1$ across the outer transducer, the transducer and the entire structure are excited into a shear motion. The shell can be made from either a metal or a dielectric. When the shell is metallic, a very thin insulating layer is assumed between the transducer electrodes and the shell. The insulating layer is very thin, its thickness and mechanical effects are neglected. $2V_1$ is a known, time-harmonic driving voltage.

$2V_2$ is the unknown output voltage. I_1 and I_2 are the input and output currents. Z is the impedance of the output circuit in time-harmonic motions. A cylindrical coordinate system is defined by $x = r\cos\theta$, $y = r\sin\theta$ and $z = z$. In the index notation below, (x, y, z) correspond to $(1, 2, 3)$.

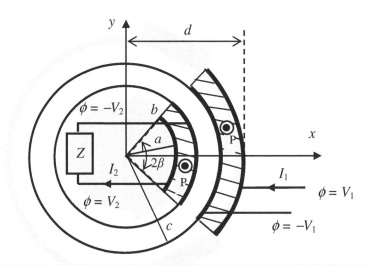

Fig. 10.18. An elastic shell with two piezoelectric transducers.

10.5.1. *Governing equations*

For the elastic shell, the governing equation is

$$c_2^2(\frac{\partial^2 u}{\partial r^2} + \frac{1}{r}\frac{\partial u}{\partial r} + \frac{1}{r^2}\frac{\partial^2 u}{\partial \theta^2}) = \ddot{u}.\qquad(10.80)$$

where $c_2^2 = \mu / \rho_1$, ρ_1 is the mass density, and μ is the shear elastic constant. For the transducers, we have

$$v_T^2(\frac{\partial^2 u}{\partial r^2} + \frac{1}{r}\frac{\partial u}{\partial r} + \frac{1}{r^2}\frac{\partial^2 u}{\partial \theta^2}) = \ddot{u},$$

$$\frac{\partial^2 \psi}{\partial r^2} + \frac{1}{r}\frac{\partial \psi}{\partial r} + \frac{1}{r^2}\frac{\partial^2 \psi}{\partial \theta^2} = 0.\qquad(10.81)$$

To calculate the charge and current on the outer electrode of the outer transducer, we need

$$D_r = e2S_{rz} + \varepsilon E_r,$$

$$Q_1 = \int_{-\beta}^{\beta} -D_r d(d\theta), \qquad (10.82)$$

$$I_1 = \dot{Q}_1 = i\omega Q_1.$$

For the charge and current on the inner electrode of the inner transducer, we have

$$D_r = e2S_{rz} + \varepsilon E_r,$$

$$Q_2 = \int_{-\beta}^{\beta} D_r\big|_{r=a} a(d\theta), \qquad (10.83)$$

$$I_2 = -\dot{Q}_2 = -i\omega Q_2.$$

We also have the following circuit condition for time-harmonic motions:

$$I_2 = 2V_2 / Z. \qquad (10.84)$$

10.5.2. Fields in different regions

Due to the symmetry in the problem, the fields in the elastic shell are given by even function of θ

$$u = A_0 J_0(\eta r) + B_0 Y_0(\eta r)$$
$$+ \sum_{n=1}^{\infty} [A_n J_n(\eta r) + B_n Y_n(\eta r)] \cos n\theta, \qquad (10.85)$$

$$T_{rz} = -A_0 \mu\eta J_1(\eta r) - B_0 \mu\eta Y_1(\eta r)$$
$$+ \sum_{n=1}^{\infty} [A_n \mu\eta J_n'(\eta r) + B_n \mu\eta Y_n'(\eta r)] \cos n\theta, \qquad (10.86)$$

where J_n and Y_n are the nth order Bessel functions of the first and second kind. A superimposed prime indicates differentiation with respect to the whole argument of a function. A_n and B_n are undetermined constants, and

$$\eta = \frac{\omega}{c_2}. \qquad (10.87)$$

The above fields satisfy the governing equation and periodic condition in θ.

For the inner transducer, the fields can be written as

$$u = C_0 J_0(\xi r) + D_0 Y_0(\xi r)$$

$$+ \sum_{m=1}^{\infty} [C_m J_{v_m}(\xi r) + D_m Y_{v_m}(\xi r)] \cos v_m \theta, \qquad (10.88)$$

$$\psi = F_0 \ln r + G_0 + \sum_{m=1}^{\infty} [F_m r^{v_m} + G_m r^{-v_m}] \cos v_m \theta,$$

$$\phi = \psi + \frac{e}{\varepsilon} u$$

$$= \frac{e}{\varepsilon} C_0 J_0(\xi r) + \frac{e}{\varepsilon} D_0 Y_0(\xi r) + F_0 \ln r + G_0$$

$$+ \sum_{m=1}^{\infty} [\frac{e}{\varepsilon} C_m J_{v_m}(\xi r) + \frac{e}{\varepsilon} D_m Y_{v_m}(\xi r) + F_m r^{v_m} + G_m r^{-v_m}] \qquad (10.89)$$

$$\times \cos v_m \theta,$$

$$T_{rz} = \bar{c} u_{,2} + e \psi_{,2}$$

$$= -\bar{c} \xi C_0 J_1(\xi r) - \bar{c} \xi D_0 Y_1(\xi r) + e F_0 \frac{1}{r}$$

$$+ \sum_{m=1}^{\infty} [\bar{c} \xi C_m J'_{v_m}(\xi r) + \bar{c} \xi D_m Y'_{v_m}(\xi r) \qquad (10.90)$$

$$+ e v_m F_m r^{v_m - 1} - e v_m G_m r^{-v_m - 1}] \cos v_m \theta,$$

$$D_r = -\varepsilon \psi_{,r} = -\varepsilon F_0 \frac{1}{r} + \sum_{m=1}^{\infty} [-\varepsilon v_m F_m r^{v_m - 1} + \varepsilon v_m G_m r^{-v_m - 1}] \cos v_m \theta,$$

$$(10.91)$$

where J_v and Y_v are the vth order Bessel functions of the first and second kind. J_m, D_m. F_m, and G_m are undetermined constants, and

$$\xi = \frac{\omega}{v_T}, \quad v_m \beta = m\pi, \quad m = 1, 2, \cdots. \qquad (10.92)$$

The above fields satisfy the governing equations and the boundary conditions at $\theta = \pm \beta$. We consider the case when β assumes a set of special values of $\beta = \pi / k$, where $k = 1, 2, 3,...$ is a positive integer, and then $v_m = m\pi / \beta = mk$ is a positive integer. In this case $\cos v_m \theta$ is a subset of $\cos n\theta$ and some simplifications will result from this.

Similarly, for the outer transducer, the general solution satisfying the governing equations and the transducer edge conditions can be written as:

$$u = P_0 J_0(\xi r) + Q_0 Y_0(\xi r)$$

$$+ \sum_{m=1}^{\infty} [P_m J_{\nu_m}(\xi r) + Q_m Y_{\nu_m}(\xi r)] \cos \nu_m \theta, \qquad (10.93)$$

$$\psi = R_0 \ln r + N_0 + \sum_{m=1}^{\infty} [R_m r^{\nu_m} + N_m r^{-\nu_m}] \cos \nu_m \theta,$$

$$\phi = \psi + \frac{e}{\varepsilon} u$$

$$= \frac{e}{\varepsilon} P_0 J_0(\xi r) + \frac{e}{\varepsilon} Q_0 Y_0(\xi r) + R_0 \ln r + N_0 \qquad (10.94)$$

$$+ \sum_{m=1}^{\infty} [\frac{e}{\varepsilon} P_m J_{\nu_m}(\xi r) + \frac{e}{\varepsilon} Q_m Y_{\nu_m}(\xi r) + R_m r^{\nu_m} + N_m r^{-\nu_m}]$$

$$\times \cos \nu_m \theta,$$

$$T_{rz} = \bar{c} u_{,2} + e \psi_{,2}$$

$$= -\bar{c} \xi P_0 J_1(\xi r) - \bar{c} \xi Q_0 Y_1(\xi r) + e R_0 \frac{1}{r} \qquad (10.95)$$

$$+ \sum_{m=1}^{\infty} [\bar{c} \xi P_m J'_{\nu_m}(\xi r) + \bar{c} \xi Q_m Y'_{\nu_m}(\xi r)$$

$$+ e \nu_m R_m r^{\nu_m - 1} - e \nu_m N_m r^{-\nu_m - 1}] \cos \nu_m \theta,$$

$$D_r = -\varepsilon \psi_{,r} = -\varepsilon R_0 \frac{1}{r} + \sum_{m=1}^{\infty} [-\varepsilon \nu_m R_m r^{\nu_m - 1} + \varepsilon \nu_m N_m r^{-\nu_m - 1}] \cos \nu_m \theta,$$

$$(10.96)$$

where P_m, Q_m, R_m, and N_m are undetermined constants.

10.5.3. *Boundary, continuity and circuit conditions*

The outer surface of the outer transducer is traction-free and the electric potential is the applied V_1. At the inner surface of the outer transducer, the potential is $-V_1$, and the displacement and the traction are the same as those of the outer surface of the elastic shell (continuity). The rest of the outer surface of the elastic shell that is not in contact with the outer transducer is traction-free. Similar boundary and continuity conditions also exist for the inner transducer. Therefore, we have

$$\phi(a) = V_2, \quad T_{rz}(a) = 0, \quad |\theta| < \beta,$$
$$\phi(b^-) = -V_2, \quad u(b^-) = u(b^+), \quad |\theta| < \beta,$$

(10.97)

$$\phi(c^+) = -V_1, \quad u(c^-) = u(c^+), \quad |\theta| < \beta,$$
$$\phi(d) = V_2, \quad T_{rz}(d) = 0, \quad |\theta| < \beta.$$

(10.98)

$$T_{rz}(b^+) = \begin{cases} T_{rz}(b^-), & |\theta| < \beta, \\ 0, & |\theta| > \beta, \end{cases}$$

$$T_{rz}(c^-) = \begin{cases} T_{rz}(c^+), & |\theta| < \beta, \\ 0, & |\theta| > \beta. \end{cases}$$

(10.99)

Substitution of the fields into the above boundary and continuity conditions yields

$$\frac{e}{\varepsilon} C_0 J_0(\xi a) + \frac{e}{\varepsilon} D_0 Y_0(\xi a) + F_0 \ln a + G_0$$

$$+ \sum_{m=1}^{\infty} [\frac{e}{\varepsilon} C_m J_{\nu_m}(\xi a) + \frac{e}{\varepsilon} D_m Y_{\nu_m}(\xi a) + F_m a^{\nu_m} + G_m a^{-\nu_m}] \quad (10.100)$$

$$\times \cos \nu_m \theta = V_2, \quad |\theta| < \beta,$$

$$-\bar{c} \xi C_0 J_1(\xi a) - \bar{c} \xi D_0 Y_1(\xi a) + e F_0 \frac{1}{a} + \sum_{m=1}^{\infty} [\bar{c} \xi C_m J_{\nu_m}'(\xi a)$$

$$+ \bar{c} \xi D_m Y_{\nu_m}'(\xi a) + e \nu_m F_m a^{\nu_m - 1} - e \nu_m G_m a^{-\nu_m - 1}] \quad (10.101)$$

$$\times \cos \nu_m \theta = 0, \quad |\theta| < \beta,$$

$$\frac{e}{\varepsilon} C_0 J_0(\xi b) + \frac{e}{\varepsilon} D_0 Y_0(\xi b) + F_0 \ln b + G_0$$

$$+ \sum_{m=1}^{\infty} [\frac{e}{\varepsilon} C_m J_{\nu_m}(\xi b) + \frac{e}{\varepsilon} D_m Y_{\nu_m}(\xi b) + F_m b^{\nu_m} + G_m b^{-\nu_m}] \quad (10.102)$$

$$\times \cos \nu_m \theta = -V_2, \quad |\theta| < \beta,$$

$$C_0 J_0(\xi b) + D_0 Y_0(\xi b) + \sum_{m=1}^{\infty} [C_m J_{\nu_m}(\xi b) + D_m Y_{\nu_m}(\xi b)] \cos \nu_m \theta$$

$$= A_0 J_0(\eta b) + B_0 Y_0(\eta b) \tag{10.103}$$

$$+ \sum_{n=1}^{\infty} [A_n J_n(\eta b) + B_n Y_n(\eta b)] \cos n\theta, \quad |\theta| < \beta,$$

$$\frac{e}{\varepsilon} P_0 J_0(\xi c) + \frac{e}{\varepsilon} Q_0 Y_0(\xi c) + R_0 \ln c + N_0$$

$$+ \sum_{m=1}^{\infty} [\frac{e}{\varepsilon} P_m J_{\nu_m}(\xi c) + \frac{e}{\varepsilon} Q_m Y_{\nu_m}(\xi c) + R_m c^{\nu_m} + N_m c^{-\nu_m}] \tag{10.104}$$

$$\times \cos \nu_m \theta = -V_1, \quad |\theta| < \beta,$$

$$A_0 J_0(\eta c) + B_0 Y_0(\eta c) + \sum_{n=1}^{\infty} [A_n J_n(\eta c) + B_n Y_n(\eta c)] \cos n\theta$$

$$= P_0 J_0(\xi c) + Q_0 Y_0(\xi c) + \sum_{m=1}^{\infty} [P_m J_{\nu_m}(\xi c) + Q_m Y_{\nu_m}(\xi c)] \tag{10.105}$$

$$\times \cos \nu_m \theta, \quad |\theta| < \beta,$$

$$\frac{e}{\varepsilon} P_0 J_0(\xi d) + \frac{e}{\varepsilon} Q_0 Y_0(\xi d) + R_0 \ln d + N_0$$

$$+ \sum_{m=0}^{\infty} [\frac{e}{\varepsilon} P_m J_{\nu_m}(\xi d) + \frac{e}{\varepsilon} Q_m Y_{\nu_m}(\xi d) + R_m d^{\nu_m} + N_m d^{-\nu_m}] \tag{10.106}$$

$$\cos \nu_m \theta = V_1, \quad |\theta| < \beta,$$

$$-\overline{c} \xi P_0 J_1(\xi d) - \overline{c} \xi Q_0 Y_1(\xi d) + e R_0 \frac{1}{r}$$

$$+ \sum_{m=1}^{\infty} [\overline{c} \xi P_m J'_{\nu_m}(\xi d) + \overline{c} \xi Q_m Y'_{\nu_m}(\xi d) + e \nu_m R_m d^{\nu_m - 1} \tag{10.107}$$

$$- e \nu_m N_m d^{-\nu_m - 1}] \cos \nu_m \theta = 0, \quad |\theta| < \beta,$$

$$-A_0\mu\eta J_1(\eta b) - B_0\mu\eta Y_1(\eta b)$$

$$+\sum_{n=1}^{\infty} [A_n\mu\eta J_n'(\eta b) + B_n\mu\eta Y_n'(\eta b)]\cos n\theta$$

$$=\begin{cases} -\overline{c}\xi C_0 J_1(\xi b) - \overline{c}\xi D_0 Y_1(\xi b) + eF_0\dfrac{1}{b} \\ \\ \quad +\sum_{m=1}^{\infty} [\overline{c}\xi C_m J_{v_m}'(\xi b) + \overline{c}\xi D_m Y_{v_m}'(\xi b) \\ \\ \quad + ev_m F_m b^{v_m-1} - ev_m G_m b^{-v_m-1}]\cos v_m\theta, \quad |\theta| < \beta, \\ \\ \qquad\qquad 0, \quad |\theta| > \beta, \end{cases} \tag{10.108}$$

$$-A_0\mu\eta J_1(\eta c) - B_0\mu\eta Y_1(\eta c)$$

$$+\sum_{n=1}^{\infty} [A_n\mu\eta J_n'(\eta c) + B_n\mu\eta Y_n'(\eta c)]\cos n\theta$$

$$=\begin{cases} -\overline{c}\xi P_0 J_1(\xi c) - \overline{c}\xi Q_0 Y_1(\xi c) + eR_0\dfrac{1}{c} \\ \\ \quad +\sum_{m=1}^{\infty} [\overline{c}\xi P_m J_{v_m}'(\xi c) + \overline{c}\xi Q_m Y_{v_m}'(\xi c) \\ \\ \quad + ev_m R_m c^{v_m-1} - ev_m N_m c^{-v_m-1}]\cos v_m\theta, \quad |\theta| < \beta, \\ \\ \qquad\qquad 0, \quad |\theta| > \beta. \end{cases} \tag{10.109}$$

The above equations still depend on θ. To obtain algebraic equations for the undetermined constants, we need to multiply Eqs. (10.100) to (10.107) by $\cos v_p\theta$ and integrate them from $-\beta$ to β for $p = 0, 1,$ 2, 3,.... Equations (10.108) and (10.109) need to be multiplied by $\cos q\theta$ and integrated from $-\pi$ to π for $q = 0, 1, 2, 3,....$ In addition, the circuit equation in Eq. (10.84) takes the following form:

$$-i\omega\int_{-\beta}^{\beta} \{-\varepsilon F_0\frac{1}{a} + \sum_{m=1}^{\infty} [-\varepsilon v_m F_m a^{v_m-1} + \varepsilon v_m G_m a^{-v_m-1}]$$

$$\times \cos v_m\theta\} a(d\theta) = \frac{2V_2}{Z}. \tag{10.110}$$

10.5.4. Numerical results and discussion

As a numerical example, consider transducers made from polarized ceramics PZT-5H. Damping is introduced by allowing the elastic constant of the ceramics to assume complex values. In our calculations c_{44} is replaced by $c_{44}(1 + iQ^{-1})$ where Q is a real number. Similarly, for the elastic shell, μ is replaced by $\mu(1 + iQ^{-1})$. We fix $Q = 100$ (except in Fig. 10.22). We introduce a Z_0 as a unit for Z:

$$Z_0 = \frac{1}{i\omega C_0}, \quad C_0 = \frac{\varepsilon_{11}\beta(b+c)}{d-c}. \tag{10.111}$$

For the middle elastic layer we consider steel with $\rho_1 = 7850\,\text{kg/m}^3$ and $\mu = 80 \times 10^9\,\text{N/m}^2$. We fix $b = 0.495$ m, $c = 0.505$ m, $d = 0.515$ m, $\beta = 10°$ (except in Fig. 10.23), and $V_1 = 220$ volt. We use the following ω_0 as a normalizing frequency:

$$\omega_0^2 = \frac{\mu\pi^2}{\rho_1(c-b)^2}. \tag{10.112}$$

In the calculations below, 15 terms are used in the series solution. The difference between using 15 and 16 terms is 0.97% when the maximum displacement is concerned.

Figure 10.19 shows the output voltage when the driving frequency sweeps through the fundamental thickness-twist frequency of the elastic shell. Near resonances the output is more pronounced. Therefore significant power transfer can be achieved near resonances. The resonant frequencies do not seem to be sensitive to Z but the output voltage is. Numerical experiments show that a larger Z implies a higher output voltage only in a certain range of Z. For very large Z the output voltage saturates when the output circuit is essentially open. Since the output voltage is of the same order of magnitude as the input voltage, effectively it is like that the input voltage is directly applied to the load circuit.

Figure 10.20 shows the displacement distribution of the outer surface ($r = c$) of the elastic shell versus θ for different inner transducer thickness. Clearly, vibration is pronounced only under and close to the piezoelectric transducers. Outside the transducers the vibration decays rapidly. This is the so-called energy trapping phenomenon of thickness modes. When energy is trapped, the shell does not experience global vibration which is desirable in this application. The figure shows that the

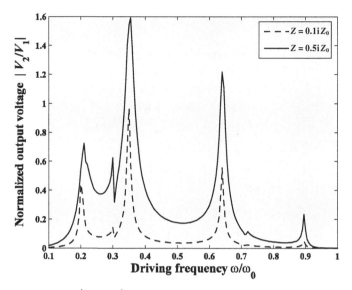

Fig. 10.19. $|V_2/V_1|$ versus ω/ω_0 ($a = 0.480$ m, $n = 15$).

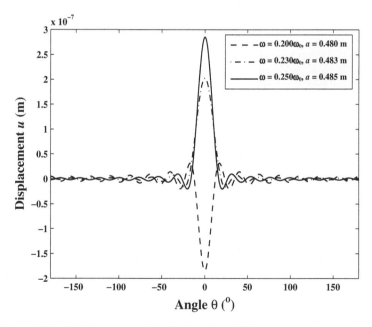

Fig. 10.20. Displacement distribution for different transducer thickness ($n = 15$, $Z = 0.1$ iZ_0).

effectiveness of trapping is sensitive to transducer thickness. The trapping is better when a = 0.485 m, i.e., when the inner and outer transducers are of the same thickness.

Figure 10.21 shows the effect of the load impedance Z on energy trapping. When Z changes, the resonant frequency varies slightly. The figure shows that energy trapping is also sensitive to Z. This is because Z affects the electric field in the inner transducer and hence the related piezoelectric stiffening effect.

Figure 10.22 shows the effect of damping. When damping increases, the vibration amplitude decreases as expected. The resonant frequency changes very little.

Figure 10.23 shows the effect of β. When β increases, the vibration is more spread as expected. It is important to note that for large β nodal points along θ begin to appear under the transducers. When there are nodal points in the transducers, material particles of the transducers do not vibrate in phase along θ. This causes cancellation of the electrical charge or voltage produced at the output transducer and lowers the power transmission effectiveness.

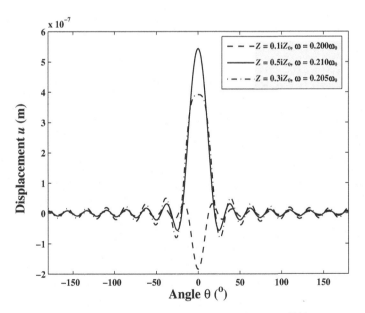

Fig. 10.21. Displacement distribution for different Z (n = 15, a = 0.480 m).

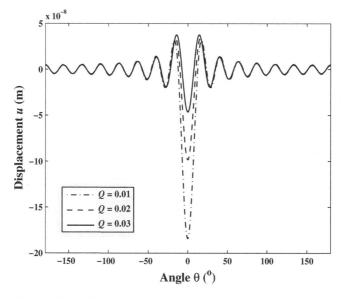

Fig. 10.22. Effect of damping on displacement distribution
($a = 0.480$ m, $n = 15$).

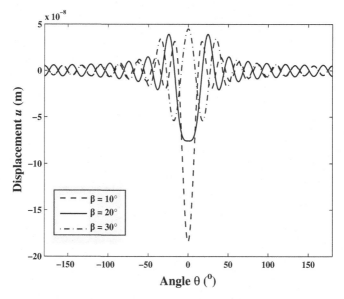

Fig. 10.23. Effect of β on displacement distribution
($a = 0.480$ m, $n = 15$).

10.6. A Transducer on an Elastic Shell

In this section we analyze vibrations of a circular cylindrical elastic shell electrically driven by a piezoelectric actuator. Consider the structure in Fig. 10.24 [68]. The circular cylindrical elastic shell is isotropic. The finite piezoelectric transducer is represented by the shaded area with an angular span of $-\beta < \theta < \beta$ in the polar coordinate system defined by $x = r\cos\theta$ and $y = r\sin\theta$. The z axis is determined from the x and y axes by the right-hand rule. The transducer is electroded at $r = b$ and a. The electrodes are very thin. The transducer is under a driving voltage $2V$ which will be assumed time-harmonic. The elastic shell is a dielectric. If a metal shell is considered, a very thin insulating layer is assumed between the bottom electrode of the transducer and the elastic shell.

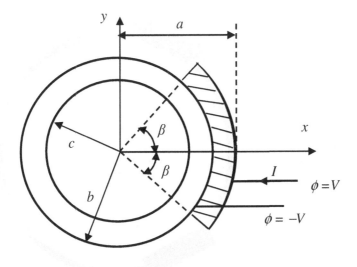

Fig. 10.24. A transducer on a shell.

10.6.1. Governing equations and fields

For the elastic shell, the governing equation is

$$c_2^2 \left(\frac{\partial^2 u}{\partial r^2} + \frac{1}{r}\frac{\partial u}{\partial r} + \frac{1}{r^2}\frac{\partial^2 u}{\partial \theta^2} \right) = \ddot{u} .$$ (10.113)

where $c_2^2 = \mu / \rho_1$, ρ_1 is the mass density, μ is the shear elastic constant. The fields that satisfy Eq. (10.113) and the periodic condition are the same as Eqs. (10.85) and (10.86). For the transducers, we have

$$v_T^2 \left(\frac{\partial^2 u}{\partial r^2} + \frac{1}{r} \frac{\partial u}{\partial r} + \frac{1}{r^2} \frac{\partial^2 u}{\partial \theta^2} \right) = \ddot{u},$$

$$\frac{\partial^2 \psi}{\partial r^2} + \frac{1}{r} \frac{\partial \psi}{\partial r} + \frac{1}{r^2} \frac{\partial^2 \psi}{\partial \theta^2} = 0. \tag{10.114}$$

The fields that satisfy Eq. (10.114) and the transducer edge conditions are the same as Eqs. (10.88) to (10.91). To calculate the charge Q_e and current I on the outer electrode of the transducer as well as the admittance Y of the structure, we need

$$D_r = e2S_{rz} + \varepsilon E_r,$$

$$Q_e = \int_{-\beta}^{\beta} -D_r a(d\theta),$$

$$I = \dot{Q}_e = i\omega Q_e, \tag{10.115}$$

$$I = Y2V.$$

10.6.2. Boundary and continuity conditions

We have the following boundary and continuity conditions:

$$\phi(a) = V, \quad T_{rz}(a) = 0, \quad |\theta| < \beta,$$

$$\phi(b^+) = -V, \quad u(b^-) = u(b^+), \quad |\theta| < \beta. \tag{10.116}$$

$$T_{rz}(b^-) = \begin{cases} T_{rz}(b^+), & |\theta| < \beta, \\ 0, & |\theta| > \beta, \end{cases} \tag{10.117}$$

$$T_{rz}(c) = 0.$$

Substituting Eqs. (10.85), (10.86) and (10.88)–(10.91) into Eqs. (10.116) and (10.117), we obtain

$$\frac{e}{\varepsilon} C_0 J_0(\xi a) + \frac{e}{\varepsilon} D_0 Y_0(\xi a) + F_0 \ln a + G_0$$

$$+ \sum_{m=1}^{\infty} \left[\frac{e}{\varepsilon} C_m J_{\nu_m}(\xi a) + \frac{e}{\varepsilon} D_m Y_{\nu_m}(\xi a) + F_m a^{\nu_m} + G_m a^{-\nu_m} \right] \tag{10.118}$$

$$\times \cos \nu_m \theta = V, \quad |\theta| < \beta,$$

$$-\overline{c}\xi C_0 J_1(\xi a) - \overline{c}\xi D_0 Y_1(\xi a) + eF_0\frac{1}{a} + \sum_{m=1}^{\infty}\ [\overline{c}\xi C_m J'_{\nu_m}(\xi a)$$

$$+\overline{c}\xi D_m Y'_{\nu_m}(\xi a) + e\nu_m F_m a^{\nu_m-1} - e\nu_m G_m a^{-\nu_m-1}] \qquad (10.119)$$

$$\times \cos\nu_m\theta = 0, \quad |\theta| < \beta,$$

$$\frac{e}{\varepsilon}C_0 J_0(\xi b) + \frac{e}{\varepsilon}D_0 Y_0(\xi b) + F_0\ln b + G_0$$

$$+\sum_{m=1}^{\infty}\ [\frac{e}{\varepsilon}C_m J_{\nu_m}(\xi b) + \frac{e}{\varepsilon}D_m Y_{\nu_m}(\xi b) + F_m b^{\nu_m} + G_m b^{-\nu_m}] \qquad (10.120)$$

$$\times \cos\nu_m\theta = -V, \quad |\theta| < \beta,$$

$$C_0 J_0(\xi b) + D_0 Y_0(\xi b)$$

$$+\sum_{m=1}^{\infty}\ [C_m J_{\nu_m}(\xi b) + D_m Y_{\nu_m}(\xi b)]\cos\nu_m\theta$$

$$= A_0 J_0(\eta b) + B_0 Y_0(\eta b) \qquad (10.121)$$

$$+\sum_{n=1}^{\infty}\ [A_n J_n(\eta b) + B_n Y_n(\eta b)]\cos n\theta, \quad |\theta| < \beta,$$

$$-A_0\mu\eta J_1(\eta b) - B_0\mu\eta Y_1(\eta b)$$

$$+\sum_{n=1}^{\infty}\ [A_n\mu\eta J'_n(\eta b) + B_n\mu\eta Y'_n(\eta b)]\cos n\theta \qquad (10.122)$$

$$= \begin{cases} -\overline{c}\xi C_0 J_1(\xi b) - \overline{c}\xi D_0 Y_1(\xi b) + eF_0\dfrac{1}{b} \\[2mm] +\displaystyle\sum_{m=1}^{\infty}\ [\overline{c}\xi C_m J'_{\nu_m}(\xi b) + \overline{c}\xi D_m Y'_{\nu_m}(\xi b) \\[2mm] + e\nu_m F_m b^{\nu_m-1} - e\nu_m G_m b^{-\nu_m-1}]\cos\nu_m\theta, \quad |\theta| < \beta, \\[2mm] \qquad\qquad 0, \quad |\theta| > \beta, \end{cases}$$

$$-A_0\mu\eta J_1(\eta c) - B_0\mu\eta Y_1(\eta c)$$

$$+\sum_{n=1}^{\infty}\ [A_n\mu\eta J'_n(\eta c) + B_n\mu\eta Y'_n(\eta c)]\cos n\theta = 0. \qquad (10.123)$$

The above equations still depend on θ. To obtain algebraic equations for the undetermined constants, we need to multiply Eqs. (10.118) to (10.121) by $\cos\nu_p\theta$ and integrate them from $-\beta$ to β for $p = 0, 1,$

2, 3,.... Equations (10.122) and (10.123) need to be multiplied by $\cos q\theta$ and integrated from $-\pi$ to π for $q = 0, 1, 2, 3,....$

10.6.3. Numerical results

As a numerical example, consider a transducer made from polarized ceramics PZT-5H. Damping is introduced by allowing the elastic constant of the ceramics c_{44} to assume complex values. In our calculations c_{44} is replaced by $c_{44}(1 + iQ^{-1})$, where Q is a large and real number. Similarly, for the elastic shell, μ is replaced by $\mu(1 + iQ^{-1})$. We fix $Q = 100$. For the elastic layer, we consider steel with $\rho_1 = 7850\,\text{kg/m}^3$ and $\mu = 80\times10^9\,\text{N/m}^2$. We fix $b = 1.005$ m, $c = 0.995$ m, $\beta = 10°$, and $V = 220$ volt. We use the following frequency ω_0 as a normalizing frequency:

$$\omega_0^2 = \frac{\mu\pi^2}{\rho_1(b-c)^2} . \tag{10.124}$$

In our calculations, eleven terms are used in the series solution. The difference between using eleven and twelve terms is 2% when the displacement $u(b, 0)$ at the first resonance is concerned.

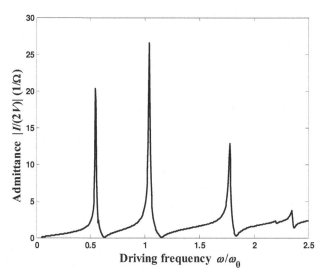

Fig. 10.25. Admittance versus driving frequency ($a = 1.010$ m).

In Fig. 10.26 we show the effect of different actuator thickness on the vibration distribution at the first resonance. A larger a represents a thicker actuator. The resonant frequency is sensitive to the actuator thickness, with lower frequencies for thicker actuators as expected. The qualitative behavior of the vibration distribution is the same for different actuator thickness.

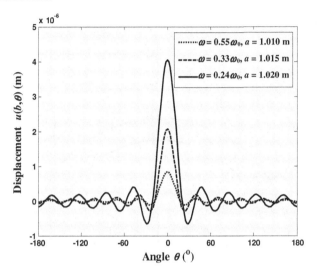

Fig. 10.26. $u(b,\theta)$ at the first resonance for different actuator thickness.

10.7. Two Transducers on an Elastic Shell

In this section we analyze the situation of two finite transducers on a circular cylindrical elastic shell, one as an actuator and the other as a sensor. Consider the cross section of a cylindrical structure shown in Fig. 10.27 [69]. The circular elastic shell is isotropic. Two finite piezoelectric transducers are located at $|\theta| < \beta$ and $|\theta - \pi| < \beta$ in the polar coordinate system defined by $x_1 = x = r\cos\theta$ and $x_2 = y = r\sin\theta$. The x_3 or z axis is determined from the x and y axes by the right-hand rule. The transducers are electroded at $r = b$ and a. The electrodes are very thin. Their mechanical effects due to inertia and stiffness are negligible. The transducer on the right with $|\theta| < \beta$ is under a driving voltage $2V_1$ which will be assumed time-harmonic. The corresponding driving current is I_1 which is unknown. This transducer acts as an actuator. The transducer on the left with $|\theta - \pi| < \beta$ is a sensor for collecting electrical output. Its voltage $2V_2$ and current I_2 are both unknown. The output electrodes are connected by a circuit whose impedance is denoted by Z in time-harmonic motions. The elastic shell is a dielectric. If a metal shell is considered, a very thin insulating layer is assumed between the bottom electrodes of the transducers and the elastic shell.

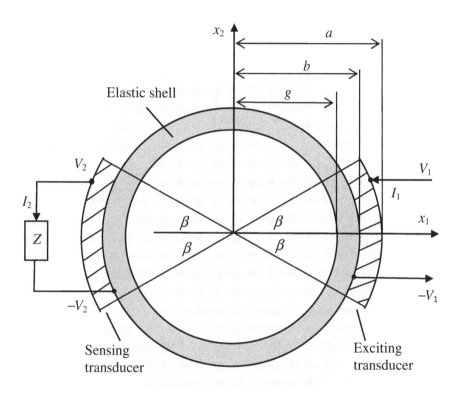

Fig. 10.27. An elastic shell with two piezoelectric transducers.

10.7.1. *Governing equations and fields*

For the elastic shell, the governing equation is

$$c_2^2 \left(\frac{\partial^2 u}{\partial r^2} + \frac{1}{r} \frac{\partial u}{\partial r} + \frac{1}{r^2} \frac{\partial^2 u}{\partial \theta^2} \right) = \ddot{u} . \tag{10.125}$$

where $c_2^2 = \mu / \rho_1$, ρ_1 is the mass density, μ is the shear elastic constant. The fields that satisfy Eq. (10.125) and the periodic condition are the same as Eqs. (10.85) and (10.86). For the transducers, we have

$$v_T^2 \left(\frac{\partial^2 u}{\partial r^2} + \frac{1}{r} \frac{\partial u}{\partial r} + \frac{1}{r^2} \frac{\partial^2 u}{\partial \theta^2} \right) = \ddot{u},$$

$$\frac{\partial^2 \psi}{\partial r^2} + \frac{1}{r} \frac{\partial \psi}{\partial r} + \frac{1}{r^2} \frac{\partial^2 \psi}{\partial \theta^2} = 0.$$

(10.126)

For the right transducer, the fields that satisfy Eq. (10.126) and the transducer edge conditions are the same as Eqs. (10.88) to (10.91). For the left transducer, the fields that satisfy Eq. (10.126) and the transducer edge conditions can be expressed by Eqs. (10.93)–(10.96). We note that for the left transducer the natural base functions for the trigonometric series are $\cos v_m (\theta - \pi)$ which are equivalent to the base functions of $\cos v_m \theta$ when v_m is an integer. This will be the case later when β assumes certain specific values. The charge and current on the outer electrode of the driving transducer on the right are given by

$$D_r = e2S_{rz} + \varepsilon E_r,$$

$$Q_1 = \int_{-\beta}^{\beta} -D_r a d\theta,$$

(10.127)

$$I_1 = \dot{Q}_1 = i\omega Q_1.$$

Similarly, the charge and current on the outer electrode of the sensing transducer on the left are

$$D_r = e2S_{rz} + \varepsilon E_r,$$

$$Q_2 = \int_{\pi-\beta}^{\pi+\beta} -D_r \big|_{r=a} a d\theta,$$

(10.128)

$$I_2 = -\dot{Q}_2 = -i\omega Q_2.$$

10.7.2. *Boundary, continuity and circuit conditions*

For the driving transducer on the right, we have

$$\phi(a) = V_1, \quad |\theta| < \beta,$$

$$T_{rz}(a) = 0, \quad |\theta| < \beta,$$

$$\phi(b^+) = -V_1, \quad |\theta| < \beta,$$

$$u(b^+) = u(b^-), \quad |\theta| < \beta.$$

(10.129)

Similarly, for the sensing transducer on the left,

$$\phi(a) = V_2, \quad |\theta - \pi| < \beta,$$
$$T_{rz}(a) = 0, \quad |\theta - \pi| < \beta,$$
$$\phi(b^+) = -V_2, \quad |\theta - \pi| < \beta,$$
$$u(b^+) = u(b^-), \quad |\theta - \pi| < \beta.$$

(10.130)

For the elastic shell, we have

$$
T_{rz}(b^-) = \begin{cases}
T_{rz}(b^+), & |\theta| < \beta. \\
T_{rz}(b^+), & |\theta - \pi| < \beta, \\
0, & \beta < \theta < \pi - \beta \text{ and } \pi + \beta < \theta < 2\pi - \beta,
\end{cases}
$$
$$T_{rz}(g) = 0, \quad -\pi < \theta < \pi.$$

(10.131)

The circuit condition for the output transducer is

$$I_2 = 2V_2 / Z.$$

(10.132)

Substituting the relevant fields into Eqs. (10.129) through (10.131), we obtain

$$
\frac{e}{\varepsilon} C_0 J_0(\xi a) + \frac{e}{\varepsilon} D_0 Y_0(\xi a) + F_0 \ln a + G_0
$$
$$
+ \sum_{m=1}^{\infty} \left[\frac{e}{\varepsilon} C_m J_{\nu_m}(\xi a) + \frac{e}{\varepsilon} D_m Y_{\nu_m}(\xi a) + F_m a^{\nu_m} + G_m a^{-\nu_m} \right]
$$
$$
\times \cos \nu_m \theta = V_1, \quad |\theta| < \beta,
$$

(10.133)

$$
-\bar{c}\xi C_0 J_1(\xi a) - \bar{c}\xi D_0 Y_1(\xi a) + eF_0 \frac{1}{a} + \sum_{m=1}^{\infty} \left[\bar{c}\xi C_m J'_{\nu_m}(\xi a) \right.
$$
$$
\left. + \bar{c}\xi D_m Y'_{\nu_m}(\xi a) + e\nu_m F_m a^{\nu_m - 1} - e\nu_m G_m a^{-\nu_m - 1} \right]
$$
$$
\times \cos \nu_m \theta = 0, \quad |\theta| < \beta,
$$

(10.134)

$$
\frac{e}{\varepsilon} C_0 J_0(\xi b) + \frac{e}{\varepsilon} D_0 Y_0(\xi b) + F_0 \ln b + G_0
$$
$$
+ \sum_{m=1}^{\infty} \left[\frac{e}{\varepsilon} C_m J_{\nu_m}(\xi b) + \frac{e}{\varepsilon} D_m Y_{\nu_m}(\xi b) + F_m b^{\nu_m} + G_m b^{-\nu_m} \right]
$$
$$
\times \cos \nu_m \theta = -V_1, \quad |\theta| < \beta,
$$

(10.135)

$$C_0 J_0(\xi b) + D_0 Y_0(\xi b)$$

$$+ \sum_{m=1}^{\infty} [C_m J_{v_m}(\xi b) + D_m Y_{v_m}(\xi b)] \cos v_m \theta$$

$$= A_0 J_0(\eta b) + B_0 Y_0(\eta b)$$ (10.136)

$$+ \sum_{n=1}^{\infty} [A_n J_n(\eta b) + B_n Y_n(\eta b)] \cos n\theta, \quad |\theta| < \beta,$$

$$\frac{e}{\varepsilon} P_0 J_0(\xi a) + \frac{e}{\varepsilon} Q_0 Y_0(\xi a) + R_0 \ln a + N_0$$

$$+ \sum_{m=1}^{\infty} [\frac{e}{\varepsilon} P_m J_{v_m}(\xi a) + \frac{e}{\varepsilon} Q_m Y_{v_m}(\xi a) + R_m a^{v_m} + N_m a^{-v_m}]$$ (10.137)

$$\times \cos v_m \theta = V_2, \quad |\theta - \pi| < \beta,$$

$$-\overline{c}\,\xi P_0 J_1(\xi a) - \overline{c}\,\xi Q_0 Y_1(\xi a) + eR_0 \frac{1}{a}$$

$$+ \sum_{m=1}^{\infty} [\overline{c}\,\xi P_m J'_{v_m}(\xi a) + \overline{c}\,\xi Q_m Y'_{v_m}(\xi a)$$ (10.138)

$$+ ev_m R_m a^{v_m-1} - ev_m N_m a^{-v_m-1}]$$

$$\times \cos v_m \theta = 0, \quad |\theta - \pi| < \beta,$$

$$\frac{e}{\varepsilon} P_0 J_0(\xi b) + \frac{e}{\varepsilon} Q_0 Y_0(\xi b) + R_0 \ln b + N_0$$

$$+ \sum_{m=1}^{\infty} [\frac{e}{\varepsilon} P_m J_{v_m}(\xi b) + \frac{e}{\varepsilon} Q_m Y_{v_m}(\xi b) + R_m b^{v_m} + N_m b^{-v_m}]$$ (10.139)

$$\times \cos v_m \theta = -V_2, \quad |\theta - \pi| < \beta,$$

$$P_0 J_0(\xi b) + Q_0 Y_0(\xi b)$$

$$+ \sum_{m=1}^{\infty} [P_m J_{v_m}(\xi b) + Q_m Y_{v_m}(\xi b)] \cos v_m \theta$$

$$= A_0 J_0(\eta b) + B_0 Y_0(\eta b)$$ (10.140)

$$+ \sum_{n=1}^{\infty} [A_n J_n(\eta b) + B_n Y_n(\eta b)] \cos n\theta, \quad |\theta - \pi| < \beta,$$

$$-A_0 \mu \eta J_1(\eta b) - B_0 \mu \eta Y_1(\eta b)$$

$$+ \sum_{n=1}^{\infty} [A_n \mu \eta J_n'(\eta b) + B_n \mu \eta Y_n'(\eta b)] \cos n\theta$$

$$= \begin{cases} -\bar{c}\xi C_0 J_1(\xi b) - \bar{c}\xi D_0 Y_1(\xi b) + eF_0 \dfrac{1}{b} \\ \quad + \displaystyle\sum_{m=1}^{\infty} [\bar{c}\xi C_m J_{\nu_m}'(\xi b) + \bar{c}\xi D_m Y_{\nu_m}'(\xi b) \\ \quad + e\nu_m F_m b^{\nu_m - 1} - e\nu_m G_m b^{-\nu_m - 1}] \cos \nu_m \theta, \quad |\theta| < \beta, \\[2em] -\bar{c}\xi P_0 J_1(\xi b) - \bar{c}\xi Q_0 Y_1(\xi b) + eR_0 \dfrac{1}{b} \\ \quad + \displaystyle\sum_{m=1}^{\infty} [\bar{c}\xi P_m J_{\nu_m}'(\xi b) + \bar{c}\xi Q_m Y_{\nu_m}'(\xi b) \\ \quad + e\nu_m R_m b^{\nu_m - 1} - e\nu_m N_m b^{-\nu_m - 1}] \cos \nu_m \theta, \quad |\theta - \pi| < \beta, \\[2em] 0, \qquad \beta < \theta < \pi - \beta \ \text{and} \ \pi + \beta < \theta < 2\pi - \beta, \end{cases} \tag{10.141}$$

$$- A_0 \mu \eta J_1(\eta g) - B_0 \mu \eta Y_1(\eta g)$$

$$+ \sum_{n=1}^{\infty} [A_n \mu \eta J_n'(\eta g) + B_n \mu \eta Y_n'(\eta g)] \cos n\theta = 0. \tag{10.142}$$

Equations (10.133)–(10.140) need to be multiplied by $\cos \nu_p \theta$ and integrated from $-\beta$ to β for $p = 0, 1, 2, 3,\ldots$ to obtain algebraic equations for the undetermined constants. Similarly, Eqs. (10.141) and (10.142) need to be multiplied by $\cos q\theta$ and integrated from $-\pi$ to π for $q = 0, 1, 2, 3,\ldots.$ In addition, with the series solution, the circuit equation given by Eq. (10.132) takes the following form:

$$-i\omega \int_{-\beta}^{\beta} -\{-\varepsilon R_0 \frac{1}{a} + \sum_{m=1}^{\infty} [-\varepsilon \nu_m R_m a^{\nu_m - 1} + \varepsilon \nu_m N_m a^{-\nu_m - 1}]$$

$$\times \cos \nu_m \theta\} a d\theta = \frac{2V_2}{Z}. \tag{10.143}$$

10.7.3. Numerical results

As a numerical example, consider transducers made from polarized ceramics PZT-5H. Damping is introduced by allowing the elastic constant of the ceramics c_{44} to assume complex values. In our calculations c_{44} is replaced by $c_{44}(1 + iQ^{-1})$ where Q is a large and real number. Similarly, for the elastic shell, μ is replaced by $\mu(1 + iQ^{-1})$. We fix $Q = 100$. We introduce a Z_0 as a unit for Z:

$$Z_0 = \frac{1}{i\omega C_0}, \quad C_0 = \frac{\varepsilon_{11}\beta(a+b)}{a-b}. \tag{10.144}$$

For the elastic layer we consider steel with $\rho_1 = 7850$ kg/m^3 and $\mu = 80 \times 10^9$ N/m^2. We fix $\beta = 10°$, $V_1 = 220$ volt, $g = 0.495$ m and $b = 0.505$ m. $a = 0.515$ m is fixed except in Fig. 10.30. We use the following frequency ω_0 as a normalizing frequency:

$$\omega_0^2 = \frac{\mu\pi^2}{\rho_1(b-c)^2}, \tag{10.145}$$

where μ is kept real.

In Fig. 10.28, we plot the normalized output voltage $|\overline{V}_2/\overline{V}_1|$ versus the driving frequency ω. A pure resistor load is used in the calculation. The output voltage assumes sharp maxima at resonant frequencies. The resonance peaks are very narrow due to small damping. 450 data points are used so that the maxima of the peaks are captured. The output voltage at resonance is of the same order of magnitude as the driving voltage.

In Fig. 10.29 we plot the displacement distribution along the outer surface of the elastic shell. The trigonometric series used can approximate the displacement distribution shown very well. The error between using 15 and 16 terms in the series of the shell is only 1.55%. Since the transducers are over smaller angular regions ($\beta = 10°$), 7 to 8 terms are used for the series of the transducers.

Figure 10.30 shows the effect of different actuator thickness on the vibration distribution. The qualitative behavior of the vibration distribution is similar for different actuator thickness. A thicker actuator with a larger a excites stronger vibrations.

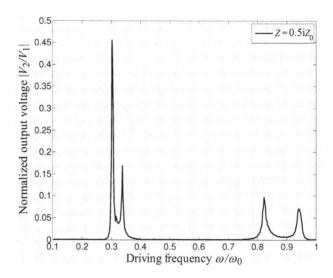

Fig. 10.28. Output voltage versus driving frequency.

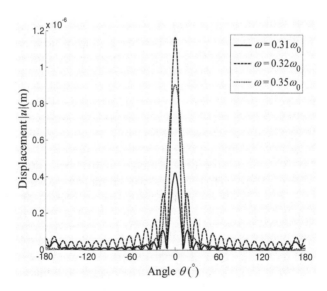

Fig. 10.29. Displacement distribution for different driving frequency
($r = b$).

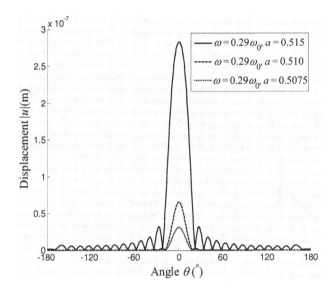

Fig. 10.30. Displacement distribution for different transducer thickness
($r = b$).

10.8. A Circular Cylinder with Unattached Electrodes

Next we analyze the electrically forced vibration of a cylinder with
unattached electrodes and air gaps whose free vibration was treated in
Sec. 8.2 (see Fig. 10.31) [51].

The fields are formally the same as those in Sec. 8.2. In the cylinder,

$$u = A_1 J_0(\xi r) + A_2 Y_0(\xi r), \quad a < r < c,$$
$$\psi = A_3 \ln r + A_4, \quad a < r < c,$$

$$(10.146)$$

$$\phi = \frac{e}{\varepsilon}[A_1 J_0(\xi r) + A_2 Y_0(\xi r)] + A_3 \ln r + A_4,$$

$$T_{rz} = -\overline{c}\,\xi[A_1 J_1(\xi r) + A_2 Y_1(\xi r)] + e\frac{A_3}{r},$$

$$(10.147)$$

$$D_r = -\varepsilon \frac{A_3}{r},$$

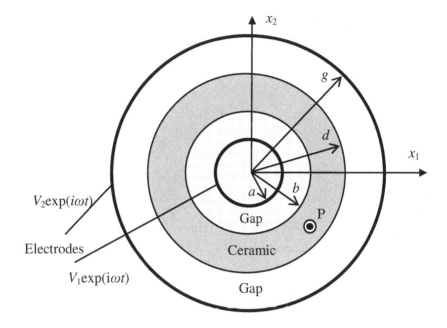

Fig. 10.31. A cylindrical shell with unattached electrodes.

where A_1 through A_4 are undetermined constants. J_0 and Y_0 are the zero-order Bessel functions of the first and second kind, and

$$\xi^2 = \frac{\rho\omega^2}{\bar{c}}. \tag{10.148}$$

In the inner air gap, we have

$$\phi = F_1 \ln r + G_1,$$
$$D_r = \varepsilon_0 E_r = -\varepsilon_0 \frac{F_1}{r}, \tag{10.149}$$

where F_1 and G_1 are undetermined constants. For the outer air gap,

$$\phi = F_2 \ln r + G_2,$$
$$D_r = -\varepsilon_0 \frac{F_2}{r}, \tag{10.150}$$

where F_2 and G_2 are undetermined constants. The above fields need to satisfy the following boundary and continuity conditions:

$$\phi = V_1 \exp(i\omega t), \quad r = a,$$
$$\phi = V_2 \exp(i\omega t), \quad r = g,$$
$$T_{rz}(r = b) = T_{rz}(r = d) = 0,$$
$$\phi(r = b^+) = \phi(r = b^-), \qquad (10.151)$$
$$\phi(r = d^+) = \phi(r = d^-),$$
$$D_r(r = b^+) = D_r(r = b^-),$$
$$D_r(r = d^+) = D_r(r = d^-),$$

which leads to the following linear equations for the undetermined constants:

$$F_1 \ln a + G_1 = V_1,$$
$$F_2 \ln g + G_2 = V_2,$$
$$-\bar{c}\xi[A_1 J_1(\xi b) + A_2 Y_1(\xi b)] + e\frac{A_3}{b} = 0,$$
$$-\bar{c}\xi[A_1 J_1(\xi d) + A_2 Y_1(\xi d)] + e\frac{A_3}{d} = 0,$$
$$F_1 \ln b + G_1 = \frac{e}{\varepsilon}[A_1 J_0(\xi b) + A_2 Y_0(\xi b)] + A_3 \ln b + A_4, \qquad (10.152)$$
$$\frac{e}{\varepsilon}[A_1 J_0(\xi d) + A_2 Y_0(\xi d)] + A_3 \ln d + A_4 = F_2 \ln d + G_2,$$
$$-\varepsilon_0 \frac{F_1}{b} = -\varepsilon \frac{A_3}{b},$$
$$-\varepsilon \frac{A_3}{d} = -\varepsilon_0 \frac{F_2}{d}.$$

Once Eq. (10.152) is solved, the charge and the current that flows into the outer electrode per unit length in the axial direction are calculated by

$$D_r = \varepsilon_0 E_r,$$
$$Q_e = \int_0^{2\pi} -D_r g\, d\theta, \qquad (10.153)$$
$$I = \dot{Q}_e = i\omega Q_e.$$

Then the impedance of the structure is given by

$$Z = (V_2 - V_1) / I \qquad (10.154)$$

As an example, consider $b = 5$ mm, $d = 8$ mm, $V_1 = 220$ volt, and $V_2 = 0$. Let $g = d + x$ and $a = b - x$. Damping is introduced by allowing the elastic material constant c to assume complex values. In our calculations c is replaced by $c(1 + iQ^{-1})$. $Q = 100$. Equation (10.152) is solved on a computer.

We plot the impedance for small x in Fig. 10.32. The figure shows the first and third resonances. The second mode does not respond to the applied electric field much because it is essentially symmetric about the shell middle surface.

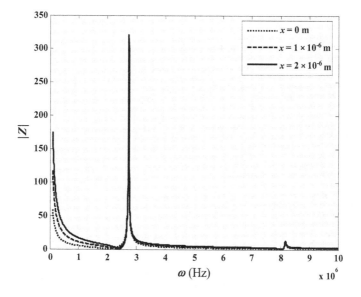

Fig. 10.32. Impedance versus driving frequency.

The resonances in Fig. 10.32 are narrow. To see the effect of the air gaps on impedance more clearly, we plot in Fig. 10.33 the maximum of the absolute value of the first resonance versus x over a relatively large range of x. The figure shows that the impedance is sensitive to x, especially for large x.

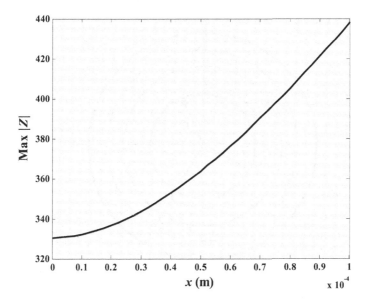

Fig. 10.33. Maximal impedance at the first resonance versus air gap thickness.

10.9. A Multilayered Shell

Consider an N-layered circular cylindrical ceramic shell as shown in Fig. 10.34 [70]. The poling directions of the layers are either along x_3 or $-x_3$. Beginning from the inner surface, the radii of the inner surface, the interfaces, and the outer surface are sequentially determined by $r = r_0$, r_1, ..., r_k, ..., r_{N-1}, and r_N. The kth layer occupies $r_{k-1} < r < r_k$, where $k = 1, 2, 3, ..., N$. The inner and outer surfaces are traction-free and are electroded. A time-harmonic voltage is applied across the two electrodes. Due to the particular material orientation, the cylinder is driven into axial TSh vibration.

10.9.1. Governing equations

In terms of u, ψ and ϕ, we have

$$v_T^2 \nabla^2 u = \ddot{u}, \quad \nabla^2 \psi = 0, \quad \phi = \psi + \frac{e}{\varepsilon} u . \tag{10.155}$$

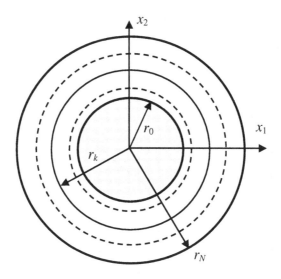

Fig. 10.34. A multilayered circular cylindrical ceramic shell.

The boundary conditions at the inner and outer surfaces are:

$$T_{rz} = 0, \quad \phi = V_1 \exp(i\omega t), \quad r = r_0,$$
$$T_{rz} = 0, \quad \phi = V_2 \exp(i\omega t), \quad r = r_N. \tag{10.156}$$

At the $N-1$ interfaces, the mechanical jump conditions according to the shear-lag model and the electric continuity conditions are

$$T_{rz}(r = r_k^+) = T_{rz}(r = r_k^-) = K[u(r = r_k^+) - u(r = r_k^-)],$$
$$\phi(r = r_k^+) = \varphi(r = r_k^-),$$
$$D_r(r = r_k^+) = D_r(r = r_k^-), \tag{10.157}$$
$$k = 1, 2, 3, \cdots, N - 1,$$

where K is an interface elastic constant. The free charge on and the current flows into the outer electrode per unit length in the axial direction of the shell are determined by

$$Q_e = \int_0^{2\pi} -D_r r_N d\theta,$$
$$I = \dot{Q}_e = i\omega Q_e. \tag{10.158}$$

Then the admittance Y of the structure per unit length in the axial direction is given by

$$I = (V_2 - V_1)Y. \tag{10.159}$$

10.9.2. Solution

In polar coordinates defined by $x_1 = r\cos\theta$ and $x_2 = r\sin\theta$, for time-harmonic and axisymmetric motions, the general solution to Eq. (10.155) is (see Eq. (10.35))

$$
\begin{aligned}
u &= A_1 J_0(\xi r) + A_2 Y_0(\xi r), \\
\psi &= A_3 \ln r + A_4,
\end{aligned}
\tag{10.160}
$$

where A_1 through A_4 are undetermined constants. J_0 and Y_0 are the zero-order Bessel functions of the first and second kind, and (see Eq. (10.37))

$$\xi^2 = \frac{\rho\omega^2}{\overline{c}}. \tag{10.161}$$

The other relevant fields are, correspondingly (see Eq. (10.38)),

$$
\begin{aligned}
\phi &= \frac{e}{\varepsilon}[A_1 J_0(\xi r) + A_2 Y_0(\xi r)] + A_3 \ln r + A_4, \\
T_{rz} &= -\overline{c}\,\xi[A_1 J_1(\xi r) + A_2 Y_1(\xi r)] + e\frac{A_3}{r}, \\
D_r &= -\varepsilon\frac{A_3}{r}.
\end{aligned}
\tag{10.162}
$$

The above solution is valid for every layer of the shell. For the kth layer we write Eqs. (10.160)$_1$ and (10.162) as

$$\mathbf{U}^{(k)}(r) = \mathbf{B}^{(k)}(r)\mathbf{A}^{(k)}, \quad k = 1, 2, 3, \dots, N, \tag{10.163}$$

where

$$
\mathbf{U}^{(k)} = \begin{bmatrix} u \\ \phi \\ T_{rz} \\ D_r \end{bmatrix}, \quad
\mathbf{A}^{(k)} = \begin{bmatrix} A_1 \\ A_2 \\ A_3 \\ A_4 \end{bmatrix},
\tag{10.164}
$$

$$\mathbf{B}^{(k)}(r) = \begin{bmatrix} J_0(\xi r) & Y_0(\xi r) & 0 & 0 \\ \dfrac{e}{\varepsilon} J_0(\xi r) & \dfrac{e}{\varepsilon} Y_0(\xi r) & \ln r & 1 \\ -\bar{c}\xi J_1(\xi r) & -\bar{c}\xi Y_1(\xi r) & \dfrac{e}{r} & 0 \\ 0 & 0 & -\dfrac{\varepsilon}{r} & 0 \end{bmatrix}. \quad (10.165)$$

At r_k, the jump and continuity conditions in Eq. (10.157) take the following form:

$$\mathbf{U}^{(k+1)}(r_k) = \mathbf{U}^{(k)}(r_k) + \mathbf{C}^{(k)}(r_k)\mathbf{A}^{(k)}, \quad k = 1, 2,\ldots, N-1, \quad (10.166)$$

where

$$\mathbf{C}^{(k)}(r) = \frac{1}{K} \begin{bmatrix} -\bar{c}\xi J_1(\xi r) & -\bar{c}\xi Y_1(\xi r) & \dfrac{e}{r} & 0 \\ 0 & 0 & 0 & 0 \\ 0 & 0 & 0 & 0 \\ 0 & 0 & 0 & 0 \end{bmatrix}.$$

$$(10.167)$$

With substitution from Eq. (10.163), Eq. (10.166) can be written as

$$\mathbf{B}^{(k+1)}(r_k)\mathbf{A}^{(k+1)} = \mathbf{B}^{(k)}(r_k)\mathbf{A}^{(k)} + \mathbf{C}^{(k)}(r_k)\mathbf{A}^{(k)}, \quad (10.168)$$

or

$$\mathbf{A}^{(k+1)} = \mathbf{T}^{(k)}(r_k)\mathbf{A}^{(k)}, \quad (10.169)$$

where

$$\mathbf{T}^{(k)}(r_k) = [\mathbf{B}^{k+1}(r_k)]^{-1}[(\mathbf{B}^k(r_k) + \mathbf{C}^k(r_k)]. \quad (10.170)$$

With successive substitutions, we have

$$\mathbf{A}^{(N)} = \mathbf{T}^{(N-1)}(r_{N-1})\cdots\mathbf{T}^{(1)}(r_1)\mathbf{A}^{(1)}. \quad (10.171)$$

Equation (10.171) and the boundary conditions in Eq. (10.156) are eight equations for the eight undetermined constants of the first and the last layers. Once these eight constants are obtained, the rest of the undetermined constants for other layers can be calculated systematically using Eq. (10.169).

10.9.3. Numerical results

For numerical results, consider a cylindrical ceramic shell made from polarized ceramics PZT-5A and PZT-6B. First we consider the case of a two-layered shell. The inner layer is PZT-5A. The outer layer is PZT-6B. Both layers are poled in the x_3 direction. The interface is perfectly bonded. Figure 10.35(a) shows the absolute value of the admittance versus the driving frequency. At resonances the admittance assumes maxima. Figures 10.35(b) and 10.35(c) show the displacement and shear stress distributions at the first two resonances. The first resonance has the largest admittance. This is because at the first resonance the shell is vibrating at its fundamental TSh mode. In this mode the shear strain does not change sign along the shell thickness. At higher resonances the shear strain changes sign and has nodal point(s) along the shell thickness.

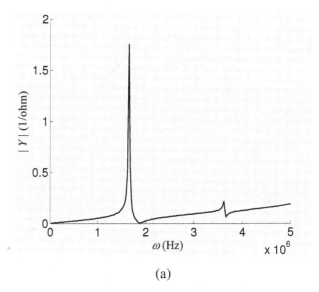

(a)

Fig. 10.35. Admittance (a), displacement (b), and stress (c) of a two-layered shell with a perfectly bonded interface and the same poling direction.

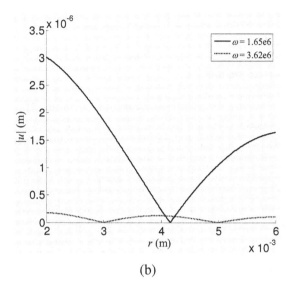

(b)

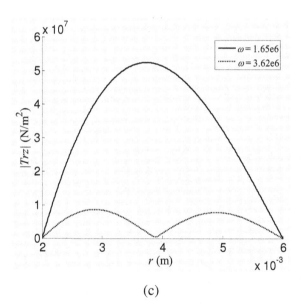

(c)

Fig. 10.35. (*Continued*)

For the same two-layered shell, if we reverse the poling direction of the inner layer of PZT-5A to the $-x_3$ direction, the corresponding results are shown in Fig. 10.36. A major difference between Figs. 10.35(a) and 10.36(a) is that in Fig. 10.36(a) the admittance at the second resonance is larger than that at the first resonance. This is a combined effect of the reversed poling and the change of sign of the shear strain along the shell thickness at the second resonance.

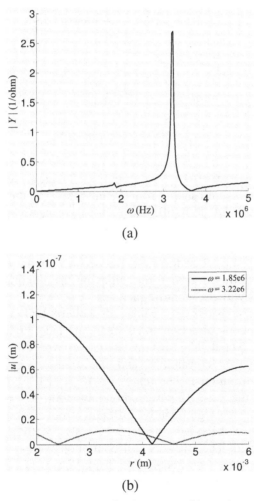

(a)

(b)

Fig. 10.36. Admittance (a), displacement (b), and stress (c) of a two-layered shell with a perfectly bonded interface and opposite poling directions.

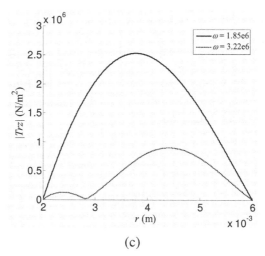

(c)

Fig. 10.36. (*Continued*)

To examine the effect of the poling direction further, next we consider a three-layered shell. From the inner layer to the outer layer, the materials are PZT-5A, PZT-6B, and PZT-5A, respectively. Figure 10.37 shows the admittance versus the driving frequency for different combinations of the poling directions. It can be seen that the maximal admittance may appear at the first, the second, or the third resonance. This offers flexibility in design for achieving the maximal admittance at different modes with different frequencies.

Finally, we consider a two-layered shell of PZT-5A and PZT-6B with the same poling direction and a weak interface. Figure 10.38 shows the admittance versus the driving frequency for different values of the interface elastic constant K and the corresponding displacement and stress distributions. The case of $K = 10^{14}$ N/m^3 is slightly different from the case of perfect bonding with $K = \infty$. The case of $K = 10^{13}$ N/m^3 is significantly different from the case of perfect bonding. When K becomes smaller, the resonant frequencies become lower as expected. The stress is still continuous across the interface as dictated by the shear-lad model. The displacement has a finite jump at the interface.

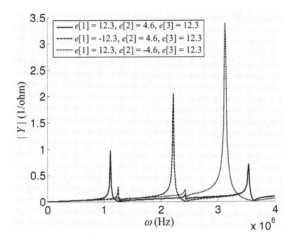

Fig. 10.37. Admittance versus driving frequency for different combinations of poling directions in a three-layered shell with perfectly bonded interfaces.

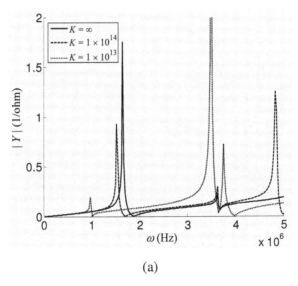

(a)

Fig. 10.38. Admittance (a), displacement (b), and stress (c) of a two-layered shell with a weak interface and the same poling direction.

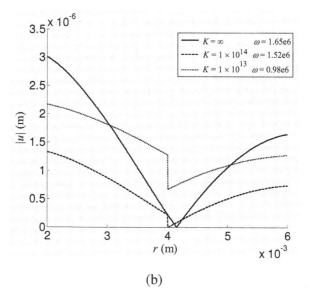

(b)

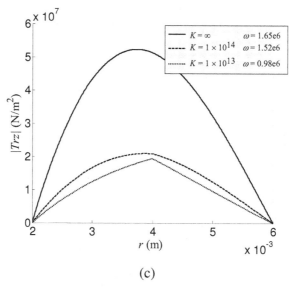

(c)

Fig. 10.38. (*Continued*)

References

[1]   A. H. Meitzler, D. Berlincourt, F. S. Welsh, III, H. F. Tiersten, G. A. Coquin and
      A. W. Warner, IEEE Standard on Piezoelectricity, IEEE, New York, 1988.
[2]   H. F. Tiersten, *Linear Piezoelectric Plate Vibrations*, Plenum, New York, 1969.
[3]   J. S. Yang, *An Introduction to the Theory of Piezoelectricity*, Springer, New York,
      2005.
[4]   H. F. Tiersten, On the nonlinear equations of thermoelectroelasticity, *Int. J. Eng.
      Sci.*, **9**, 587–604, 1971.
[5]   D. F. Nelson, *Electric, Optic and Acoustic Interactions in Crystals*, John Wiley and
      Sons, New York, 1979, pp. 481–513.
[6]   V. V. Varadan, J.-H. Jeng and V. K. Varadan, Form invariant constitutive relations
      for transversely isotropic piezoelectric materials, *J. Acoust. Soc. Am.*, **82**, 337–341,
      1987.
[7]   B. A. Auld, *Acoustic Fields and Waves in Solids*, Vol. 1, John Wiley and Sons,
      New York, 1973.
[8]   J. L. Bleustein, A new surface wave in piezoelectric materials, *Appl. Phys. Lett.*,
      13, 412–413, 1968.
[9]   X. Wang and Z. Zhong, Two-dimensional time-harmonic dynamic Green's
      functions in transversely isotropic piezoelectric solids, *Mech. Res. Commun.*, **30**,
      589–593, 2003.
[10] X. Wang and Z. Zhong, A moving piezoelectric screw dislocation, *Mech. Res.
      Commun.*, **29**, 425–429, 2002.
[11] Y. V. Gulyaev, Electroacoustic surface waves in solids, *Soviet Phys. JETP Lett.*, **9**,
      37–38, 1969.
[12] F. S. Hickernell, Shear horizontal BG surface acoustic waves on piezoelectrics: A
      historical note, *IEEE Trans. Ultrasonics, Ferroelectrics, and Frequency Control*,
      52, 809–811, 2005.
[13] J. S. Yang, *Analysis of Piezoelectric Devices*, World Scientific, Singapore, 2006.
[14] Z. H. Qian, F. Jin, T. J. Lu and K. Kishimoto, Transverse surface waves in
      functionally graded piezoelectric materials with experimental variation, *Smart
      Materials and Structures*, **17**, #065005, 2008.
[15] F. L. Guo and R. Sun, Propagation of Bleustein-Gulyaev wave in 6mm
      piezoelectric materials loaded with viscous fluid, *Int. J. Solids and Structures*, **45**,
      3699–3710, 2008.
[16] C. Maerfeld and P. Tournois, Pure shear elastic surface wave guided by the
      interface of two semi-infinite media, *Appl. Phys. Lett.*, **19**, 117–118, 1971.
[17] H. Fan, J. S. Yang and L. M. Xu, Piezoelectric waves near an imperfectly bonded
      interface between two half-spaces, *Appl. Phys. Lett.*, **88**, #203509, 2006.
[18] N. Liu, J. S. Yang, Z. H. Qian and K. Kishimoto, Interface waves in functionally
      graded piezoelectric materials, *Int. J. Eng. Sci.*, **48**, 151–159, 2010.

[19] Y. V. Gulyaev and V. P. Plesskii, Acoustic gap waves in piezoelectric materials, *Soviet Phys.-Acoust.*, **23**, 410–413, 1977.

[20] C. L. Chen, On the electroacoustic waves guided by a cylindrical piezoelectric surface, *J. Appl. Phys.*, **44**, 3841–3847, 1973.

[21] J. L. Bleustein, Some simple modes of wave propagation in an infinite piezoelectric plate, *J. Acoust. Soc. Am.*, **45**, 614–620, 1969.

[22] J. S. Yang and H. G. Zhou, On the effect of the electric field in the free space surrounding a finite piezoelectric body, *IEEE Trans. Ultrasonics, Ferroelectrics, and Frequency Control*, **53**, 1557–1559, 2006.

[23] Z. H. Qian, J. S. Yang and K. Kishimoto, Propagation of thickness-twist waves in a piezoelectric ceramic plate with unattached electrodes, *Ultrasonics*, **49**, 501–504, 2009.

[24] J. S. Yang, Z. G. Chen and Y. T. Hu, Mass sensitivity of thickness-twist modes in a rectangular piezoelectric plate of hexagonal crystals, *IEEE Trans. Ultrasonics, Ferroelectrics, and Frequency Control*, **54**, 882–887, 2007.

[25] Z. H. Qian, K. Kishimoto, N. Liu and J. S. Yang, Propagation of thickness-twist waves in a piezoelectric ceramic plate in contact with viscous fluids, *Acta Mechanica*, submitted.

[26] J. S. Yang, Z. G. Chen and Y. T. Hu, Propagation of thickness-twist waves through a joint between two semi-infinite piezoelectric plates, *IEEE Trans. Ultrasonics, Ferroelectrics, and Frequency Control*, **54**, 888–891, 2007.

[27] J. S. Yang, Z. G. Chen and Y. T. Hu, Trapped thickness-twist modes in an inhomogeneous piezoelectric plate, *Phil. Mag. Lett.*, **86**, 699–705, 2006.

[28] J. S. Yang and S. H. Guo, Effects of piezoelectric coupling on Bechmann's number for thickness-twist waves in a plate of hexagonal crystals, *IEEE Trans. Ultrasonics, Ferroelectrics, and Frequency Control*, **53**, 1960–1962, 2006.

[29] F. M. Li and Y. S. Wang, Study on wave localization in disordered periodic layered piezoelectric composite structures, *Int. J. Solids and Structures*, **42**, 6457–6474, 2005.

[30] F. M. Li, Y. S. Wang, C. Hu, and W. H. Huang, Wave localization in randomly disordered periodic layered piezoelectric structures, *Acta Mechanica Sinica*, **22**, 559–567, 2006.

[31] F. M. Li, M. Q. Xu and Y. S. Wang, Frequency-dependent localization length of SH-waves in randomly disordered piezoelectric phononic crystals, *Solid State Commun.*, **141**, 296–301, 2007.

[32] Z. H. Qian, F. Jin, Z. K. Wang and K. Kishimoto, Dispersion relations for SH-wave propagation in periodic piezoelectric composite layered structures, *Int. J. Eng. Sci.*, **42**, 673–689, 2004.

[33] J. S. Yang, Z. G. Chen, Y. T. Hu, S. N. Jiang and S. H. Guo, Propagation of thickness-twist waves in a multi-sectioned piezoelectric plate of 6mm crystals, *Arch. Appl. Mech.*, **77**, 689–696, 2007.

[34] R. G. Curtis and M. Redwood, Transverse surface waves on a piezoelectric material carrying a metal layer of finite thickness, *J. Appl. Phys.*, **44**, 2002–2007, 1973.

[35] X. J. Li, Effect of domain structure on surface wave, *Ferroelectrics*, **47**, 3–7, 1983.

[36] H. H. Feng and X. J. Li, Shear-horizontal surface waves in a layered structure of piezoelectric ceramics, *IEEE Trans. Ultrasonics, Ferroelectrics, and Frequency Control*, **40**, 167–170, 1993.

[37] Z. K. Wang, F. Jin, Z. Zong and T. J. Wang, The propagation of a layer-confined Love wave in layered piezoelectric structures, *Key Eng. Mater.*, **183–187**, 725–730, 2000.

[38] F. Jin, Z. K. Wang and T. J. Wang, The Bleustein–Gulyaev (B–G) wave in a piezoelectric layered half-space, *Int. J. Eng. Sci.*, **39**, 1271–1285, 2001.

[39] J. K. Du, X. Y. Jin, J. Wang and K. Xian, Love wave propagation in functionally graded piezoelectric material layer, *Ultrasonics*, **46**, 13–22, 2007.

[40] Z. H. Qian, F. Jin, K. Kishimoto and T. J. Lu, Surface Love waves in a functionally graded piezoelectric substrate covered with a dielectric layer, *Int. J. Solids and Structures*, submitted.

[41] F. Jin, K. Kishimoto, H. Inoue and T. Tateno, Experimental investigation on the interface properties evaluation in piezoelectric layered structures by Loves waves propagation, *Key Eng. Mater.*, **297–300**, 807–812, 2005.

[42] H. Fan, J. S. Yang and L. M. Xu, Antiplane piezoelectric surface waves over a ceramic half-space with an imperfectly bonded layer, *IEEE Trans. Ultrasonics, Ferroelectrics, and Frequency Control*, **53**, 1695–1698, 2006.

[43] Z. G. Chen, Y. T. Hu and J. S. Yang, Shear horizontal piezoelectric waves in a ceramic plate imperfectly bonded to two ceramic half-spaces, *J. Mech.*, **24**, 229–239, 2008.

[44] X. F. Li and J. S. Yang, Piezoelectric gap waves between a piezoceramic half-space and a piezoceramic plate, *Sensors Actuators Phys.*, **132**, 472–479, 2006.

[45] F. Peng, H. Liu and S. Y. Hu, Love wave propagation in a layered piezoelectric structure immersed in a viscous fluid, *Key Eng. Mater.*, **306–308**, 1211–1216, 1006.

[46] J. K. Chen, W. C. Wang, J. Wang, Z. T. Yang and J. S. Yang, A thickness mode acoustic wave sensor for measuring interface stiffness between two elastic materials, *IEEE Trans. Ultrasonics, Ferroelectrics, and Frequency Control*, **55**, 1678–1681, 2008.

[47] J. S. Yang, Thickness-twist edge modes in a semi-infinite piezoelectric plate of crystals with 6mm symmetry, *IEEE Trans. Ultrasonics, Ferroelectrics, and Frequency Control*, **54**, 220–221, 2007.

[48] J. S. Yang and A.-K. Soh, A new mass sensor based on thickness-twist edge modes in a piezoelectric plate, *Europhys. Lett.*, **77**, #28003, 2007.

[49] J. S. Yang and S. H. Guo, Thickness-twist modes in a rectangular piezoelectric resonator of hexagonal crystals, *Appl. Phys. Lett.*, **88**, #153506, 2006.

[50] N. T. Adelman and Y. Stavsky, Radial vibrations of axially polarized piezoelectric ceramic cylinders, *J. Acoust. Soc. Am.*, **57**, 356–360, 1975.

[51] Z. T. Yang, J. S. Yang, S. H. Guo and L. Cao, Thickness-shear vibration of a circular cylindrical ceramic cylinder with unattached electrodes and air gaps, *Science in China-G*, **52**, 1423–1427, 2009.

[52] J. S. Yang, Shear horizontal vibrations of a piezoelectric/ ferroelectric wedge, *Acta Mechanica*, **173**, 13–17, 2004.

[53] Z. T. Yang, J. S. Yang, Y. T. Hu and Q.-M. Wang, Vibration characteristics of a circular cylindrical panel piezoelectric transducer, *IEEE Trans. Ultrasonics, Ferroelectrics, and Frequency Control*, **55**, 2327–2335, 2008.

[54] J. S. Yang, Z. G. Chen and Y. T. Hu, An exact analysis of a rectangular plate piezoelectric generator, *IEEE Trans. Ultrasonics, Ferroelectrics, and Frequency Control*, **54**, 190–195, 2007.

[55] J. S. Yang, J. J. Liu and J. Y. Li, Analysis of a rectangular ceramic plate in electrically forced thickness-twist vibration as a piezoelectric transformer, *IEEE Trans. Ultrasonics, Ferroelectrics, and Frequency Control*, **54**, 830–835, 2007.

[56] J. S. Yang, Z. G. Chen and Y. T. Hu, Electrically forced vibration of a thickness-twist mode piezoelectric resonator with non-uniform electrodes, *Acta Mechanica Solida Sinica*, **20**, 266–274, 2007.

[57] J. S. Yang, Z. G. Chen and Y. T. Hu, Vibration of a thickness-twist mode piezoelectric resonator with asymmetric, non-uniform electrodes, *IEEE Trans. Ultrasonics, Ferroelectrics, and Frequency Control*, **55**, 841–848, 2008.

[58] H. P. Hu, Z. G. Chen, J. S. Yang and Y. T. Hu, An exact analysis of forced thickness-twist vibrations of multi-layered piezoelectric plates, *Acta Mechannica Solida Sinica*, **20**, 211–218, 2007.

[59] Z. T. Yang, J. S. Yang and Y. T. Hu, Energy trapping in power transmission through an elastic plate by finite piezoelectric transducers, *IEEE Trans. Ultrasonics, Ferroelectrics, and Frequency Control*, **55**, 2493–2501, 2008.

[60] Z. T. Yang, S. H. Guo, J. S. Yang and Y. T. Hu, Electrically forced vibration of an elastic plate with a finite piezoelectric actuator, *J. Sound Vib.*, **321**, 242–253, 2009.

[61] F. Yang, B. B. Tang, Y. T. Hu and J. S. Yang, Excitation and detection of shear horizontal waves in an elastic plate using two piezoelectric transducers, to be submitted.

[62] B. X. Zhang, A. Bostrom and A. J. Niklasson, Antiplane shear waves from a piezoelectric strip actuator: exact versus effective boundary condition solutions, *Smart Materials and Structures*, **13**, 161–168, 2004.

[63] B. X. Zhang, C. H. Wang and A. Bostrom, Study of SH acoustic radiation field excited by a piezoelectric strip, *Acta Physica Sinica*, **54**, 2111–2117, 2005.

[64] S. N. Jiang, Q. Jiang, Y. T. Hu, X. F. Li, S. H. Guo and J. S. Yang, Analysis of a piezoelectric ceramic shell in thickness-shear vibration as a power harvester, *Int. J. Appl. Electromagn. Mech.*, **24**, 25–31, 2006.

[65] J. S. Yang, Z. G. Chen and Y. T. Hu, Theoretical modeling of a thickness-shear circular cylinder piezoelectric transformer, *IEEE Trans. Ultrasonics, Ferroelectrics, and Frequency Control*, **54**, 621–626, 2007.

[66] Z. T. Yang, S. H. Guo and J. S. Yang, Transmitting electric energy through a closed elastic wall by acoustic waves and piezoelectric transducers, *IEEE Trans. Ultrasonics, Ferroelectrics, and Frequency Control*, **55**, 1380–1386, 2008.

[67] Z. T. Yang and S. H. Guo, Energy trapping in power transmission through a circular cylindrical elastic shell by finite piezoelectric transducers, *Ultrasonics*, **48**, 716–723, 2008.

[68] Z. T. Yang, J. S. Yang and Y. T. Hu, Electrically forced shear horizontal vibration of a circular cylindrical elastic shell with a finite piezoelectric actuator, *Arch. Appl. Mech.*, **79**, 955–964, 2009.

[69] F. Yang, Y. T. Hu and J. S. Yang, Shear-horizontal vibration of an elastic circular cylindrical shell with two piezoelectric transducers, *IEEE Trans. Ultrasonics, Ferroelectrics, and Frequency Control*, **56**, 1708–1715, 2009.

[70] F. Yang, Y. T. Hu and J. S. Yang, Electrically forced antiplane vibrations of a multilayered circular cylindrical piezoelectric shell of polarized ceramics with weak interfaces, to be submitted.

[71] H. Jaffe and D. A. Berlincourt, Piezoelectric transducer materials, *Proc. IEEE*, **53**, 1372–1386, 1965.

[72] G. Feuillard, M. Lethiecq, Y. Janin, C. Alemany, B. Jimenez, C. Miller and W. Wolny, Experimental determination of SAW properties of 5 standard piezoceramics, *Proc. IEEE Ultrasonics Symp.*, 1994, pp. 447–451.

[73] J. G. Gualtieri, J. A. Kosinski and A. Ballato, Piezoelectric materials for acoustic wave applications, *IEEE Trans. Ultrasonics, Ferroelectrics, and Frequency Control*, **41**, 53–59, 1994.

[74] K. Tsubouchi, K. Sugai and N. Mikoshiba, AlN material constants evaluation and SAW properties on AlN/Al$_2$O$_3$ and AlN/Si, *Proc. IEEE Ultrasonics Symp.*, 1981, pp. 375–380.

Appendix 1
Notation

E_i	–	Electric field
D_i	–	Electric displacement
T_{kl}	–	Stress tensor
S_{kl}	–	Strain tensor
$u = u_3(x_1, x_2, t)$	–	Antiplane displacement
ϕ	–	Electric potential
ρ	–	Mass density
$c = c_{44}$	–	Shear elastic constant
$e = e_{15}$	–	Piezoelectric constant
$\varepsilon = \varepsilon_{11}$	–	Dielectric constant

$$\psi(x_1, x_2, t) = \phi - \frac{e}{\varepsilon}u$$

ε_0	–	Permittivity of free space
$n = \sqrt{\dfrac{\varepsilon}{\varepsilon_0}}$	–	Refractive index
$k^2 = \dfrac{e^2}{\varepsilon c}$		
$\bar{c} = c(1 + k^2)$	–	Piezoelectrically stiffened elastic constant
$\bar{\varepsilon} = \varepsilon(1 + k^2)$	–	Effective dielectric constant including piezoelectric effect
$\bar{k}^2 = \dfrac{e^2}{\varepsilon\bar{c}} = \dfrac{k^2}{1 + k^2}$	–	Electromechanical coupling coefficient
$v_T = \sqrt{\dfrac{\bar{c}}{\rho}}$	–	Shear wave speed
μ	–	Shear elastic constant or fluid viscosity
q	–	Free charge density

371

Q_e	–	Free charge (a scalar)
σ	–	Surface free charge per unit area
I	–	Current
V	–	Voltage
Z	–	Impedance
Y	–	Admittance
i	–	Imaginary unit
Q	–	Material quality factor
FS	–	Face-shear
TT	–	Thickness-twist
TSh	–	Thickness-shear
TSt	–	Thickness-stretch
FGM	–	Functionally graded (gradient) material

Appendix 2
Material Constants

Material constants for a few common ceramics and crystals of 6mm symmetry are summarized below. The permittivity of free space is $\varepsilon_0 = 8.854 \times 10^{-12}$ F/m.

For PZT-5H $\rho = 7500 \ kg/m^3$. The material matrices are [71]

$$[c_{pq}] = \begin{pmatrix} 12.6 & 7.95 & 8.41 & 0 & 0 & 0 \\ 7.95 & 12.6 & 8.41 & 0 & 0 & 0 \\ 8.41 & 8.41 & 11.7 & 0 & 0 & 0 \\ 0 & 0 & 0 & 2.3 & 0 & 0 \\ 0 & 0 & 0 & 0 & 2.3 & 0 \\ 0 & 0 & 0 & 0 & 0 & 2.325 \end{pmatrix} \times 10^{10} \, \text{N/m}^2,$$

$$[e_{ip}] = \begin{pmatrix} 0 & 0 & 0 & 0 & 17 & 0 \\ 0 & 0 & 0 & 17 & 0 & 0 \\ -6.5 & -6.5 & 23.3 & 0 & 0 & 0 \end{pmatrix} \text{C/m}^2,$$

$$[\varepsilon_{ij}] = \begin{pmatrix} 1700\varepsilon_0 & 0 & 0 \\ 0 & 1700\varepsilon_0 & 0 \\ 0 & 0 & 1470\varepsilon_0 \end{pmatrix} =$$

$$\begin{pmatrix} 1.505 & 0 & 0 \\ 0 & 1.505 & 0 \\ 0 & 0 & 1.302 \end{pmatrix} \times 10^{-8} \, \text{C/(V} - \text{m)}.$$

For PZT-5H, an equivalent set of material constants are

$$s_{11} = 16.5, \quad s_{33} = 20.7, \quad s_{44} = 43.5,$$

$$s_{12} = -4.78, \quad s_{13} = -8.45 \times 10^{-12}\,\text{m}^2/\text{N},$$

$$d_{31} = -274, \quad d_{15} = 741, \quad d_{33} = 593 \times 10^{-12}\,\text{C/N},$$

$$\varepsilon_{11} = 3130\varepsilon_0, \quad \varepsilon_{33} = 3400\varepsilon_0.$$

When poling is along other directions, the material matrices can be obtained by tensor transformations. For PZT-5H, when poling is along the x_1 axis, we have

$$[c_{pq}] = \begin{pmatrix} 11.7 & 8.41 & 8.41 & 0 & 0 & 0 \\ 8.41 & 12.6 & 7.95 & 0 & 0 & 0 \\ 8.41 & 7.95 & 12.6 & 0 & 0 & 0 \\ 0 & 0 & 0 & 2.325 & 0 & 0 \\ 0 & 0 & 0 & 0 & 2.3 & 0 \\ 0 & 0 & 0 & 0 & 0 & 2.3 \end{pmatrix} \times 10^{10}\,\text{N/m}^2,$$

$$[e_{ip}] = \begin{pmatrix} 23.3 & -6.5 & -6.5 & 0 & 0 & 0 \\ 0 & 0 & 0 & 0 & 0 & 17 \\ 0 & 0 & 0 & 0 & 17 & 0 \end{pmatrix} \text{C/m}^2,$$

$$[\varepsilon_{ij}] = \begin{pmatrix} 1.302 & 0 & 0 \\ 0 & 1.505 & 0 \\ 0 & 0 & 1.505 \end{pmatrix} \times 10^{-8}\,\text{C/Vm}.$$

When poling is along the x_2 axis,

$$[c_{pq}] = \begin{pmatrix} 12.6 & 8.41 & 7.95 & 0 & 0 & 0 \\ 8.41 & 11.7 & 8.41 & 0 & 0 & 0 \\ 7.95 & 8.41 & 12.6 & 0 & 0 & 0 \\ 0 & 0 & 0 & 2.3 & 0 & 0 \\ 0 & 0 & 0 & 0 & 2.325 & 0 \\ 0 & 0 & 0 & 0 & 0 & 2.3 \end{pmatrix} \times 10^{10}\,\text{N/m}^2,$$

$$[e_{ip}] = \begin{pmatrix} 0 & 0 & 0 & 0 & 0 & 17 \\ -6.5 & 23.3 & -6.5 & 0 & 0 & 0 \\ 0 & 0 & 0 & 17 & 0 & 0 \end{pmatrix} C/m^2,$$

$$[\varepsilon_{ij}] = \begin{pmatrix} 1.505 & 0 & 0 \\ 0 & 1.302 & 0 \\ 0 & 0 & 1.505 \end{pmatrix} \times 10^{-8} C/Vm.$$

Material constants of a few other polarized ceramics are given in the following tables [71]:

Material	c_{11}	c_{12}	c_{13}	c_{33}	c_{44}	c_{66}
PZT-4	13.9	7.78	7.40	11.5	2.56	3.06
PZT-5A	12.1	7.59	7.54	11.1	2.11	2.26
PZT-6B	16.8	8.47	8.42	16.3	3.55	4.17
PZT-5H	12.6	7.91	8.39	11.7	2.30	2.35
PZT-7A	14.8	7.61	8.13	13.1	2.53	3.60
PZT-8	13.7	6.99	7.11	12.3	3.13	3.36
BaTiO$_3$	15.0	6.53	6.62	14.6	4.39	4.24
	$\times 10^{10}$ N/m^2					

Material	e_{31}	e_{33}	e_{15}	ε_{11}	ε_{33}
PZT-4	−5.2	15.1	12.7	0.646	0.562
PZT-5A	−5.4	15.8	12.3	0.811	0.735
PZT-6B	−0.9	7.1	4.6	0.360	0.342
PZT-5H	−6.5	23.3	17.0	1.505	1.302
PZT-7A	−2.1	9.5	9.2	0.407	0.208
PZT-8	−4.0	13.2	10.4	0.797	0.514
BaTiO$_3$	−4.3	17.5	11.4	0.987	1.116
	C/m^2			$\times 10^{-8}$ C/Vm	

Density	PZT-5H	PZT-5A	PZT-6B	PZT-4
kg/m^3	7500	7750	7550	7500

Density	PZT-7A	PZT-8	BaTiO$_3$
kg/m^3	7600	7600	5700

Material constants for a few other piezoceramics are [72]:

Material	Pz24	Pz26	Pz27	Pz28	Pz34
ρ (kg/m^3)	7.65	7.65	7.75	7.65	7.6

Material	Pz24	Pz26	Pz27	Pz28	Pz34
c_{11}^E (GPa)	177.3	171.3	142.8	142.2	188.4
c_{12}^E (GPa)	106.5	15.5	99.8	88.1	80.2
c_{13}^E (GPa)	98.2	107.4	92.2	82.1	82.3
c_{33}^E (GPa)	127.2	133.2	114.5	108.1	127.2
c_{44}^E (GPa)	33.5	33.3	22.8	29.8	58.9

Material	Pz24	Pz26	Pz27	Pz28	Pz34
e_{15} (C/m^2)	8.10	8.84	11.74	9.26	2.31
e_{31} (C/m^2)	-0.26	-1.16	-3.71	-2.80	3.81
e_{33} (C/m^2)	9.73	15.67	15.77	12.1	6.87

Material	Pz24	Pz26	Pz27	Pz28	Pz34
$\varepsilon_{11}^S / \varepsilon_0$	477	890	1120	910	181
$\varepsilon_{33}^S / \varepsilon_0$	239	790	880	520	154

Material constants for a few crystals of 6mm symmetry are [73]:

	Zinc Oxide ZnO	Aluminum Nitride AlN	α - Silicon Carbide α - Sic
$\rho\,(10^3\,\text{kg/m}^3)$	5.665		3.211
c_{11} (GPa)	209	345	500
c_{12} (GPa)		125	92
c_{13} (GPa)	104.6	120	
c_{33} (GPa)	210.6	395	564
c_{44} (GPa)	42.3	118	168
c_{66} (GPa)	44.25	110	204
e_{15} (C/m^2)	-0.48	-0.48	0.08
e_{31} (C/m^2)	-0.57	-0.58	
e_{33} (C/m^2)	1.32	1.55	0.2
$\varepsilon_{11}\,(\varepsilon_0)$			6.7
$\varepsilon_{33}\,(\varepsilon_0)$			6.52

	Cadium Selenide SdSe	Cadium Selenide SdSe	Zinc Sulfide ZnS
$\rho\,(10^3\,\text{kg/m}^3)$	5.684	4.820	3.980
c_{11} (GPa)	74.1	90.7	120.4
c_{12} (GPa)	45.2	58.1	69.2
c_{13} (GPa)	39.3	51.0	62.0
c_{33} (GPa)	83.6	93.8	127.6
c_{44} (GPa)	13.17	15.04	22.8
c_{66} (GPa)	14.45	16.3	25.6
e_{15} (C/m^2)		-0.21	-0.0683
e_{31} (C/m^2)		-0.24	-0.0140
e_{33} (C/m^2)		0.44	0.272
$\varepsilon_{11}\,(\varepsilon_0)$		9.02	
$\varepsilon_{33}\,(\varepsilon_0)$		9.53	

Material constants for AlN are also given in [74]: $\rho = 3512 \, \text{kg/m}^3$,

$$[c_{pq}] = \begin{pmatrix} 345 & 125 & 120 & 0 & 0 & 0 \\ 125 & 345 & 120 & 0 & 0 & 0 \\ 120 & 120 & 395 & 0 & 0 & 0 \\ 0 & 0 & 0 & 118 & 0 & 0 \\ 0 & 0 & 0 & 0 & 118 & 0 \\ 0 & 0 & 0 & 0 & 0 & 110 \end{pmatrix} \times 10^9 \, \text{N/m}^2,$$

$$[e_{ip}] = \begin{pmatrix} 0 & 0 & 0 & 0 & -0.48 & 0 \\ 0 & 0 & 0 & -0.48 & 0 & 0 \\ -0.58 & -0.58 & 1.55 & 0 & 0 & 0 \end{pmatrix} \text{C/m}^2,$$

$$[\varepsilon_{ij}] = \begin{pmatrix} 8.0 & 0 & 0 \\ 0 & 8.0 & 0 \\ 0 & 0 & 9.5 \end{pmatrix} \times 10^{-11} \, \text{F/m}.$$

Material constants of a few 4mm crystals are [73]:

	Dilithium Tetraborate $Li_2B_4O_7$	Strontium Niobate $Sr_{.5}Ba_{.5}Nb_2O_6$
$\rho \, (10^3 \text{kg/m}^3)$	2.439	5.4
c_{11} (GPa)	135	210.1
c_{12} (GPa)	3.57	65.7
c_{13} (GPa)	33.5	35.5
c_{33} (GPa)	56.8	116.6
c_{44} (GPa)	58.5	66.3
c_{66} (GPa)	46.7	68.9
e_{15} (C/m^2)	0.472	5.19
e_{31} (C/m^2)	0.290	−1.89
e_{33} (C/m^2)	0.928	9.81
ε_{11} (ε_0)	8.90	370
ε_{33} (ε_0)	8.07	282

Index

379